Environmental Pollution and Assessment in Mining Areas

Environmental Pollution and Assessment in Mining Areas

Guest Editors

Mari Luz García Lorenzo
José María Esbrí
Oscar Andreu Sánchez

MDPI

Basel • Beijing • Wuhan • Barcelona • Belgrade • Novi Sad • Cluj • Manchester

Guest Editors
Mari Luz García Lorenzo
Departamento de
Mineralogía y Petrología
Universidad Complutense
de Madrid
Madrid
Spain

José María Esbrí
Departamento de
Mineralogía y Petrología
Universidad Complutense
de Madrid
Madrid
Spain

Oscar Andreu Sánchez
Departamento de Biología
Vegetal
Universitat de València
Valencia
Spain

Editorial Office
MDPI AG
Grosspeteranlage 5
4052 Basel, Switzerland

This is a reprint of the Special Issue, published open access by the journal *Minerals* (ISSN 2075-163X), freely accessible at: https://www.mdpi.com/journal/minerals/special_issues/FI5E0PS186.

For citation purposes, cite each article independently as indicated on the article page online and as indicated below:

Lastname, A.A.; Lastname, B.B. Article Title. *Journal Name* **Year**, *Volume Number*, Page Range.

ISBN 978-3-7258-5299-4 (Hbk)
ISBN 978-3-7258-5300-7 (PDF)
https://doi.org/10.3390/books978-3-7258-5300-7

Contents

About the Editors

Mari Luz García Lorenzo

Mari Luz García Lorenzo is affiliated with the Department of Mineralogy and Petrology at Universidad Complutense de Madrid, Spain. Her research interests center on environmental geochemistry and soil contamination, with particular emphasis on mining-impacted areas. She has developed extensive work on the mobility and toxicity of potentially toxic elements in contaminated soils, as well as the environmental implications of acid mine drainage. Throughout her career, she has led and participated in national and international research projects focused on pollution monitoring, ecological risk assessment, and the design of soil remediation strategies.

José María Esbrí

José María Esbrí is affiliated with the Department of Mineralogy and Petrology at Universidad Complutense de Madrid, Spain. His scientific work focuses on the environmental consequences of historical mining, the study of mobility, immobilization processes, and the transfer of potentially toxic elements between environmental compartments. He has contributed to multiple research initiatives related to the geochemical characterization of mining residues, waste remediation, and risk-based approaches, participating in several metal mining restoration projects.

Oscar Andreu Sánchez

Oscar Andreu Sánchez is affiliated with the Department of Plant Biology, Faculty of Biology, University of Valencia. His research integrates soil science, environmental chemistry, and ecotoxicology to address contamination in mining and industrial areas. He is particularly involved in the study of bioavailability and the ecological effects of potentially toxic elements in soils. His current projects include the development of pollution indices and bioassay-based tools for evaluating and categorizing mining wastes, with the goal of supporting environmental risk assessment and land restoration efforts.

Preface

This Reprint, derived from the Special Issue 'Environmental Pollution and Assessment in Mining Areas', compiles a diverse range of scientific contributions that explore the complex challenges posed by mining-related contamination. This volume brings together original research and review articles from multidisciplinary teams spanning several continents, offering new perspectives on the mobility, toxicity, and environmental behavior of potentially harmful elements in soils, waters, and sediments. The Special Issue was designed to improve understanding of environmental risks and promote evidence-based assessment and remediation of mining-impacted ecosystems. In today's society, where the need to secure the supply of many critical raw materials has risen in importance, developing knowledge of these aspects related to mining waste management will be essential to promoting more sustainable mining that will allow mining projects to begin in countries with restrictive environmental legislation. We hope this compilation will serve as a valuable resource for researchers, environmental professionals, and policymakers not only in restoration/remediation projects for abandoned mines, but also as a contribution to the development of new, more sustainable mining techniques. We would like to express our gratitude to all the contributing authors, reviewers, and *Minerals* editorial staff for their collaboration and support in preparing this Special Issue.

Mari Luz García Lorenzo, José María Esbrí, and Oscar Andreu Sánchez
Guest Editors

minerals

MDPI

Article

Advancing the Understanding of Sediment Contamination Dynamics in the Iron Quadrangle (Brazil): A Comparative Analysis of Pollution Indices for PTE Assessment

Raphael Vicq [1,*], Mariangela G. P. Leite [2], Lucas P. Leão [2], Hermínio A. Nallini Júnior [2], Patricia Gomes [1], Rita Fonseca [3] and Teresa Valente [1,*]

1 Earth Sciences Department, Institute of Earth Sciences, University of Minho, 4710-057 Braga, Portugal; patriciagomes@dct.uminho.pt

2 Department of Geology, Federal University of Ouro Preto, Morro do Cruzeiro Campus, Ouro Preto 35400-000, MG, Brazil; mgpleite@gmail.com (M.G.P.L.); lucas.leao@ufop.edu.br (L.P.L.); nalini@ufop.edu.br (H.A.N.J.)

3 Earth Science Institute, University of Évora, Largo Dos Colegiais 2, 7004-516 Évora, Portugal; rfonseca@uevora.pt

* Correspondence: raphaelcosta@dct.uminho.pt (R.V.); teresav@dct.uminho.pt (T.V.)

Abstract: The assessment of sediment contamination is a critical component in understanding the dynamics of potentially toxic elements (PTEs) in aquatic ecosystems, particularly in regions with intensive mining activities. This study focuses on the Rio das Velhas basin, located in the Iron Quadrangle (IQ), one of the most important mining provinces in the world, characterized by extensive anthropogenic pressures and rich geological diversity. A comprehensive evaluation of sediment contamination in this region was conducted, applying multiple univariate and multielement indices, including the contamination factor (CF), enrichment factor (EF), modified contamination degree (mCd), pollution index (PI), modified pollution index (MPI), and ecological risk index (RI). A high sampling density (1 sample per 15 km^2) enabled the creation of geochemical maps and the identification of contamination hotspots. The results revealed that As and Cd are the most concerning elements, with concentrations exceeding regional background levels. While EF provided a more sensitive and comprehensive spatial distribution of contamination, MPI emerged as a robust index for capturing geochemical trends in complex environments. The study also highlighted that over 20% of the samples exceeded guideline values for sediment quality, posing ecological risks. Elevated concentrations of PTEs, particularly As and Cd, raise concerns about their potential mobilization and bioaccumulation, threatening aquatic ecosystems. These findings underscore the urgent need for enhanced monitoring and targeted management strategies in mining-impacted basins. This work not only advances the understanding of sediment contamination dynamics in the IQ but also establishes a methodological framework for evaluating sediment quality in heavily impacted mining regions worldwide.

Keywords: environmental risk assessment; pollution indices; mining impact; geochemical mapping; Velhas River basin

1. Introduction

Sediments are among the most sensitive environmental compartments to contamination by potentially toxic elements (PTEs), mainly due to their high adsorption/desorption capacity. This property is primarily attributed to clay minerals and iron and aluminum oxides–hydroxides in the clay size fraction, which facilitate the accumulation and binding of PTEs [1]. As a result, sediments act as critical geochemical reservoirs for toxic elements and play a key role in their distribution and bioavailability in aquatic ecosystems [2].

Studies show that more than 90% of the total PTE concentration in river environments is typically sequestered in sediments [3,4]. Consequently, sediments serve as a controlling matrix for the mobility, availability, and eventual transfer of these substances to the hydrosphere and/or biota. This dynamic poses potential risks to ecosystems and human health, as many metals and metalloids are known to be toxic and carcinogenic [4–6].

Beyond contamination, sediments serve as natural archives of recent environmental change [7], providing a historical record of pollution and allowing spatial analyses to determine the patterns of contamination distribution patterns in aquatic systems [8–10]. This role is crucial in heavily industrialized or anthropogenically affected regions. For example, studies in mining-affected basins show significantly elevated PTE concentrations in river sediments compared to regional backgrounds or crustal averages [9–12]. This trend is demonstrated by the Velhas River basin, located in Brazil's Iron Quadrangle (IQ), one of the most important mineral provinces in the world.

Sediment Quality Control Guidelines (SQCQs) are widely employed to assess sediment pollution. These guidelines estimate the degree of contamination and the potential adverse effects of PTE on aquatic organisms [13,14]. However, sediment quality indicators can often achieve a more comprehensive understanding of geochemical trends [10,15]. These tools classify sediment quality based on specific PTE concentrations and offer valuable insight into contamination levels. Several methods have been developed to quantify metal enrichment and accumulation in sediments, allowing the classification of contamination levels and their associated impacts [16–19]. Typically, these methods involve comparing observed concentrations with regional background values, as well as average crustal concentrations. Ecological indices further complement this analysis by integrating contamination levels with ecological risks, providing a holistic evaluation of the status of aquatic systems [17,20].

The upper course of the Velhas River basin, within the IQ, has been the subject of numerous geochemical studies due to its extensive mining legacy, present activity, and diverse geological framework [11,21–26]. Despite substantial research efforts, existing studies often lack sufficient sampling density and fail to incorporate comprehensive multielement sediment quality assessments. As a result, the broader environmental and ecological impacts of sediment contamination in this region remain inadequately explored.

To address these limitations, this study undertakes a detailed evaluation of the environmental quality and ecological health of the sediments in the upper course of the Velhas River. A set of ecological indicators—including the contamination factor (CF), enrichment factor (EF), modified contamination degree (mCd), pollution index (PI), and modified pollution index (MPI)—is applied to characterize the contamination levels and risks.

Additionally, high-density geochemical maps are generated, providing spatially explicit information on sediment contamination patterns and intensity. This comprehensive approach aims to enhance the understanding of contamination dynamics in the Iron Quadrangle while establishing a robust framework for monitoring and managing sediment quality in heavily impacted mining regions.

2. Materials and Methods

2.1. Study Area

The upper Velhas River region is located in the Iron Quadrangle, a major mining region in Minas Gerais, Brazil, covering an area of 3200 km^2. This area includes nine towns: Rio Acima, Itabirito, Nova Lima, Ouro Preto, Raposos, Caeté, Belo Horizonte, Sabará, and

Santa Luzia. It supplies water to the metropolitan area of Belo Horizonte, which has more than 5.7 million inhabitants, making it the third-largest urban center in Brazil [27]. The river basin is bordered by Belo Horizonte to the north, Ouro Preto to the south–southeast, Serra da Moeda to the west, and Serra da Piedade to the east (Figure 1). The Velhas River and its tributaries cross areas of varying land use, from forested areas to highly urbanized and mined regions.

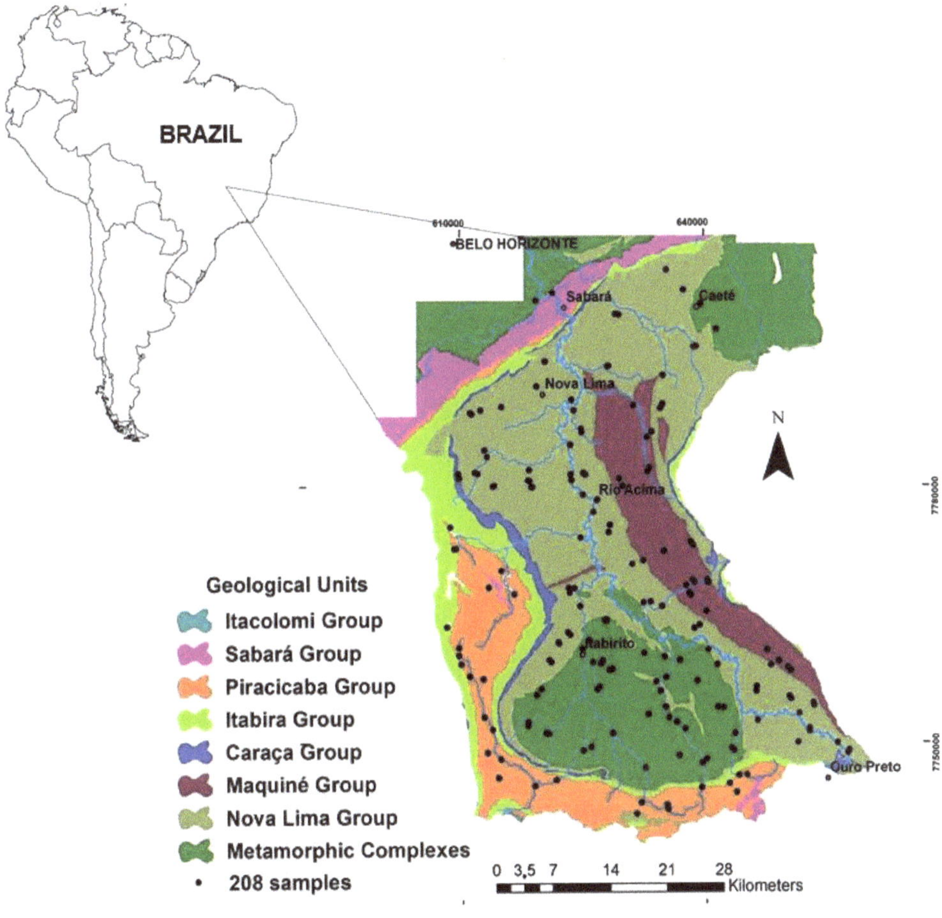

Figure 1. Simplified geological map of the upper Velhas River basin.

The presence of such economically significant deposits is the result of the region's geological diversity, which includes Archean and Proterozoic rocks (Figure 1). The IQ is recognized for its geological diversity and complexity, encompassing a variety of lithostratigraphic units that offer insights into the region's tectonic, sedimentary, and metamorphic history.

The geology of the basin is thought to be organized into four major lithostratigraphic units, from oldest to youngest: (i) Metamorphic complexes, which constitute the crystalline basement, composed of poly-deformed gneissic rocks of tonalitic composition and granite, granodiorite, mafic, and ultramafic intrusions forming the Bonfim, Santa Rita, Caeté, Belo Horizonte, Santa Bárbara, and Bação complexes [28,29]; (ii) the Rio das Velhas supergroup, composed of metasedimentary and metavolcanic rocks [30,31], subdivided into two groups: Nova Lima and Maquiné. The Nova Lima group predominantly features volcano–sedimentary rocks such as carbonate schists, metacherts, banded iron formations, and phyllites. The Maquiné group includes a basal unit of metaconglomerates, overlain by massive and sericitic quartzites, sericite–quartz–schists, and phyllites [31]; (iii) the Minas supergroup is up to 6000 m thick, composed mainly of pelitic and quartzose metasediments, placed discordant above the Rio das Velhas green belt, and subdivided into four groups that record a progression of depositional environments, including the Caraça group (basal metaconglomerates and metarenites, indicating fluvial to shallow marine conditions), Itabira group (composed of chemical sediments, particularly BIFs, marking a period of significant iron deposition), the Piracicaba group (metapelites and additional chemical sedimentary sequences), and the Sabará group (the youngest unit, consisting mainly of terrigenous sediments, with a basal conglomeratic phyllite, indicating a more dynamic sedimentary regime during the late stages of basin evolution) [31]; (iv) the Itacolomi group, which comprises quartzites and metaconglomerates that represent fluvial-deltaic to shallow marine environments; (v) and Tertiary and Quaternary deposits, representing the ongoing sedimentary processes within the region, thus typically associated with modern fluvial systems and weathering profiles [30].

The economy of the region is mainly based on mining activities (iron, gold, limestone, dolomite, bauxite, manganese, and topaz), as well as the steel industry and tourism [32].

Mining activities, particularly for iron and gold ores, have had a significant impact on the basin, contributing to widespread sediment contamination and environmental problems. The combination of mining, urbanization, and industrial activities highlights the need for comprehensive environmental monitoring in this region.

Gold is the primary mineral resource extracted in the Alto Rio das Velhas region, predominantly associated with the volcanic–sedimentary sequences of the Nova Lima group, which hosts Brazil's most significant gold district. The Cuiabá, Morro Velho, and Raposos mines contain the largest gold reserves in the region, with estimated resources of 11.20, 6.60, and 4.30 million tonnes, respectively, and average gold grades of 8.49, 10.00, and 6.27 g per tonne [33]. As by-products of gold extraction, all active mines produce silver, sulfur, and arsenic. Arsenic concentrations range between 3.07 and 8.87 mg/kg, while sulfur concentrations vary between 6.61% and 9.06%. Additionally, gold mineralization is associated with the conglomerate lenses of the Maquiné group. Beyond gold, the region also hosts significant deposits of serpentinites from the Córrego dos Boiadeiros complex and ultramafic rocks from the Quebra-Osso group, both widely used as smelting materials [34].

Regarding iron ore, the region contains high-grade iron ore bodies, typically with iron (Fe) content above 64%. Enriched itabirites, with Fe concentrations ranging from 30% to 60%, are also mined. Major active deposits include the Água Limpa (Itabirito), Vargem Grande (Nova Lima), Fábrica and Fábrica Nova (Ouro Preto and Congonhas), and Casa de Pedra (Congonhas) mines [35]. Over the past 30 years, these deposits have been extensively exploited, with an estimated annual production of 190 million tonnes (Mt). In terms of mining waste, it was reported that, in 2016, approximately 295 Mt of tailings were stored in dams, while 183 Mt of waste was deposited in open-air storage sites [36].

2.2. Sampling Procedures

To achieve a high sampling density, sediments were collected at the mouth of all 3rd-order basins within the region [37], as defined using ArcGIS 10.8 software. The selection process involved overlaying the hypsometric, topographic, and hydrographic maps at a 1: 25,000 scale, provided by the Institute of Water Management (IGAM) and the Company of Mineral Resources Production (CPRM). This approach results in the collection of 208 stream sediment samples, achieving a density of approximately one sample per 15 km^2.

To account for the channel geomorphological variability, nine subsamples were collected along a stretch of 250 to 500 m of each sampling site, representing distinct geomorphological patterns, such as riffles, pools, and transitional areas. Samples were taken from the right margin, left margin, and center of the channel. Margin samples were collected at a depth of 0.50 m from the riverbed. The subsamples were mixed in the field to create a composite sample representative of the site. The homogenized material was quartered to obtain a 500 g composite sample, which was stored in labeled plastic bags and transported to the laboratory under refrigerated conditions to preserve its integrity [38,39].

2.3. Chemical Analyses and Quality Control

The sediment samples were air-dried and sieved, and 1 g of the fraction smaller than 63 μm was subjected to aqua regia digestion (HCl with HNO_3, in a 3:1 ratio) at the Laboratory of Geochemistry of the Federal University of Ouro Preto. The process was begun by adding a small volume of Milli-Q water to form a sediment "pulp". Subsequently, 7.0 mL of concentrated HCl and 2.3 mL of concentrated HNO_3 (Merck p.a.) were added to each sample. The samples were left to react in a fume hood for 16 h, followed by heating on a hot plate at 100 ± 5 °C for two hours to reduce the volume. After cooling, the samples were filtered through 0.45 μm pore filter paper, and the final volume was adjusted to 100 mL with Milli-Q water in a volumetric flask [39].

The digested samples were analyzed using a spectrophotometer of atomic emission with an inductively coupled plasma source (ICP-OES), a brand called Spectro/Model Ciros CCD. The instrument was used to quantify the concentrations of elements, such as Al, As, Ca, Cd, Co, Cr, Cu, Fe, K, Mg, Mn, Na, Ni, Pb, Ti, and Zn.

The digestion with aqua regia is known as pseudo-total digestion, extracting elements from the non-silicate fraction. This method captures trace elements of significant environmental interest, including those associated with oxides, sulfides, clay minerals, and organic matter [40–42]. To ensure robust and reliable data, quality assurance measures included the preparation of a blank for every ten samples and the duplicate analysis of 10% of the samples. Accuracy was verified by comparing results against certified reference material LKSD-01 (CCNRP-Ottawa, Canada). Recovery rates ranged between 93.3% and 102.5% (Table 1), indicating satisfactory precision and reliability.

Table 1. Percentage recovery rates for elements in certified reference material (LKSD-01—CCNRP-Ottawa, Canada).

Element	Measured Concentrations ($mg{\cdot}kg^{-1}$)	Certified Value LKSD 01 ($mg{\cdot}kg^{-1}$)	Recovery Rates (%)
As	29.2 ± 0.8	30	97.3
Cd	1.14 ± 0.02	1.2	95
Cr	11.2 ± 0.3	12	93.3
Cu	45.1 ± 1.2	44	102.5
Pb	83.4 ± 2.6	84	99.3
Zn	330.3 ± 9.3	337	98

2.4. Pollution Quantification Indices

2.4.1. Contamination Factor (CF)

The CF calculation was performed using Equation (1) [17]:

$$CF = \frac{C_{PTE}}{C_{background}} \quad (1)$$

where C_{PTE} represents the concentration of PTE in the sediments of the Velhas River basin, and $C_{background}$ corresponds to the background value of the same element in the basin. The background values adopted as reference were As = 20.6 mg·kg^{-1}; Cd = 1.02 mg·kg^{-1}; Pb = 28.4 mg·kg^{-1}; Cu = 37.9 mg·kg^{-1}; Cr = 151.3 mg·kg^{-1}; Ni = 56.9 mg·kg^{-1}; and Zn = 63.1 mg·kg^{-1} [43]. The adoption of regional reference values is especially recommended in complex geological environments [44].

This index is among the simplest methods for evaluating sediment contamination. It categorizes contamination into four qualitative levels based on the CF value:

- CF < 1: low contamination;
- 1 < CF < 3: moderate contamination;
- 3 < CF < 6: considerable contamination;
- CF > 6: very high contamination.

2.4.2. Enrichment Factor (EF)

This index normalizes PTE concentrations by comparing the region's reference concentrations with a reference metal, called a normalizing element, allowing for more effective comparison, especially in regions with older soils and sediments and with great geological diversity [44]. Although iron (Fe) is commonly used as a normalizing element, its high background value in the Rio Velhas basin (166,000 mg·kg^{-1}) [40]—four times higher than the average value in the upper crust [45]—may lead to the underestimation of FE values [46]. In this study, aluminum (Al) was selected as the normalizing element [46,47] due to its low natural mobility and role as a primary constituent of clay minerals found in river sediments [48–50]. EF for each element was calculated according to Equation (2) [51]:

$$EF = \frac{\frac{Ci}{C_{Al}}}{\frac{Bi}{B_{Al}}} \quad (2)$$

where,

- Ci is the concentration of each element in the sediment sample;
- C_{Al} is the concentration of the normalization element, in this case aluminum (Al), in the same sediment sample;
- Bi is the reference geochemical background value of each element;
- B_{Al} is the reference geochemical background value of the reference element, aluminum (Al).

Five qualitative categories are used to describe the enrichment levels, as summarized in Table 2.

Table 2. Enrichment factor (EF) classes.

Class	Value	Description—Sediment Enrichment
1	EF < 2	No enrichment/natural influence
2	2 < EF < 5	Moderate enrichment
3	5 < EF < 20	Severe enrichment
4	20 < EF < 40	Very severe enrichment
5	EF > 40	Extremely severe enrichment

The enrichment factor (EF) can provide insight into how to differentiate an anthropogenic source from a natural process. EF values close to 1 indicate a crustal source, while values greater than 10 are related to anthropogenic sources and/or processes [51].

2.4.3. Modified Contamination Degree (mCd) and Modified Pollution Index (MPI)

Multielement indices were developed to address the limitations of pollution indices based on single elements. The modified contamination degree (mCd) and pollution index (PI) provide more comprehensive assessments by integrating data from multiple PTEs [17,52]. Building on these, the modified pollution index (MPI) was introduced [53], incorporating the enrichment factor (EF) into its calculation for a more nuanced evaluation of contamination levels. These indices are calculated using Equations (3)–(5). Table 3 presents the quality classification ranges for the multielement indices.

$$\mathrm{mCd} = \frac{\sum_{i=1}^{n} \mathrm{cf}^{i}}{n} \tag{3}$$

$$\mathrm{PI} = \frac{\sqrt{(\mathrm{Cf\ average})^2 + (\mathrm{Cfmax})^2}}{2} \tag{4}$$

$$\mathrm{MPI} = \frac{\sqrt{(\mathrm{Ef\ average})^2 + (\mathrm{Efmax})^2}}{2} \tag{5}$$

Table 3. Thresholds for sediment quality classification for multielement indices.

Class	Sediment Qualification	mCd	PI	MPI
0	Unpolluted	mCd < 1.5	PI < 0.7	MPI < 1
1	Slightly polluted	1.5 < mCd < 2	0.7 < PI < 1	1 < MPI < 2
2	Moderately polluted	2 < mCd < 4	1 < PI < 2	2 < MPI < 3
3	Moderately–heavily polluted	4 < mCd < 8	----	3 < MPI < 5
4	Severely polluted	8 < mCd < 16	2 < PI < 3	5 < MPI < 10
5	Heavily polluted	16 < mCd < 32	PI > 3	MPI > 10
6	Extremely polluted	mCd > 32	---	----

2.4.4. Ecological Risk and Potential Ecological Risk

The potential ecological risk index (RI) evaluates the sensitivity of the biological community at a given site by considering the toxicity and environmental effects of individual elements [15]. This index compares the concentration of the element in the local samples with regional reference values, adjusting for a factor known as the biological toxic response. This factor reflects the relative toxicity of each element and varies as follows: As = 10; Cu = Pb = 5; Zn = 1; Cr = Ni = 2; Cd = 30 [15,17]. The formula for the potential ecological risk of an individual element is expressed in Equation (6).

$$\mathrm{RI} = \sum_{i=1}^{n} \mathrm{Er}^{i} = \sum_{i=1}^{n} \mathrm{Tr}^{i} \times \mathrm{Cf}^{i} \tag{6}$$

where,

- Er^i is the potential ecological risk index of an individual element;
- Tr^i is the biological toxic response factor of an individual element;
- Cf^i is the contamination factor for each single element.

The overall RI aggregates the individual risks to provide a comprehensive assessment of ecological vulnerability. This index categorizes ecological risk into four levels, ranging from low to very high, as summarized in Table 4.

Table 4. Potential and modified ecological risk index classification grades.

Er^i	Ecological Grade (Er^i)	RI	Ecological Grade (RI)
$Er^i < 40$	Low risk	RI < 150	Low risk
$40 < Er^i < 80$	Moderate risk	150 < RI < 300	Moderate risk
$80 < Er^i < 160$	Considerable risk	300 < RI < 600	Considerable risk
$160 < Er^i < 320$	High risk	----	----
$Er^i > 320$	Very high risk	RI > 600	Very high risk

2.4.5. Mapping of Pollution Indices

Maps with several indices were prepared using ArcGIS 10.8 software, adopting the World Geodetic System 1984 (WGS84) datum and the IDW (inverse distance weighted) geostatistical interpolation tool.

3. Results and Discussion

Table 5 presents the statistical parameters for element concentrations in 208 sediment samples from the upper Velhas River basin and compares these concentrations with similar studies conducted in the IQ. Table 6 provides additional context by comparing these concentrations to those observed in rivers worldwide, along with the threshold effect concentration level (TEL) and probable effect level (PEL).

Table 5. Descriptive statistics of sediment samples from the upper Velhas River and comparison of PTEs concentration with similar studies in the IQ, Minas Gerais, Brazil.

Element	Unit	Min	Max	Mean	Median	APA Sul (2005) [24] Min–Max	Pereira et al. (2007) [23] Min–Max	Gonçalves (2010) [25] Min–Max	Mendonça (2012) [26] Min–Max
As	$mg \cdot kg^{-1}$	1.6	1691	32.3	1.6	4–873	2–580	1.6–167	1.6–69
Cd		0.4	14.7	1.13	0.4	1.6–12.2	----	----	----
Cr		6.5	572	115	92.8	44–1077	30–510	197–632	8–198
Cu		0.3	234	27.7	22.3	17–841	20–110	8–97	0.3–180
Mn		41	10,053	1317	756		----	----	----
Ni		1.2	157	36.3	30.5	10–332	----	65–220	0.6–47
Pb		0.4	70.2	22.5	20.2	7–47	----	1–50	1–47
Zn		18.6	181	53.8	48.7	28–175	----	50–131	30–119

All analyzed elements showed skewed distributions, with a median consistently lower than the mean. The disparity between these two values was particularly pronounced for As, with the mean approximately 20 times higher than the median. This highlights the probable influence of localized enrichment in the dataset.

The elements with the highest variability were As, Ni, Cu, and Pb, with maximum concentrations of two to three orders of magnitude greater than their respective minimum values. In contrast, Cd and Zn displayed more uniform concentrations.

The abundance sequence of PTEs in the basin followed the order Cr > Zn > Ni > As > Cu > Pb > Cd, which differs from the typical sequence in the Earth's crust (Cr > Ni > Zn > Cu > Pb > As > Cd). As and Cd stand out, with average concentrations 16 and 11 times higher than their crustal averages, respectively [54]. Pb and Cu were also slightly elevated, with concentrations 1.5 times above their crustal abundances.

Compared to global rivers affected by mining and other anthropogenic activities (Table 6) the Velhas River basin displayed alarming PTEs concentrations. The maximum As concentration (1691 mg/kg) was the highest among all datasets considered, while the average As concentration (32.3 mg/kg) exceeded the maximum values reported for rivers such as the Liaohe, Luanhe, Luan, and Yangtze Rivers (China), the South Plate River (United States), and the Tigris River (Turkey). Similarly, the maximum concentrations of Cd and Cr in the Velhas River sediments surpassed the extreme values observed in most other river systems. This study revealed maximum concentrations higher than those documented in previous research within the same basin, likely due to the higher sampling density. Therefore, this comprehensive sampling approach allowed for a more detailed characterization of localized contamination hotspots.

Table 6. PTEs concentrations in sediment samples from various rivers around the world and TEL (threshold effect concentration level) and PEL (probable effect level) values.

River/Location	As	Cd	Cr	Cu	Pb	Zn
Liaoche River, China [55]	9.9	1.2	35.1	17.8	10.6	50.2
Luanhe River, China [1]	3.4–13.6	0.02–0.24	11.6–76.2	9.6–35.6	23–43.7	12.9–94.7
Yangtze River, China [56]	9.1	0.2	79.1	24.7	23.8	82.9
Tigris River, Turkey [57]	2–85	0.7–3	28.4–163.4	11.2–297	62–392.4	60–247
Danubio River, Europe [58]	8.1–388	1.1–32.9	26.5–556.5	31.1–8088	14.7–542	78–2010
Axios River, Greece [59]	1–40	1–11	39–180	14–93	11–140	42–271
South Plate River, USA [60]	2.8–31	0.1–22	33–71	18–480	19–270	82–3700
Rimac River, Peru [61]	21–1543	0.5–31	24–71	51–796	62–2281	160–8076
Luan River, China [62]	2–12.9	0.03–0.37	28.7–152.7	6.5–179	8.6–38.3	21–25.7
S. Domingos, Portugal [63]	-	1.2	78.9	331.2	3307	168.8
TEL [64]	5.9	0.6	37.3	35.7	35.8	123
PEL [64]	17	3.5	90	197	91.3	315

Comparisons with the Guideline Values for Sediment Quality (GVSQs) [64] revealed important exceedances for several elements. Cr, As, Cu and Cd concentrations exceeded the TEL and PEL values in 57.9%, 33.3%, 32,7%, and 1.6% of the samples, respectively, suggesting a high potential for harmful effects on aquatic organisms interacting with the river sediments. For Pb, none of the samples exceeded the PEL value, but 16.9% of the samples exceeded the TEL. The average concentrations of As and Cr were above the PEL and that of were between the TEL and PEL thresholds, indicating a substantial ecological problem. This pattern highlights the pressing need for targeted mitigation strategies in the study basin, particularly As, Cr, and Cd.

3.1. Sediment Quality Assessment by Single Pollution Indexes

3.1.1. Contamination Factor (CF)

Table 7 summarizes the results of the CF for each PTE. Except for Cr, Pb, and Zn, all the analyzed PTEs showed at least one sampling point classified as having high contamination.

Table 7. Minimum and maximum CF values for PTEs and percentage of samples in each CF class—Velhas River basin.

	As	Cd	Cr	Cu	Pb	Zn
Minimum	0.08	0.4	0.04	0.01	0.01	0.3
Maximum	84.1	14.4	3.8	6.2	2.5	2.9
Percentage of Element Samples by Class						
Absence of Contamination	71.5	70.7	71.6	72.1	72.7	71.6
Moderate Contamination	16.9	22.9	26.8	26.2	27.3	28.4
Considerable contamination	4.9	3.9	1.6	1.1	0.0	0.0
High Contamination	6.7	2.5	0.0	0.6	0.0	0.0

The highest CF values were recorded for As and Cd, with 11.6% of the samples indicating considerable contamination for As and 6.4% for Cd. For other elements, less than 2% of samples exhibited CFs in this contamination range. However, moderate contamination was prevalent in over 25% of the samples for Cu, Cr, Pb, and Cd, underscoring the widespread impact of these elements.

3.1.2. Enrichment Factor (EF)

Table 8 shows the percentages of samples classified in each class for the enrichment factor (EF). Except for Cr and Zn, all the analyzed PTEs show at least one point with very high enrichment.

Table 8. Minimum and maximum EF values for PTEs and percentage of samples in each EF class—Velhas River basin.

	As	Cd	Cr	Cu	Pb	Zn
Minimum	0.03	0.2	0.04	0.02	0.03	0.17
Maximum	226.5	21.9	18.8	24.5	25.6	13.5
Percentage of Element Samples by Class						
No enrichment	68.3	61.2	61.2	61.8	64.5	53.5
Moderate enrichment	18.0	26.8	29.0	27.9	21.9	32.9
Severe enrichment	9.3	10.4	9.8	9.8	13.1	13.6
Very severe enrichment	2.2	1.6	0.0	0.5	0.5	0.0
Extremely severe enrichment	2.2	0.0	0.0	0.0	0.0	0.0

The enrichment factor (EF) values (Table 8) reveal varying levels of enrichment for all analyzed PTEs in the Velhas River basin, with Pb, Zn, As, and Cd showing the most severe enrichments. Notably, Pb and Zn exhibited severe enrichment in more than 13% of the samples, while As and Cd showed 9.3% and 10.4%, respectively. Cr and Cu followed closely, with 9.8% of samples classified under the severe enrichment category.

A maximum EF of 226.5 was observed for As, with 4.4% of samples classified as severe to extremely severe enrichment. Cd, Cu, and Pb also showed enrichment in this range, though at much lower frequencies (1.6% for Cd and 0.5% for Cu and Pb).

The enrichment pattern indicates localized contamination hotspots within the Velhas River basin. Regional background values for EF calculations instead of global crustal averages ensure a more accurate representation of local geochemical conditions [43,47,54]. If global averages were used, the proportion of samples classified as contaminated would significantly increase.

The EF results highlight As as a major contaminant, with its severe enrichment values requiring particular attention. Pb and Zn also demonstrate notable enrichment, suggesting potential anthropogenic contributions, such as the historical mining activities in this basin.

3.2. *Spatial Distribution of Quality Indexes (EF and CF)*

Figure 2a,b illustrate the spatial distributions of univariate sediment quality indices, highlighting locations with CF > 3 and EF > 10. These thresholds correspond to considerable contamination and severe enrichment from anthropogenic sources, respectively. The maps also delineate the distribution of the Nova Lima group lithology, which is known to contribute to higher concentrations of PTEs in the basin.

Samples with EF > 10 and CF > 3 for As and Cd are concentrated in the headwater of the basin, near Ouro Preto. In this region, elevated EF values were also recorded for Cu, Zn, Pb, Cr, and Ni. However, the central portion of the basin, particularly in Rio Acima, Nova Lima, Caeté, and Sabará, presents a higher density of samples with elevated EF and CF values, primarily for As, Cd, and Cu. These areas experience considerable anthropogenic influence and are geologically dominated by the carbonate–quartz–schist of the Nova Lima group.

In Nova Lima, the highest EF and CF values for As were identified in rural communities, such as Honório Bicalho, a finding corroborated by earlier studies [22,24]. Similarly, Cd showed its peak levels (CF = 14.4 and EF = 21.9) in the Caeté region. Both elements demonstrated overlapping spatial distribution, with their highest concentrations recorded in similar areas.

The contamination by As is strongly linked to the presence of Nova Lima group rocks, rich in sulfide deposits, and historical and ongoing gold mining activities. These activities, prominent between the 17th and 19th centuries, have left a legacy of abandoned mines and tailings piles that continue to impact the environment.

Cr and Ni exhibited comparable spatial patterns, with their highest EF values concentrated in the basin's headwaters, where rocks from the Nova Lima group dominate. Cu, being a chalcophile, shares a similar spatial presence with As, reflecting its association with Nova Lima group lithology and its typical Cu-As-Ni-Cr geochemical signature.

Pb presented the highest EF levels in the southwestern part of the basin, particularly in the Mata Porcos Creek watershed. This area is heavily influenced by six active mining operations, which notably contribute to Pb release into the environment. Pb concentrations were correlated with Fe, likely due to its association with magnetite and co-precipitation processes involving Fe compounds [54,65]. Zn, while exhibiting a similar spatial distribution to Pb, has fewer points with EF > 3.

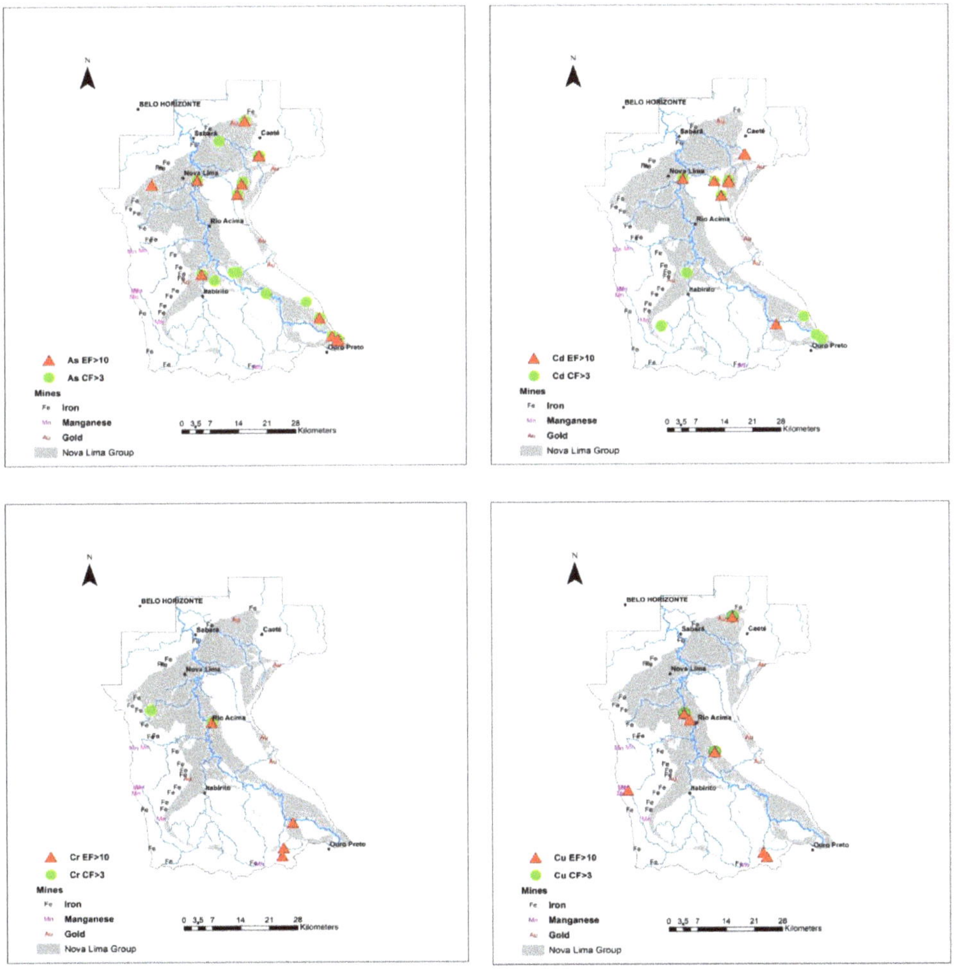

(**a**)

Figure 2. *Cont.*

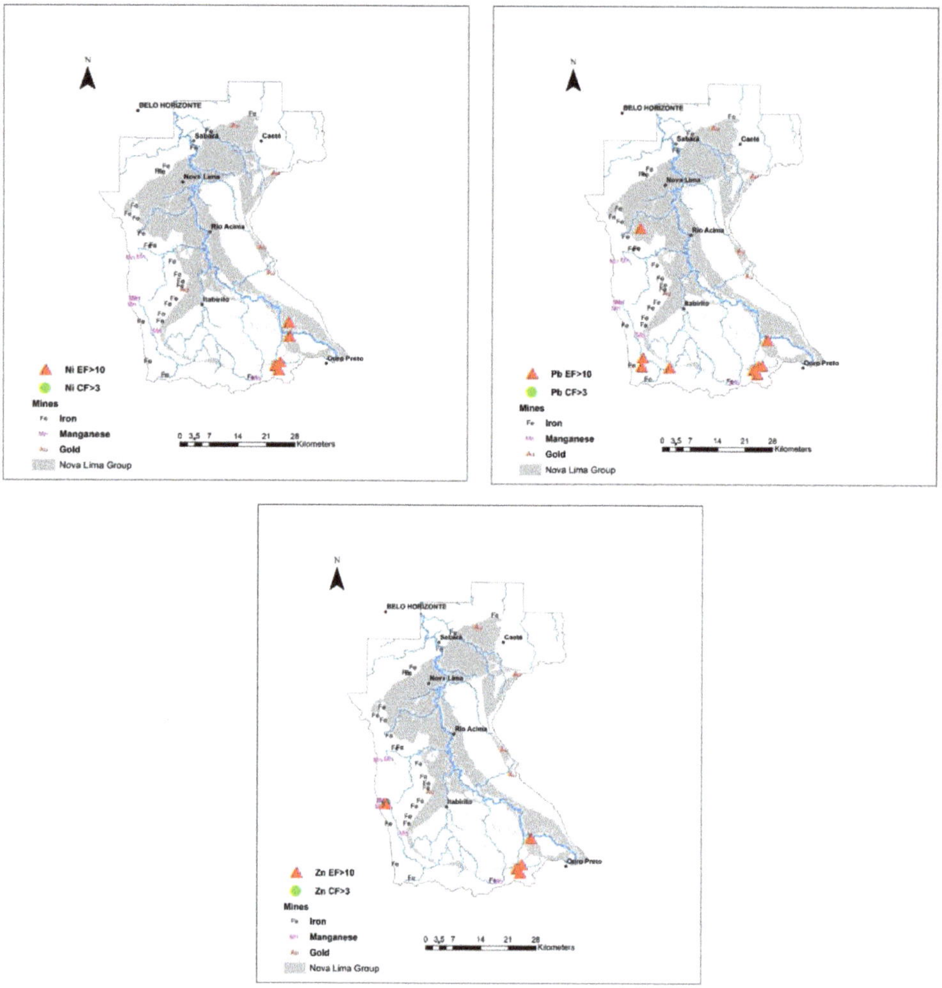

(b)

Figure 2. (**a**) Maps with spatial distributions of occurrences of CF > 3 and EF > 10 of single-element (As, Cd, Cr, Cu); (**b**) Maps with spatial distributions of occurrences of CF > 3 and EF > 10 of single-element (Ni, Pb, and Zn).

Although both CF and EF indicate contamination, CF tends to show higher proportions of uncontaminated samples while identifying more points with very high contamination levels for As, Cr, and Cu. This discrepancy suggests that CF may have lower sensitivity in evaluating contamination, particularly in complex geological and environmental settings. In contrast, EF provides a more detailed spatial representation due to its normalization with Al, which reduces the variability associated with natural geochemical conditions [47,53]. Previous studies [8,15] confirm that EF is better suited for identifying subtle contamination trends, underscoring the limitations of CF in such complex environments.

3.3. Sediment Quality Assessment by Multielemental Pollution Indexes

The multielement indices revealed distinct insights into sediment quality, as shown in Table 9. The modified contamination degree (mCd) ranged from 0.2 to 16.8, while the

PI and MPI demonstrated broader ranges, from 0.3 to 60.7 and 0.3 to 163.3, respectively. This disparity arises from the distinct methodology used to calculate each index, with PI and MPI incorporating maximum values in their formulas, thereby amplifying the final values. According to the mCd classification, 87% of the sediment samples were categorized as non-polluted, a proportion considerably higher than those determined by PI (18.5%) and MPI 1 (21.3%).

Table 9. Minimum and maximum mCD, PI, and MPI values for PTEs and percentage of samples in each mCD, PI, and MPI class—Velhas River basin.

Values	%mCD	%PI	%MPI
Minimum	0.2	0.3	0.3
Maximum	16.8	60.7	163.3
	Percentage of Element Samples by Class		
No Pollution	86.9	18.5	21.3
Slightly Polluted	6	21.8	16.9
Moderate Pollution	4.4	41.7	14.8
Moderate to Strong Pollution	2.2	---	18.6
Severe Pollution	0.50	6.5	17.5
Strong Pollution	0	11.5	10.9
Extreme Pollution	0	---	---

In contrast, when examining polluted samples, mCd classified only 0.5% of samples as severely polluted and did not identify any sample in the strong pollution range. In comparison, PI and MPI identified 18% and 28.4% of the samples, respectively, as having severe or strong pollution. These discrepancies highlight the variability in the sensitivity and classification thresholds among the indices.

The integrated analysis of these indices reveals overlapping areas of substantial pollution. Specifically, points with PI values above five and mCd values higher than four were observed in the Ouro Preto, Itabirito, Nova Lima, and Caeté regions. These locations also correspond to areas with MPI values exceeding 12, suggesting a consistent spatial pattern of contamination.

Among the indices, mCd appears to overestimate the proportion of sediment samples classified as non-polluted. This discrepancy arises from the higher thresholds required for mCd to indicate pollution, which are comparatively lower in PI and MPI [15]. Consequently, PI and MPI identified a larger fraction of polluted samples. Notably, MPI demonstrated greater sensitivity, effectively classifying sediment samples across all pollution categories. This characteristic makes MPI particularly advantageous for evaluating complex geochemical environments, as it incorporates enrichment factors normalized by Al.

In summary, while mCd provides a conservative pollution estimate, MPI is a more robust tool for assessing sediment quality in the Velhas River basin. Its ability to normalize elemental data and classify pollution across a comprehensive range of categories makes it especially valuable for understanding contamination dynamics in complex mining-impacted environments. Table 10 summarizes the ecological risk potential (Er) for each element and the ecological risk index (RI) for all analyzed sediments in the Velhas River basin.

Approximately 96% of the analyzed trace elements (excluding As and Cd) were classified as presenting a low ecological risk at all sampled locations ($Er < 40$). However, As and Cd stood out for their elevated Er values, with 4.4% and 6.6% of the samples, respectively, falling into the categories of considerable to very high contamination.

The RI values varied considerably, ranging from 7.1 to 1092.2, with 3.9% of the samples exhibiting a considerable to high ecological risk. Spatial analysis of RI highlighted clusters of sampling points with high RI values (316 to 446) located in the municipalities of Ouro

Preto, Caeté, and Itabirito. The highest RI value (1092.2) was recorded in Nova Lima, near abandoned gold mining tailings, underscoring the contribution of historical mining activities to contamination in an area with gold mining tailings.

Table 10. Minimum and maximum Er and RI values for PTEs and percentage of samples in each Er and RI class—Velhas River basin.

Values	As	Cd	Cr	Cu	Pb	Zn	RI
Minimum	0.8	1.3	0.1	0.04	0.07	0.3	7.1
Maximum	820.9	433.2	2.2	30.8	12.3	2.0	1092.2
Percentage of Element Samples by Class							
Low Contamination	91.8	83.6	100	100	100	100	92.3
Moderate Contamination	3.8	9.8	0.0	0.0	0.0	0.0	3.8
Considerable Contamination	2.8	2.2	0.0	0.0	0.0	0.0	3.4
High Contamination	1.1	3.8	0.0	0.0	0.0	0.0	0.5
Very High Contamination	0.5	0.6	0.0	0.0	0.0	0.0	0.0

The high Er values for As and Cd predominantly drove the elevated RI values. These two elements showed mean Er values of 17.7 and 33, respectively, markedly higher than those of the other analyzed elements, whose average RI values ranged between 0.90 and 4.12. Cd, recognized as a highly toxic and persistent element, is often found in elevated concentrations in sediments and soils near mining regions [66,67]. Its toxicity is of particular concern due to its potential to bioaccumulate benthic organisms and propagate through the food chain, ultimately impacting aquatic ecosystems and human health [68,69].

3.4. Spatial Distribution of Multielement Quality Indexes

The spatial distribution of the multielement indices (mCd, PI, MPI, and RI) in the Velhas River basin, overlaid with the mining concession areas for Au, Fe, and Mn, reveals critical contamination hotspots (Figure 3).

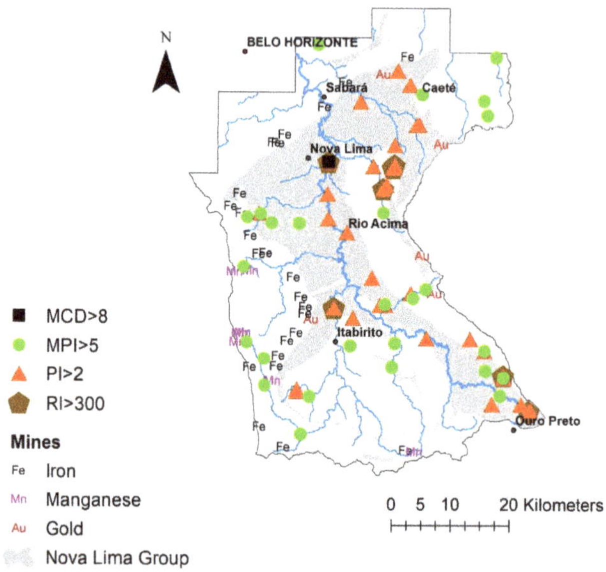

Figure 3. Map showing the distribution of mining concession areas for Fe and/or Mn and Au, suspended mining activities for Fe and Au, and occurrence of samples with mCd > 8, PI > 2, MPI > 5, and RI > 300.

The results indicate a prominent area of contamination extending from the basin's headwaters near Ouro Preto, through Itabirito, Nova Lima, and Caeté. This region has been historically impacted by mining operations, as reflected in the dense clustering of mining concession areas. The association between gold mining concessions and the Nova Lima group lithology is particularly evident, highlighting their contribution to contamination. Elevated values of mCd, PI, and MPI are concentrated in these areas, suggesting a strong relationship between mining activities and pollution levels.

The analysis focused on sites exceeding thresholds indicative of severe pollution, with mCd > 8, PI > 2, MPI > 5, and RI > 300.

4. Conclusions

The present study represents a comprehensive calculation of PTE contamination indices across the Velhas River basin, the most polluted in Minas Gerais, and located in one of the world's most important mining provinces. The high sampling density enabled a detailed evaluation of lithological controls on sediment geochemistry, revealing well-defined patterns in elemental behavior.

Intense contamination of As and Cd was identified across large portions of the basin, with two areas consistently showing high contamination across all analyzed parameters. The first is centered on the municipalities of Ouro Preto and Mariana, while the second encompasses Rio Acima, Nova Lima, Caeté, and Sabará.

The elevated concentrations of these elements in these regions are closely linked to the geology, mainly the carbonate–quartzite–schist rocks of the Nova Lima group, as well as centuries of mining activity since the 18th century and rapid urbanization over the past four decades. While many points near mining areas are monitored according to environmental legislation, this study revealed high uni- and multielement index values in areas influenced by the natural weathering of geological materials. For the first time, detailed geochemical mapping has delineated these risks, which are not currently monitored.

Previous studies in the basin highlighted high levels of As and Cd contamination, but this study advances the understanding by applying contamination indices that account for regional background values, effectively distinguishing natural from anthropogenic contributions. The geochemical maps created in this study provide a more effective environmental assessment tool, pinpointing critical contamination hotspots and identifying rural communities and urban areas exposed to severe environmental risks. These findings underscore the need for relevant authorities to intervene and take protective measures.

In conclusion, this study provides a robust foundation for targeted environmental management in the Velhas River basin, emphasizing the need for comprehensive monitoring and remediation to mitigate the risks posed by PTE contamination.

Author Contributions: Conceptualization, R.V., T.V., M.G.P.L., L.P.L. and H.A.N.J.; Methodology, R.V., M.G.P.L. and L.P.L.; Validation, R.V. and T.V.; Writing—original draft, R.V., T.V., M.G.P.L., L.P.L., H.A.N.J., P.G. and R.F.; Supervision, R.V., T.V. and M.G.P.L. All authors have read and agreed to the published version of the manuscript.

Funding: The work is funded by national funds through the FCT—Fundação para a Ciência e Tecnologia—I.P., in the framework of the UIDB/04683 and UIDP/04683—Instituto de Ciências da Terra programs. The authors highly appreciate the financial support of the institutions CNPq, FAPEMIG, and mainly CAPES for the scholarship Proc. no 10228/13-6.

Data Availability Statement: Data are contained within the article.

Acknowledgments: The authors highly appreciate the valuable comments of the anonymous reviewers.

Conflicts of Interest: The authors declare no conflicts of interest.

References

1. Liu, M.; Chen, J.; Sun, X.; Hu, Z.; Fan, D. Accumulation and transformation of heavy metals in surface sediments from the Yangtze River estuary to the East China Sea shelf. *Environ. Pollut.* **2019**, *245*, 111–121. [CrossRef] [PubMed]
2. Vicente, V.A.S.; Pinto, M.C.; Dinis, P.; Pratas, J.A.M.S. Ecological and Human Health Risk Assessment Based on Stream Sediments from Coastal Oecusse (Timor). *Water* **2024**, *16*, 3020. [CrossRef]
3. Walling, D.E. Human impact on land–ocean sediment transfer by the world's Rivers. *Geomorphology* **2006**, *79*, 192–216. [CrossRef]
4. Zhang, Z.; Lua, Y.; Li, H.; Tua, Y.; Liua, B.; Yang, Z. Assessment of heavy metal contamination, distribution and source identification in the sediments from the Zijiang River. *China Sci. Total Environ.* **2018**, *645*, 235–243. [CrossRef] [PubMed]
5. Fu, F.; Wang, Q. Removal of heavy metal ions from wastewaters: A review. *J. Environ. Manag.* **2011**, *92*, 407–418. [CrossRef]
6. Garrett, R.; Reimann, C.; Smith, D.; Xie, X. From geochemical prospecting to international geochemical mapping: A historical overview. *Geochem. Explor. Environ. Anal.* **2008**, *8*, 205–217. [CrossRef]
7. Cabral Pinto, M.M.; Ferreira da Silva, E.A. Heavy Metals of Santiago Island (Cape Verde) alluvial deposits: Baseline value maps and human health risk assessment. *Int. J. Environ. Res. Public Health* **2019**, *16*, 2. [CrossRef]
8. Çevik, F.; Göksu, M.Z.L.; Derici, O.B.; Fındık, Ö. An assessment of metal pollution in surface sediments of Seyhan dam by using enrichment factor, geoaccumulation index and statistical analyses. *Environ. Monit. Assess.* **2009**, *152*, 309–317. [CrossRef]
9. de Vicq, R.; Matschullat, J.; Leite, M.G.P.; Junior, H.A.N.; Mendonça, F.P.C. Iron Quadrangle stream sediments, Brazil: Geochemical maps and reference values. *Environ. Earth Sci.* **2015**, *74*, 4407–4417. [CrossRef]
10. Jahan, S.; Strezov, V. Comparison of pollution indices for the assessment of heavy metals in the sediments of seaports of NSW, Australia. *Mar. Pollut. Bull.* **2019**, *128*, 295–306. [CrossRef]
11. Matschullat, J.; Borba, R.P.; Deschamps, E.; Figueiredo, B.F.; Ga-Brio, T.; Schwenk, M. Human and environmental contamination in the Quadrilátero Ferrífero, Brazil. *Appl. Geochem.* **2000**, *15*, 181–190. [CrossRef]
12. Leão, L.P. Mapeamento Geoquímico de Sedimentos Fluviais: Diferentes Métodos e suas Aplicações Ambientais. Ph.D. Thesis, Programa de Pós-Graduação em Evolução Crustal e Recursos Naturais, Departamento de Geologia, Escola de Minas, Universidade Federal de Ouro Preto, Ouro Preto, Brazil, 2019.
13. MacDonald, D.D.; Ingersoll, C.G.; Berger, T.A. Development and evaluation of consensus-based sediment quality guidelines for freshwater ecosystems. *Arch. Environ. Contam. Toxicol.* **2000**, *39*, 20–31. [CrossRef] [PubMed]
14. Deckere, E.; De Cooman, W.; Leloup, V.; Meire, P.; Schmitt, C.; von der Ohe, P.C. Development of sediment quality guidelines for freshwater ecosystems. *J. Soils Sediments* **2011**, *11*, 504–517. [CrossRef]
15. Duodu, G.O.; Goonetilleke, A.; Ayoko, G.A. Comparison of pollution indices for the assessment of heavy metal in Brisbane River sediment. *Environ. Pollut.* **2016**, *219*, 1077–1091. [CrossRef]
16. Müller, G. Index of geoaccumulation in the sediments of the Rhine River. *Geojournal* **1969**, *2*, 108–118.
17. Håkanson, L. An ecological risk index for aquatic pollution control. A sedimentological approach. *Water Res.* **1980**, *14*, 975–1001. [CrossRef]
18. Ridgway, J.; Shimmield, G. Estuaries as repositories of historical contamination and their impact on shelf seas. *Estuar. Coast. Shelf Sci.* **2002**, *55*, 903–928. [CrossRef]
19. Abrahim, G.M.S.; Parker, R.J. Assessment of heavy metal enrichment factors and the degree of contamination in marine sediments from Tamaki Estuary, Auckland, New Zealand. *Environ. Monit. Assess.* **2008**, *136*, 227–238. [CrossRef]
20. Hilton, J.; Davison, W.; Ochsenbein, U. A mathematical model for analysis of sediment coke data. *Chem. Geol.* **1985**, *48*, 281–291. [CrossRef]
21. Borba, R.P.; Figueiredo, B.R.; Rawlins, B.; Matschullat, J. Arsenic in Water and Sediment in the Iron Quadrangle, State of Minas Gerais, Brazil. *Appl. Geochem.* **2000**, *15*, 181–190. [CrossRef]
22. Deschamps, E.; Matschullat, J. *Arsênio Antropogênico e Natural: Um Estudo em Regiões do Quadrilátero Ferrífero*; Fundação Estadual do Meio Ambiente—FEAM, Belo Horizonte: Minas Gerais, Brazil, 2007; p. 330.
23. Pereira, J.C.; Guimarães Silva, A.K.; Nalini, H.A., Jr.; Pacheco Silva, E.; De Lena, J.C. Distribuição, fracionamento e mobilidade de elementos traço em sedimentos superficiais. *Quim. Nova* **2007**, *30*, 1249–1255. [CrossRef]
24. APA Sul RMBH. 2005. Projeto de Geoquímica Ambiental, Mapas Geoquímicos Escala 1:225.000. Fernanda G. da Cunha, Gilberto J. Machado—Belo Horizonte: Secretaria Estadual Meio Ambiente e Companhia de Pesquisa de Recursos Minerais. 80p., v. 7: 17 Mapas (Série Programa Informações Básicas para a Gestão Territorial—GATE). Available online: https://rigeo.sgb.gov.br/bitstream/doc/10218/16/rel_apa_sulrmbhv1.pdf (accessed on 10 February 2025).
25. Gonçalves, G.H.T. Avaliação Geoambiental de Bacias Contíguas Situadas na Área de Proteção Ambiental Cachoeira das Andorinhas e Floresta Estadual do Uaimii, Ouro Preto-MG: Diagnóstico e Percepção Ambiental. Master's Thesis, Escola de Minas, Departamento de Geologia, Programa de Pós-graduação em Evolução Crustal e Recursos Naturais, Universidade Federal de Ouro Preto, Preto, Brazil, 2010.

26. Mendonça, F.P.C. Influência da Mineração na Geoquímica das Águas Superficiais e nos Sedimentos no Alto Curso da Bacia do Ribeirão Mata Porcos, Quadrilátero Ferrífero—Minas Gerais. Master's Thesis, Escola de Minas, Departamento de Geologia, Programa de Pós-Graduação em Evolução Crustal e Recursos Naturais, Universidade Federal de Ouro Preto, Preto, Brazil, 2012.
27. Instituto Brasileiro de Geografia e Estatística (IBGE) Censo Demográfico do Brasil. 2021. Available online: https://censo2022.ibge.gov.br/ (accessed on 15 December 2024).
28. Ladeira, E.A.; Roeser HM, P.; Tobschall, H.J. Evolução Petrogenética do Cinturão de Rochas Verdes, Rio das Velhas, Quadrilátero Ferrífero, Minas Gerais. In Proceedings of the Simpósio Geológico, Belo Horizonte, Brazil, 10–14 August 1983; pp. 149–165.
29. Teixeira, W.; Sabaté, P.; Barbosa, J.; Noce, C.M.; Carneiro, M.A. Archean and paleoproterozoic tectonic evolution of the São Francisco Craton. In *Proceedings of International Geological Congress, Rio de Janeiro, 2000. Tectonic Evolution of South America*; Cordani, U.G., Milani, E.J., Thomaz Filho, A., Campos, D.A., Eds.; 31st International Geological Congress, Rio de Janeiro; 2000; Volume 31, pp. 101–137. Available online: https://rigeo.sgb.gov.br/handle/doc/19419 (accessed on 10 February 2025).
30. Dorr, J.N. Physiographic, Stratigraphic, and Structural Development of the Quadrilátero Ferrífero, Minas Gerais. U.S. Geological Survey. 1969. Professional. Paper. Available online: https://pubs.usgs.gov/pp/0641a/report.pdf (accessed on 12 December 2024).
31. Alkmim, F.F.; Marshak, S. Transamazonian Orogeny in the Southern São Francisco Craton Region, Minas Gerais, Brazil: Evidence for Paleoproterozoic collision and collapse in the Quadrilátero Ferrífero. *Precambrian Res.* **1998**, *90*, 29–58. [CrossRef]
32. Nalini, H.A., Jr. *Estudosgeoambientais no Quadrilátero Ferrífero: Mineração e Sustentabilidade*; School of Mines, Department of Geology, Escola de Minas, Universidade Federal de Ouro Preto: Minas Gerais, Brazil, 2009; p. 52.
33. Zucchetti, M.; Baltazar, O.F. Projeto Rio das Velhas: Texto Explicativo do Mapa Geológico Integrado—Escala 1:100.000/Organizado por DNPM/CPRM—Belo Horizonte, 2a Reimpressão. 2000. Available online: https://rigeo.sgb.gov.br/bitstream/doc/22798/1/rio%20da%20velhas_2000.pdf (accessed on 7 February 2025).
34. Vieira, F.W.R.; Oliveira, G.A.I. Geologia do Distrito Aurífero de Nova Lima, Minas Gerais. In *Principais Depósitos Minerais do Brasil*; Schobbenhaus, C., Coelho, C.E.S., Eds.; DNPM/CVRD: Brasília, Brazil, 1988; Volume 3, pp. 377–391.
35. Roeser, H.M.P.; Roeser, P.A. O Quadrilátero Ferrífero-MG, Brasil: Aspectos Sobre sua História, Seus Recursos Minerais e Problemas Ambientais Relacionados. Geonomos. 2010. Available online: https://periodicos.ufmg.br/index.php/revistageonomos/article/view/11598 (accessed on 25 January 2025).
36. Anuário Mineral Brasileiro: Principais Substâncias Metálicas/Agência Nacional de Mineração; Coordenação Técnica de Marina Dalla Costa—Brasília: ANM. 2022; 30p. Available online: https://www.gov.br/anm/pt-br/assuntos/economia-mineral/publicacoes/anuario-mineral/anuario-mineral-brasileiro/PreviaAMB2022.pdf (accessed on 26 January 2025).
37. Bølviken, B.; Bogen, J.; Jartun, M.; Langedal, M.; Ottesen, R.; Volden, T. Overbank sediments: A natural bed blending sampling medium for large-scale geochemical mapping. *Chemom. Intell. Lab. Syst.* **2004**, *74*, 183–199. [CrossRef]
38. Salminen, R.; Tarvainen, T.; Demetriades, A.; Duris, M.; Fordyce, F.M.; Gregorauskiene, V.; Kahelin, H.; Kivisilla, J.; Klaver, G.; Klein, H.; et al. *FOREGS Geochemical Mapping Field Manual. Guide 47*; Geological Survey of Finland: Espoo, Finland, 1998; p. 36.
39. United States Environmental Protection Agency (USEPA) Method 3005A—Acid Digestion of Waters for Total Recoverable or Dissolved Metals for Analysis by FLAA or ICP Spectroscopy. 2001; p. 5. Available online: https://www.epa.gov/sites/default/files/2015-12/documents/3005a.pdf (accessed on 16 December 2024).
40. Calmano, W.; Förstner, U. *Sediments and Toxic Substances: Environmental Effects and Ecotoxicity, 1st ed*; Springer: Berlin/Heidelberg, Germany, 1996; p. 332.
41. Ferreira, A.; Inácio, M.; Morgado, P.; Batista, M.; Ferreira, L.; Pereira, V.; Pinto, M. Low-density geochemical mapping in Portugal. *Appl. Geochem.* **2001**, *16*, 1323–1331. [CrossRef]
42. Albanese, S.S.; De Vivo, B.; Lima, A.; Cicchella, D. Geochemical background and baseline values of toxic elements in stream sediments of Campania region (Italy). *J. Geochem. Explor.* **2006**, *93*, 21–34. [CrossRef]
43. Vicq, R.; Leite, M.G.P.; Leão, L.P.; Júnior, H.A.N.; Valente, T. Geochemical Mapping and Reference Values of Potentially Toxic Elements in a Contaminated Mining Region: Upper Velhas River Basin Stream Sediments, Iron Quadrangle, Brazil. *Minerals* **2023**, *13*, 1545. [CrossRef]
44. Reimann, C.; de Caritat, P. Distinguishing between natural and anthropogenic sources for elements in the environment: Regional geochemical surveys versus enrichment factors. *Sci. Total. Environ.* **2005**, *337*, 91–107. [CrossRef]
45. Wedepohl, K.H. The composition of the continental crust. *Geochim. Cosmochim. Acta* **1995**, *59*, 1217–1232. [CrossRef]
46. Birch, G. An assessment of aluminum and iron in normalisation and enrichment procedures for environmental assessment of marine sediment. *Sci. Total. Environ.* **2020**, *727*, 138123. [CrossRef]
47. Birch, G. A review and critical assessment of sedimentary metal indices used in determining the magnitude of anthropogenic change in coastal environments. *Sci. Total. Environ.* **2023**, *854*, 158129. [CrossRef]
48. Balls, P.; Hull, S.; Miller, B.; Pirie, J.; Proctor, W. Trace metal in Scottish estuarine and coastal sediments. *Mar. Pollut. Bull.* **1997**, *34*, 42–50. [CrossRef]

49. Muñoz-Barbosa, A.; Huerta-Diaz, M.A. Trace metal enrichments in nearshore sediments and accumulation in mussels (Modiolus capax) along the eastern coast of Baja California, Mexico: Environmental status in 1995. *Mar. Pollut. Bull.* **2013**, *77*, 71–81. [CrossRef] [PubMed]
50. Rubio, B.; Nombella, M.A.; Vilas, F. Geochemistry of major and trace elements in sediments of the Ria de Vigo (NWSpain): An assessment of metal pollution. *Mar. Pollut. Bull.* **2000**, *40*, 968–980. [CrossRef]
51. Nolting, R.F.; Ramkema, A.; Everaarts, J.M. The geochemistry of Cu, Cd, Zn, Ni and Pb in sediment cores from the continental slope of the Banc d'Arguin (Mauritania). *Cont. Shelf Res.* **1999**, *19*, 665–691. [CrossRef]
52. Nemerow, N.L. *Stream, Lake, Estuary, and Ocean Pollution*; Van Nostrand Reinhold: New York, NY, USA, 1991.
53. Brady, J.P.; Ayoko, G.A.; Martens, W.N.; Goonetilleke, A. Development of a hybrid pollution index for heavy metals in marine and estuarine sediments. *Environ. Monit. Assess.* **2015**, *187*, 306. [CrossRef]
54. Reimann, C.; Caritat, P. *Chemical Elements in the Environment: Factsheets for the Geochemist and Environmental Scientist*; Springer: Berlin/Heidelberg, Germany, 1998; p. 398.
55. Ke, X.; Gui, S.; Huang, H.; Zhan, H.; Wang, C.; Guo, W. Ecological risk assessment and source identification for heavy metals in surface sediment from the Liaohe River protected area, China. *Chemosphere* **2017**, *175*, 473–481. [CrossRef]
56. Wang, L.; Liang, T. Geochemical fractions of rare earth elements in soil around a mine tailing in Baotou, China. *Sci. Rep.* **2015**, *5*, 12483. [CrossRef]
57. Varol, M. Assessment of heavy metal contamination in sediments of the Tigris River (Turkey) using pollution índices and multivariate statistical techniques. *J. Hazard. Mater.* **2011**, *195*, 355–364. [CrossRef]
58. Woitke, P.; Wellmitz, J.; Helm, D.; Kube, P.; Lepom, P.; Litheraty, P. Analysis and assessment of heavy metal pollution in suspended solids and sediments of the river Danube. *Chemosphere* **2003**, *51*, 633–642. [CrossRef]
59. Karageorgis, A.; Nikolaidis, N.; Karamanos, H.; Skoulikidis, N. Water and sediment quality assessment of the Axios River and its coastal environment. *Cont. Shelf Res.* **2003**, *23*, 1929–1944. [CrossRef]
60. Heiny, J.S.; Tate, C.M. Concentration, distribution, and comparison of selected trace elements in bed sediment and fish tissue in the South Platte River Basin, USA, 1992–1993. *Arch. Environ. Contam. Toxicol.* **1997**, *32*, 246–259. [CrossRef] [PubMed]
61. Mendez, W. *Contamination of Rimac River Basin, Peru, due to Mining Tailings, TRITALWR, MS Dissertation, Environmental Engineering and Sustainable Infrastructure*; The Royal Institute of Technology (KTH): Stockholm, Sweden, 2005.
62. Liu, X.; Jiang, J.; Yan, Y.; Dai, Y.Y.; Deng, B.; Ding, S.; Su, S.; Sun, W.; Li, Z.; Gan, Z. Distribution and risk assessment of metals in water, sediments, and wild fish from Luan River, China. *Chemosphere* **2018**, *196*, 45–52. [CrossRef] [PubMed]
63. Gomes, P.; Valente, T. (In Press). Potential Accumulation of Strategic Elements in Mining Dams—From Remining to Rehabilitation. Special Issue "Congresso Nacional de Geologia". Comunicações Geológicas. Available online: https://portaberta.uminho.pt/display/cv-t-c5f77356040f5772681a5b4b1097ad98 (accessed on 10 February 2025).
64. Environment Canada, Guidelines Division Technical Secretariat of the CCME Task Group on Water Quality Guidelines, Ottawa. 2003. EPC-98E. CCME (Canadian Council of Ministers of the Environment) Protocol for the Derivation of Canadian Sediment Quality Guidelines for the Protection of Aquatic Life. 2003. Available online: https://ccme.ca/en/resources/sediment (accessed on 10 February 2025).
65. Merian, E.; Anke, M.; Ihnat, M.; Stoeppler, M. *Elements and Their Compounds in the Environment: Occurrence, Analysis and Biological Relevance*; WILEY-VCH: Weinheim, Germany, 2004; p. 1773.
66. Zheng-Qi, X.; Shi-Jun, N.; Xian-Guo, T.; Cheng-Jiang, Z. Calculation of heavy metals' toxicity coefficient in the evaluation of potential ecological risk index. *Environ. Sci. Technol.* **2008**, *31*, 112–115.
67. Pan, Y.; Li, H. Investigating heavy metal pollution in mining brownfield and its policy implications: A case study of the Bayan Obo rare Earth mine, Inner Mongolia. *China. Environ. Manag.* **2016**, *57*, 879–893. [CrossRef]
68. Olayinka-Olagunju, J.O.; Dosumu, A.A.; Olatunji-Ojo, A.M. Bioaccumulation of heavy metals in pelagic and benthic fishes of Ogbese River, Ondo State, South-Western Nigeria. *Water Air Soil Pollut.* **2021**, *232*, 44. [CrossRef]
69. Muhammad, S.; Ali, W.; Rehman, I.U. Potentially harmful elements accumulation and health risk assessment of edible fish tissues caught from the Phander Valley, Northern Pakistan. *Biol. Trace Elem. Res.* **2021**, *200*, 4837–4845. [CrossRef]

minerals

MDPI

Article

Geochemistry and Mineralogy of Precipitates from Passive Treatment of Acid Mine Drainage: Implications for Future Management Strategies

Joaquín Delgado [1,*], Olivia Lozano [1], Diana Ayala [2], Domingo Martín [1] and Cinta Barba-Brioso [1]

1 Department of Crystallography, Mineralogy and Agricultural Chemistry, University of Seville, 41071 Seville, Spain; olozano2@us.es (O.L.); dmartin5@us.es (D.M.); cbarba@us.es (C.B.-B.)
2 Faculty of Engineering and Applied Sciences, University of Las Americas, UDLA Park, Quito 170124, Ecuador; diana.ayala.munoz@udla.edu.ec
* Correspondence: jdelgado15@us.es; Tel.: +34-600722205

Abstract: Traditional mining activities in Zaruma-Portovelo (SE Ecuador) have led to high concentrations of pollutants in the Puyango River due to acid mine drainage (AMD) from abandoned waste. Dispersed alkaline substrate (DAS) passive treatment systems have shown efficacy in neutralizing acidity and retaining metals and sulfates in acidic waters, achieving near a 100% retention for Fe, Al, and Cu, over 70% for trace elements, and 25% for $SO_4{}^{2-}$. However, significant solid residues are generated, requiring proper geochemical and mineralogical understanding for management. This study investigates the fractionation of elements in AMD precipitates. Results indicate that Fe^{3+} and Al^{3+} predominantly precipitate as low-crystallinity oxyhydroxysulfate minerals such as schwertmannite [$Fe^{3+}{}_{16}(OHSO_4)_{12-13}O_{16}\cdot 10\text{–}12H_2O$] and jarosite [$KFe^{3+}{}_3(SO_4)_2(OH)_6$], which retain elements like As, Cr, Cu, Pb, and Zn through adsorption and co-precipitation processes. Sulfate removal occurs via salts like coquimbite [$AlFe_3(SO_4)_6(H_2O)_{12}\cdot 6H_2O$] and gypsum [$CaSO_4\cdot 2H_2O$]. Divalent metals are primarily removed through carbonate and bicarbonate phases, with minerals such as azurite [$Cu(OH)_2\cdot 2CuCO_3$], malachite [$Cu_2(CO_3)(OH)_2$], rhodochrosite [$MnCO_3$], and calcite [$CaCO_3$]. Despite the effectiveness of DAS, leachates from the precipitates exceed regulatory thresholds for aquatic life protection, classifying them as hazardous and posing environmental risks. However, these residues offer opportunities for the recovery of valuable metals.

Keywords: AMD treatment; divalent metal carbonate; jarosite–alunite group; waste management

1. Introduction

Acid mine drainage (AMD) is referred to as a water course with both high acidity and concentrations of metal(loid)s and sulfates [1] generated by the oxidative dissolution of metal sulfides in abandoned mining areas and uncontrolled tailing dumps. It is a major cause of hydrographic basin contamination [2]. AMD poses a worldwide concern in mining, creating extreme environmental impacts to be responsibly managed [3,4]. In this sense, several passive remediation techniques, including dispersed alkaline substrates (DAS) [4–7], have been designed to reduce their environmental impact. Passive treatments are currently growing for the restoration of abandoned mine areas since they do not require the continuous addition of reagents and need occasional maintenance with little energy consumption [8,9], resulting in an economical option for AMD remediation. These technologies, especially DAS, have been successfully applied in medium- and high-acidity waters with high metal loads, and they are widely developed despite problems of saturation or a decrease in effectiveness of the systems in the medium term [10–14].

While a significant amount of research has been dedicated to the occurrence and treatment of AMD, there has been limited investigation into the assessment and proper management of post-treatment residues of AMD. These treatments generate solid metal-rich residues or sludge (AMDp, acid mine drainage precipitates) that are also qualified as wastes [4,15]. These present high variability in chemical composition and physical properties depending on the AMD treated [16], making it difficult to control and predict their stability and environmental behavior. In fact, they could generate drainage themselves with negative environmental impacts if their storage conditions are non-stabilized. This demonstrates the importance of the proper management of the solid waste generated during the neutralization of AMD-polluted sources [17]. According to the literature, the recommendations to store the residues are very varied, but many authors agree that the residues are unstable and should be stored in a dry environment [15]. According to regulations, any contact with acidic or alkaline water and DAS residues must be avoided.

Standardized leaching tests are commonly used to characterize the potential release of contaminants from treatment wastes, such as the pH control leaching procedure (TCLP) or sequential extraction procedure (SEP), used increasingly by more researchers to understand the main processes in the mine waste environment: sulfide oxidation and the retention of mobilized elements by secondary minerals via sorption and/or coprecipitation processes [18–20]. In this sense, effective extractants and detailed studies of environmental conditions have been tested [21,22]. Additionally, according to Jouini et al. [17], metal bonds in Fe-rich AMD have still not been investigated enough in multi-step passive treatment methods, despite their importance in management strategies. These studies could allow the evaluation of various residues such as hazardous waste [15–23] as well as the establishment of the potential valorization of such metals from mining residues [12].

In this context, the aims of the present study were the following: (1) to gain a better understanding of trace metal cycles in environmental systems affected by AMD, with this objective being based on the detailed geochemical characterization of trace elements mobilized from the generated solid wastes (AMDp) in a laboratory passive multi-step treatment for Fe-rich AMD; and (2) to investigate mineral transformation and its influence on element fixation as a basis of future studies on waste reutilization and better management strategies for appropriate storage.

2. Study Area

Mining activities in southern Ecuador have impacted the Puyango River basin (Figure 1), a key watershed extending through southern Ecuador into northern Peru before flowing into the Pacific Ocean. The Puyango River flows through the Zaruma-Portovelo gold mining site (ZPMS), located atop the Paleozoic El Oro metamorphic complex and overlaid with Cretaceous volcanic rocks intruded by Cenozoic igneous complexes. Primary ores in the area contain gold and silver in quartz veins, along with sulfides like arsenopyrite [FeAsS], chalcopyrite [$CuFeS_2$], galena [PbS], and sphalerite [ZnS].

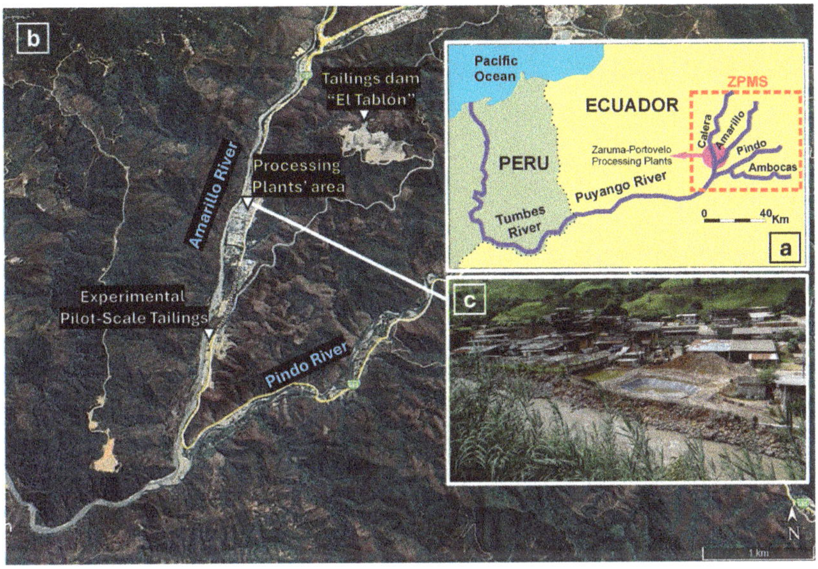

Figure 1. A geographic map of the Zaruma-Portovelo gold mining site (ZPMS). (**a**) The location of the AMD creek treated in the experimental columns, (**b**) details of the Zaruma-Portovelo processing plant, and (**c**) a photograph showing the wastes of the processing plants around the Amarillo River.

In 2008, the Peruvian government raised concerns about contamination in the lower Puyango basin, referencing the 1971 Binational Convention. Since then, Ecuadorian authorities have undertaken remediation efforts, including constructing a community tailings facility. Nevertheless, substantial, unclassified volumes of mineralogically diverse mine waste continue to pose AMD risks, necessitating ongoing mitigation or remediation [24]. Also, approximately 100 small-scale processing plants in the Zaruma-Portovelo site [25] store tailings in dams near rivers or discharge them directly, promoting oxidation and generating potential AMD sources [26]. Several studies have documented the environmental effects of these mining operations on the Amarillo and Calera Rivers within the Puyango basin [27–29]. Furthermore, the climatic conditions of the basin (dry megathermal), characterized by high temperatures and precipitation variability (24 °C annual average, 500–2000 mm rainfall), promotes the oxidation of unmanaged waste dispersed along the riverbanks.

3. Materials and Methods

3.1. Experimental Design and Sampling

A reactive substrate system (Figure 1) combining calcium and magnesium DAS according to the conceptual model developed by Rötting and others [14,30] was implemented at the National Research Institute of Geology, Mining, and Metallurgy (INIGEMM) laboratories in Quito, Ecuador. AMD sources located in the ZPMS, selected for treatment, exhibited mean pH values of 2.7 and an electrical conductivity of 4.7 mS/cm, both indicative of high acidity (4361 mg/L eq as $CaCO_3$ equivalents) and elevated concentrations of Al, Cd, Cu, Fe, Mn, Pb, Zn, and sulfates. The system operated at both low and high hydraulic flow rates for 8 months, proving high efficiency in increasing water's pH and achieving approximately an 80% retention of Fe, Al, Mn, and trace elements. A complete description of the design, results, and hydrochemical model can be found in Delgado et al. [2,24].

The dissolution of limestone sand promotes protons' decline in solution, which increases the pH and induces the formation of an "ochre" horizon in the first few centimeters of the column due to chemical shock, followed by a white horizon that may correspond to Fe and Al precipitates, respectively [2]. DAS-Ca alone is proved to be insufficient to remediate high concentrations of divalent metals such as Zn, Mn, Ni, and Cd derived from hydrochemical conditions [31–33]. For this reason, Rötting and collaborators [34] conducted laboratory experiments demonstrating that caustic magnesia [MgO] could be used to remove high concentrations of divalent metals from solution.

To evaluate the effectiveness of the acid water treatment, solids (AMDp) generated in the columns at various depths were sampled after the treatment. Five samples were taken from the DAS-Ca column located at depths of 3.5, 7.5, 8.5, 10, and 15 cm, while four samples were collected from the DAS-Mg column: two samples coinciding with superficial green–white precipitates and the other two at 1 and 12 cm depths (Figure 1).

3.2. Chemical Study

Total element concentrations were determined as the sum of 5 steps of the selected SEP [35]. Sequential extractions are widely used for exploration purposes and to study element speciation in natural environments (i.e., [36–39]) and gradually more in mine waste's environments to understand trace metal mobility. In this study, changes were implemented to the ratio of the sample to the extractant volume and to experimental settings for high concentrations of iron oxides and oxyhydroxysulfates. Due to the presumable nature of the samples, this study was performed according to Caraballo et al. [35] and the schematic SE steps are summarized in Table 1.

Table 1. Five-step sequential extraction procedure scheme. * Mineral phases dissolved in each step according to Caraballo et al. [35].

Sequential Extraction Step	* Dissolved Phases	Elements Released into Solution	Possible Dissolved Phases	Preferentially Dissolved Minerals
(1) Water soluble fraction: 200 mg of sample and 20 mL deionizied water, shake for 12 h at room temperature.	Gypsum	Ca and SO_4^{2-}; Fe, K, Mg, Mn, Zn and Cd-Co	Gypsum, secondary sulfates and other salts, hydroxides. Fe-amorphous	Secondary sulfates and other salts
(2) Sorbed and exchangeable fraction: 20 mL of 1 M NH-acetate (4.5 pH buffer). shake for 1 h at RT.	Calcite	Ca and adsorbed elements; Fe and Al- K, As, Cu, Mn, Zn	Metal-divalent carbonate type, including calcite; Fe-amorphous or very poorly ordered Fe(III) precipitates	Calcite and some clay minerals
(3) Poorly ordered Fe(III) oxyhydroxides and oxyhydroxysulfates: 20 mL of 0.2 M NHy-oxalate (3 pH buffer). 30 min shake in darkness and at RT.	Schwertmannite, hydrobasaluminite and gibbsite	Fe, Al and SO_4^{2-} and trace associated Residual Ca	schwertmannite; jarosite-alunite group	Mainly schwertmannite and 2-line ferrihydrite
(4) Highly ordered Fe (III) hydroxides and oxides: 20 mL of 0.2 M NH4-oxalate (3 pH buffer), 80 °C water bath for 1h.	Goethite	K, Fe, Al and SO_4^{2-}; also As, Cu, Zn, Mn, Mg Residual Ca	jarosite-alunite group; residual Ca-Mg-Mn oxides	Goethite, jarosite, 6-line ferrihydrite and hematite
(5) Residue digestion: 3 mL of HNO_3 + 7.5 mL of HF + 2.5 mL of $HClO_4$	Residue (wood chips)	Organic elements	Wood and residual silicates and clay assimilated	Silicates

Major and trace elements were determined in <0.2 μm extracted solutions by inductively coupled plasma–atomic emission spectroscopy (ICP-OES, model Optima 8300, Perkin Elmer, Waltham, MA, USA). The main key elements (Ag, Al, K, Mg, Mn, As, Be, Ca, Cd, Co, Cr, Cu, Mo, Na, Nb, Ni, Pb, Sn, Sr, and Zn) were reported. Calibration and accuracy controls were certified with ICP Multi-Element standard solutions from AccuStandard, SQS-01-1CRM, and SQS-02-R1-1CRM, according to EPA 200.7 and EPA 6010. Certified reference material (No. 1244) and inter-laboratory tests for trace metals (No. 586) (CRM-European Environmental Production) were measured as quality controls. Each step of the SEP was analyzed in triplicate, calculating the relative percentual difference (Equation (1)) among the replicates and the mean values. Most values were below the admissible value (5%), except for Ag, which stood at 10%.

$$\%RPD = (S - D)/((S + D)/2) \times 100 \tag{1}$$

where S = the determined value of the samples and D = the value of the duplicates.

The internal recovery of this SEP was assessed by comparing the sum of each metal in the five fractions with its acid-digested concentration, in percentages. The recovery rates for Al, Zn, Cu, Fe, and Mn ranged from 102.2% to 97.3%, 91.6% to 105.8%, 90.8% to 109.0%, 90.0% to 109.0%, and 91.4% to 107.4%, respectively. The recovery percentages for Cu and Zn in the Mg-1 sample were high, probably due to inherent errors in the SEP method, since through acid attack for the extraction of the sorbed and exchangeable fraction (F2), the elements associated with poorly ordered Fe^{3+} or hydroxide precipitates could also be released, adding excess values to the measurement.

3.3. Mineralogical Characterization

X-ray diffraction (XRD) measurements were performed with a Panalytical Empyrean diffractometer (Malvern Panalytical, Malvern, UK) with Cu-Kα radiation at 40 kV and 20 mA. A step size of 0.008° 2θ was used in the range of 5–80°, and a step time of 0.5 s/step was set. The software used for mineral identification were EVA v16.0 (Bruker-AXS) and Match! v3.15 (Crystal Impact), and the BGMN/Profex v4.8 for Rietveld refinement was employed. Structure models for Rietveld analysis were selected based on the Powder Diffraction File and from the Crystallography Open Database. The crystallinity order of the analyzed samples was calculated using the software Diffract.Eva v4.1 (Bruker-AXS). This crystallinity order represented the proportion of crystalline material in the samples, as indicated by the total integrated areas of the full pattern. In samples with higher crystallinity, the control refinement fit values (such as discrepancy values, χ^2, and agreement factors R, Rwp, Rexp and RB) were satisfactory. However, in samples with lower crystallinity and poor-quality data, these values were higher. This was why, due to the low crystallinity of some minerals, the percentages of the mineral phases were semi-quantitative.

Selected samples were studied using a Tescan-MIRA 3 (Kohoutovice, Brno, Czech Republic) scanning electron microscope (SEM) operating with an accelerating voltage of 25 kV. Compositional information was extracted by back-scattered electron (BSE) imaging and energy-dispersive X-ray (EDX) microanalysis (Link ISIS system). The samples were coated with a sputtering Quorum-Q150R ES with gold–platinum.

4. Results and Discussion

4.1. Metal Partition from AMD to Solid Phases

During AMD treatment, the gradual increase in the pH along the DAS-Ca column led to the development of an "ochre horizon" in the upper centimeters of the solid residue, followed by the formation of a "white horizon" (Figure 2a). The chemical conditions developed along the column triggered Fe depletion from the solution, probably forming low-crystalline oxyhydroxides or oxy-hydroxysulfates such as H-jarosite [$H_3O)Fe_3(SO_4)_2(OH)_6$], K-jarosite [$KFe_3(SO_4)_2(OH)_6$], schwertmannite [$Fe_8O_8(OH)_6(SO_4)$], ferrihydrite [$Fe(OH)_3$], hematite [α-Fe_2O_3], goethite [α-FeOOH], lepidocrocite [γ-FeOOH], and maghemite [γ-Fe_2O_3] [2].

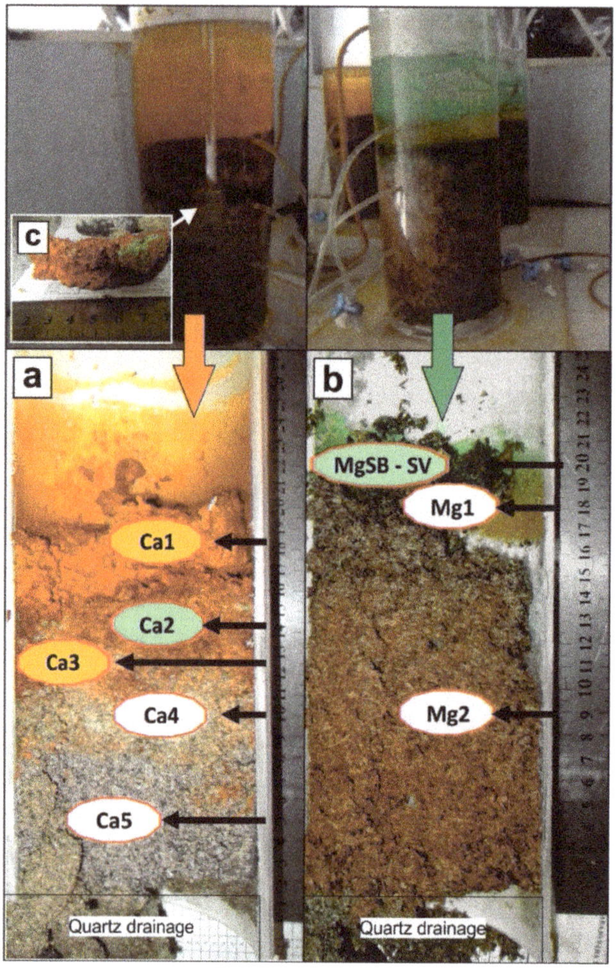

Figure 2. A passive treatment system based on the DAS concept [2]. (**a**) Photographs of the sampling points in the DAS-Ca column, (**b**) DAS-Mg column, taken after 8 months of treatment, and (**c**) detail of the inner Ca2 layer showing green precipitates.

Indeed, the "ochre horizon" (Figure 3a) exhibited the highest concentrations of Fe, with retention rates exceeding 90% (Ca1; 336 g/kg) and 80% (Ca3; 149 g/kg) of the total metal content (Figure 3a; Table 2), as confirmed by the corresponding visual observations.

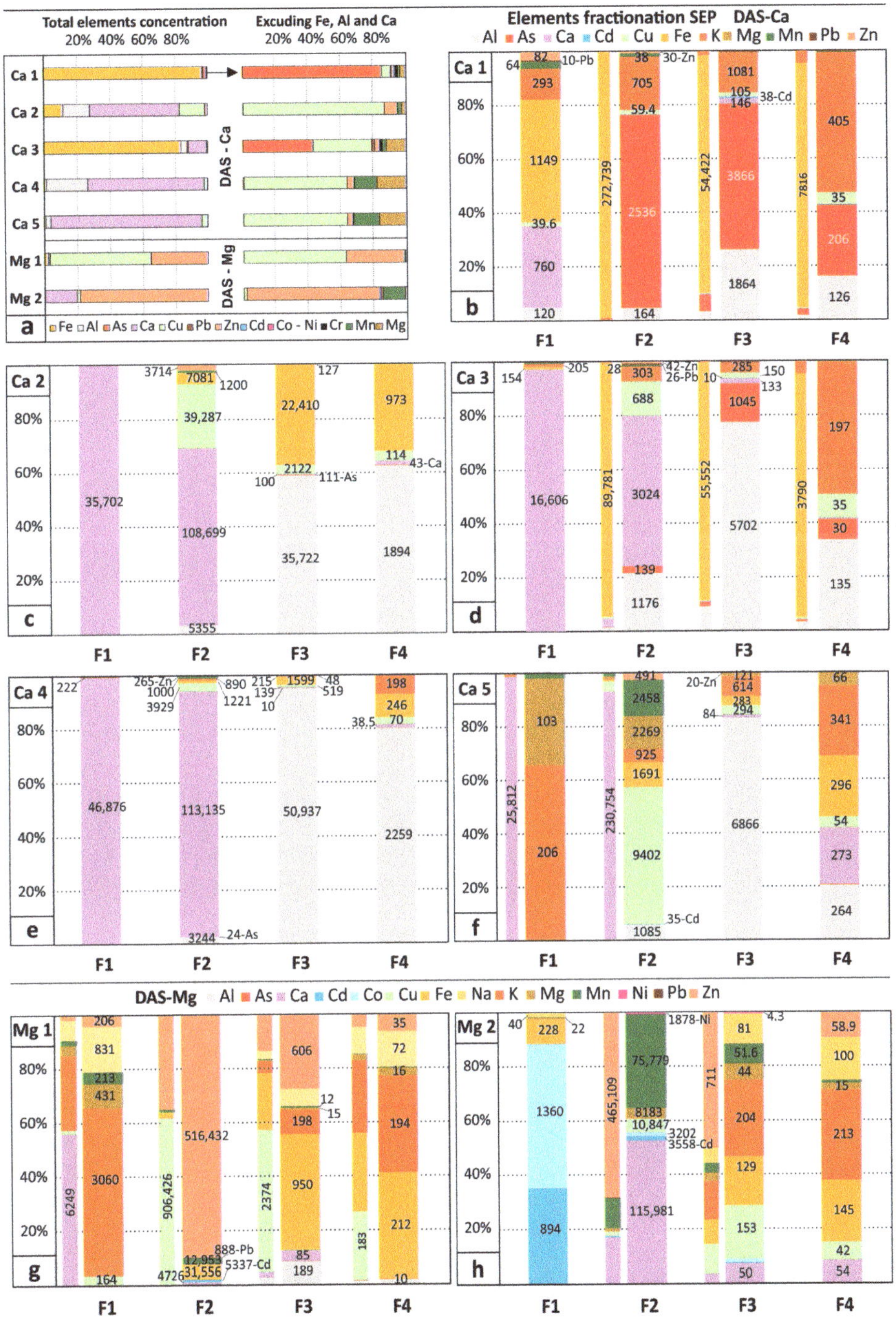

Figure 3. (**a**) The total concentration in each sample (% respect to total) of the main elements involved in the AMD neutralization process. (**b**–**h**) Sequential extraction data showing the percentage of the total extracted data in each fraction. Where necessary, elements with higher concentrations were suppressed (thin bars) to improve the visualization of trace elements. Concentrations are expressed in mg/kg.

Table 2. A summary of the solid wastes' pseudototal concentration showing the elements frequently associated with AMD environments in the Zaruma-Portovelo mine district. <d.l., below the detection limit.

		Das—Ca					Das—Mg	
		Ca1	Ca2	Ca3	Ca4	Ca5	Mg1	Mg2
Ag	mg/kg	9.74	23.23	16.33	36.15	100.76	11.82	60.0
Fe	mg/kg	336,422	30,527	149,324	3096	2324	32,842	727
Al	mg/kg	2497	43,040	7091	56,474	8288	1576	162
As	mg/kg	6616	127	1215	37.6	11.2	43.0	31.1
Ca	mg/kg	1085	145,278	19,878	160,770	257,355	11,060	116,385
Cd	mg/kg	39.6	0.71	9.75	<d.l.	34.7	5347	4453
Co	mg/kg	3.49	<d.l.	2.41	<d.l.	<d.l.	773	4565
Cr	mg/kg	118	21.4	28.6	14.0	19.2	20.2	68.4
Cu	mg/kg	423	41,688	1015	4632	9842	909,599	11,385
K	mg/kg	2626	1227	1114	1672	3603	4023	943
Mg	mg/kg	290	1305	347	1330	2597	916	8253
Mn	mg/kg	107	1223	74	1024	2471	13,181	75,841
Mo	mg/kg	<d.l.	17.7	1.27	25.3	24.9	0.86	4.87
Na	mg/kg	745	554	492	579	1174	1257	460
Ni	mg/kg	4.01	12.4	2.29	1.49	<d.l.	486	1883
Pb	mg/kg	39.0	1.80	47.3	<d.l.	<d.l.	888	41.2
Sb	mg/kg	112	62.3	70.2	68.1	68.6	92.3	112
Sr	mg/kg	21.1	574	36	588	1130	69.8	1364
Zn	mg/kg	139	3860	86	320	517	517,344	465,916

The SEP data of the surficial sample, Ca1, indicated the presence of high concentrations of Ca (760 mg/kg). In addition, Ca increased at greater depths, so Ca3 was present in 17 g/kg, associated with the labile fraction of the residues (Figure 3b and Figure 3d, respectively). Furthermore, in the exchangeable fraction (F2), the release of low-crystalline iron hydroxide seemed evident, along with weakly adsorbed trace elements onto these compounds, as previously described by other authors [40]. The presence of iron and trace metals in the F2 of sample Ca1 (Figure 3b) endorsed this interpretation.

The movement of trace elements from solution to solid Fe-Al phases involves adsorption and/or coprecipitation mechanisms. Research on ochre-like deposits rich in iron in acidic conditions indicates that jarosite and schwertmannite may have the greatest amounts of As and Cr [41], as well as other metals like Cu, Pb, and Zn [42]. Furthermore, ferrihydrite appears to be related to high concentrations of Pb and relatively high contents of Zn and Ni [43]. In this regard, the main concentration of As was related to the higher concentrations of Fe and K in the Ca1 and Ca3 samples (Figure 3b,d), presumably associated with schwertmannite and jarosite. Contrary to expectations, although significant concentrations of As, Cu, and Pb associated with fractions F3 and F4 (involving Fe(III) oxyhydroxides and oxyhydroxysulfates) were leached, concentrations of up to 2536 mg/kg of As (Ca1 sample) were extracted from F2, the fraction associated with carbonates (Figure 3b). This could be attributed to the dissolution of amorphous or disordered Fe(III) precipitates, along with the release of adsorbed elements such as As, Cu, Mn, and Zn, induced by the exposure to acetic acid at a pH of 4.5 within the F2 fraction. Additionally, excluding the samples from DAS-Mg, the trace metals Cu, Cd, Pb, and Zn showed considerable concentrations associated with this section of the iron profile (up to 105, 38.1, 27.7, and 42.2 mg/kg, respectively), confirming the link between Fe-rich mineral phases and trace elements' retention.

The deeper 'white horizon' likely corresponded to Al precipitates formed when the pH increased, as described by Rötting et al. [14]. This was confirmed by the higher Al concentration throughout the column, with the highest value (56 g/kg) present in sample

Ca4 (Table 2). According to Macías et al. [15], aluminum may be trapped as hydroxides or hydroxysulfates in white precipitates, which were mainly concentrated in sample Ca4, as shown in Figure 2a. The high Al concentrations extracted in F3 and F4 confirmed the dissolution of Al-rich sulfate phases such as alunite [$KAl_3(SO_4)_2(OH)_6$], basaluminite [$Al_4(SO_4)(OH)_{10}\cdot 5(H_2O)$], jurbanite [$Al(SO_4)(OH)\cdot 5(H_2O)$], or Al-hydroxides [$Al(OH)_{3(am)}$, gibbsite $Al(OH)_3$, boehmite γ-AlO(OH), and diaspore AlO(OH)] that would be important as the pH rises above 6 [25,44,45]. Furthermore, the presence of alunite could be elucidated by considering the relative concentration of potassium in this profile section. These precipitates frequently contained gypsum (confirmed by the high concentration of Ca associated with the F1 fraction, Figure 3c,e). Furthermore, based on the hydrochemical model developed by Delgado et al. [2], the dominant aqueous species ($AlSO_4^+$, $Al(OH)_2^+$, and $AlOH^{2+}$) and the pH range between 5.8 and 6.4 in these sections of the profile were consistent with the precipitation of Al phases.

Elevated concentrations of Al, Cu, and Ca and low Fe concentrations were measured in sample Ca2 (30 g/kg). This can be attributed to the significant dissolution of the calcium substrate and unexpected local alkaline conditions indicated by the presence of green precipitates in the "ochre horizon" of DAS-Ca (Figure 2c). AMDp are commonly associated with the precipitation of divalent metal hydroxides, as described by Cortina et al. [32], when the pH reaches neutral-to-alkaline conditions. This supports the high concentrations of Al (43 g/kg), Cu (42 g/kg), Mg (1300 mg/kg), Mn (1223 mg/kg), and Zn (3860 mg/kg) at this depth (Ca2) of the profile.

It is noted that the neutral conditions obtained in the final section of the DAS-Ca are typically insufficient for the formation of divalent metal hydroxides [32]. According to Baes and Mesler [46], divalent metals are most effectively retained by hydroxides at a pH between 8 and 10. Based on these premises, the presence of high concentration of Ca in the final section of DAS-Ca (F2 of samples Ca4 and Ca5, Figure 3) added to the addition of alkalinity by the magnesia dissolved in DAS-Mg (supported by the presence of Mg in F1 of Ca5) appear to provide an ideal scenario for divalent metal retention. However, the control over the fractionation of divalent metals appeared to be mainly exerted by carbonate phases [2]. These conditions were obtained towards the end of DAS-Ca (sample Ca5) and were accompanied by significant concentrations of Ca, Mg, Cu, or Fe (until 231, 2.3, 9.4, and 1.7 g/kg, respectively) associated with the exchangeable phase (Figure 3f). Additionally, the fractionation of divalent elements in this zone of the profile seemed to be somehow influenced by the availability of aluminum, with these phases extracted in F3 of the SEP, as previously suggested by Caraballo et al. [35].

The samples Mg1 and Mg2 exhibited the highest concentrations of Cd, Co, Cu, Ni, Pb, and Zn (Table 2 and Figure 3g,h). The DAS-Mg hydrochemical model suggested that primary carbonate species in water quickly disappeared upon entering the DAS-Mg column due to the formation of carbonate minerals like malachite, azurite, and, to a lesser extent, smithsonite [$ZnCO_3\cdot 1H_2O$] [2]. These findings align with the results observed in the experimental columns, where the highest Cu and Zn retentions (900 and 516 g/kg, respectively) were linked to sample Mg1 (exchangeable fraction, Figure 3g). This was supported by a bluish-green color change in the precipitates at the DAS-Mg surface. Additionally, most Zn partitioning from the solution to the solid phase occurred in the final section of DAS-Mg, with Mg1 (around 40%) and Mg2 (80%) being the most significant (Figure 3a). In calcareous settings with high levels of divalent metals (like the DAS-Mg column), the amount of these metals in water (including Cd and Mn) is typically controlled by the precipitation of solid metal–carbonate compounds [47–49]. In fact, Zn and Mn showed a clear association with the carbonate fraction of the AMDp. While Cd showed some association with the water-soluble fraction along with Co (894 and

1360 mg/kg, respectively), its higher concentrations (3558 and 3202 mg/kg) were linked to the F2 fraction (Figure 3h). Other divalent metals, such as Ni and Cr, exhibited similar behavior, showing higher concentrations in the F2 fraction, particularly in the final stage of the treatment (e.g., Ni reached 1878 mg/kg in F2, Figure 3h).

4.2. DAS Mineralogical Assemblages

In the diffraction pattern of the surficial sample of DAS-Ca (Ca1), the presence of amorphous and low-crystalline phases required background subtraction so that the peaks of some minerals became evident (Figure 4a).

In the "iron profile", the precipitation of the first oxyhydroxysulfates, such as schwertmannite and jarosite, could trap toxic elements like As, Cr, Cu, Pb, and Zn from the solution [41,42]. According to Hudson-Edwards et al. [50], these are the controlling phases at low pH values. Observations by SEM-EDS showed a clear association between S and Al-Fe and the dissemination of As and Pb trapped in this column's section (Figure 5a). Other sulfate species could play an important role here, depending on the availability of certain elements in the solution. The high proportion of Fe^{3+} in solution could lead to coquimbite ($Fe_3^{3+}(SO_4)_3 \cdot 9H_2O$) formation, and if the water contained more than 1500 mg/L sulfate, the presence of gypsum was expected due to the dissolution of the reactive calcium across the substrate. Based on phosphate concentrations and the hydrochemical model reported by Delgado et al. [2,26], brushite [$Ca(PO_3OH) \cdot 2H_2O$] could form in AMD neutralization processes, as suggested by the XRD data (Figure 4a and Table 3). This phase has been described by other authors under similar pH conditions during wetland restoration treatments. Furthermore, the model indicated the possible presence of other Fe (strengite, $FePO_4$) and lanthanide (monazite, La, $Ce(PO_4)$) phases at a low pH that could not be detected with the applied methods.

Sample Ca2 (Figure S1, Supplementary Data) was dominated by amorphous phases but also presented calcite and gypsum, which accounted 54% and 45% of the crystalline phases, respectively (Table 3). This sample corresponded to a zone of the profile where the nature of the alkaline reagent could be observed. However, the relatively high concentration of Cu, Zn (Figure 3c), and gypsum (rhombohedral) in this section seemed to be a consequence of the beginning neutralization processes [51,52], as evidenced by the green–white color of the precipitates. In fact, the SEM images confirmed the presence of some Al-hydroxide amorphous precipitates and discrete Cu carbonated phases. The low presence of Fe associated with S and Al could explain the commonly described substitution of Fe to Al in the jarosite–alunite group [53,54] as the pH increased (Figure 5b,c).

Likewise, in the Ca1 sample, the background of Ca3 (8.5 cm-deep profile) was subtracted to improve the identification of mineral phases. The results, consistent with the modeled hydrochemical data [2], pointed to the presence of schwertmannite (5.8%) and jarosite (6.9%), with presence of plumbojarosite [$Pb_{0.5}Fe_3^{3+}(SO_4)_2(OH)_6$] (Figure S2). Although distinguishing among these phases is challenging [41], these results were also checked with the chemical data (Table 2). In this case, a preferential retention of Pb could be observed, supporting the XRD interpretations. Regarding trace elements, the SEP results (F3 and F4 SEP extractant) and potassium distribution among the fractions could indicate a relative affinity between both As–schwertmannite and Cu–jarosite.

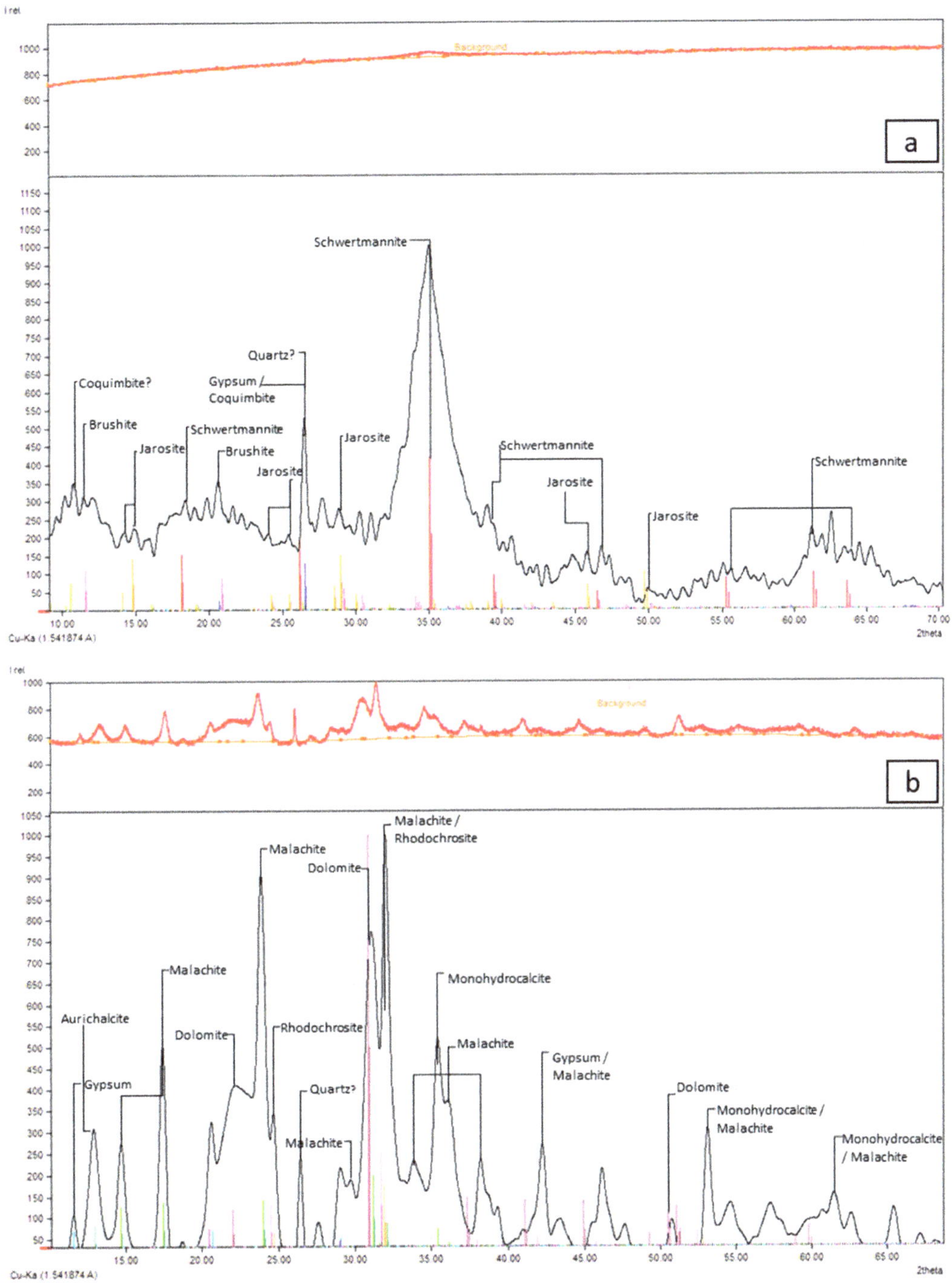

Figure 4. Representative XRD pattern of samples showing main mineral phases identified and their correlation with patterns obtained from database by Match, (**a**) DAS-Ca (Ca1), and (**b**) DAS-Mg (Mg1).

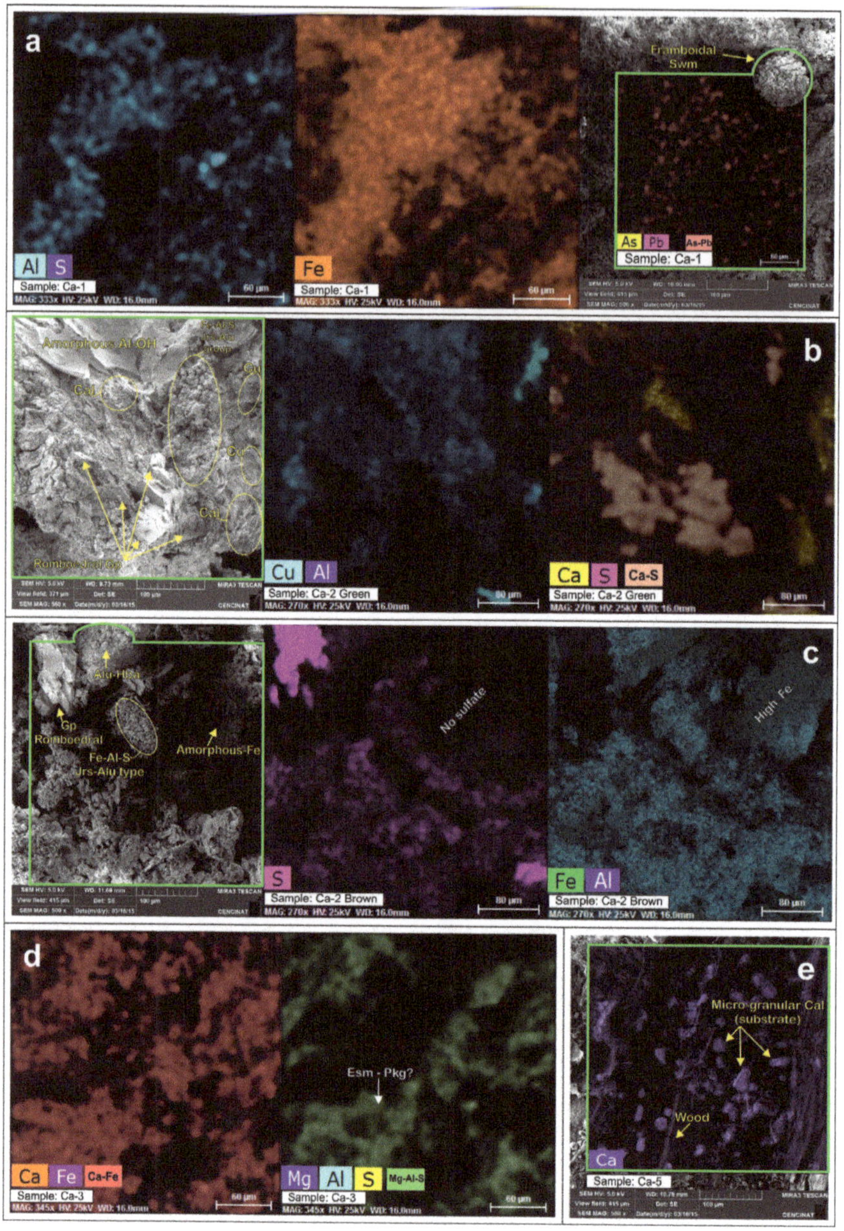

Figure 5. SEM images revealing the composition and morphology of mineralogical assemblages from DAS-Ca. The green frame highlights the correlation between the SE image and compositional mapping using Quantax. (**a**) Sample Ca1, where the relationship of Fe and Al with trace elements such as As and Pb can be observed. Framboidal aggregates of schwertmannite can also be observed. (**b**) Sample Ca2, showing the rhombohedral Gp and the formation of aluminum hydroxides and copper carbonates in local alkaline conditions and (**c**) the presence of Fe (microcrystalline aggregates (<10 μm) of Fe-Al-S, attributed to the jarosite–alunite group). The larger-sized spherical aggregates exhibited low concentrations of Fe (alunite–basaluminite). (**d**) Samples Ca-3, showing the presence of Mg-Al sulfates (identical compositional map) that could imply the formation of epsomite or pickeringite. (**e**) Microgranular calcite on the reactive substrate over wood shavings in sample Ca5.

Table 3. Semiquantitative results made by Rietveld showing the grade of crystallinity of the samples and chi^-2GOF obtained before analysis. The XRD pattern and identification peak can be seen in the Supplementary Figures S1–S6. Trace percentage in red font (<1%). Mineral symbols are according to IMA–CNMNC [55].

		DAS—Ca					DAS—Mg		Drain
Mineral Phases	**SAMPLE**	**Ca1**	**Ca2**	**Ca3**	**Ca4**	**Ca5**	**Mg1**	**Mg2**	**Pindo**
Boehmite	γ-AlO(OH) Bhm	-	-	3.4	-	-	-	-	-
Brushite	$Ca(PO_3OH){\cdot}2H_2O$ Bsh	9.5	-	5.9	-	-	-	-	-
Calcite	$CaCO_3$ Cal		54.5		51.6	85.6	-	59.2	-
Coquimbite	$AlFe_3(SO_4)_6(H_2O)_{12}{\cdot}6H_2O$ Coq	22.1	-		-	-	-	-	-
Diaspore	AlO(OH) Dsp		-		1.2	-	-	-	-
Dolomite	Ca, $Mg(CO_3)_2$ Dol		-		3.0	-	42.9	-	-
Gypsum	$CaSO_4$ Gp	2.4	45.5	74.1	44.2	13.7	2.0	-	-
Jarosite	$KFe^{3+}{}_3(SO_4)_2(OH)_6$ Jrs	14.7	-	6.9	-	-	-	-	-
Melanterite	$Fe^{2+}(H_2O)_6SO_4{\cdot}H_2O$ Mln		-	4.0	-	-	-	-	-
Quartz	SiO_2 Qtz	15.9	-		-	0.7	1.9	0.6	72.0
Schwertmannite	$Fe^{3+}{}_{16}(OH,SO_4)_{12\text{-}13}O_{16}{\cdot}10\text{-}12H_2O$ Swm	35.4	-	5.8	-	-	-	-	-
Aragonite	$CaCO_3$ Arg	-	-	-	-	-	-	34.1	-
Aurichalcite	$(Zn,Cu)_5(CO_3)_2(OH)_6$ Ach	-	-	-	-	-	11.6	-	-
Brucite	$Mg(OH)_2$ Brc	-	-	-	-	-	-	1.1	-
Azurite/Malachite	$Cu(OH)_2{\cdot}2(CuCO_3)$ Azu/$Cu_2(CO_3)(OH)_2$ Mlc	-	-	-	-	-	24.2	-	-
Monohydrocalcite	$CaCO_3{\cdot}H_2O$ Mhcal	-	-	-	-	-	1.9	-	-
Rhodochrosite	$MnCO_3$ Rds	-	-	-	-	-	9.1	2.0	-
Siderite	$FeCO_3$ Sd	-	-	-	-	-	6.5	3.1	-
Albite	$Na(AlSi_3O_8)$ Ab	-	-	-	-	-	-	-	8.7
Chlorite	$(Mg,Fe)_3(Si,Al)_4O_{10}(OH)_2(Mg,Fe)_3(OH)_6$ Chl	-	-	-	-	-	-	-	4.0
Magnetite	$Fe^{2+}Fe^{3+}{}_2O_4$ Mag	-	-	-	-	-	-	-	1.0
Microcline	$K(AlSi_3O_8)$ Mcc	-	-	-	-	-	-	-	1.5
Mica	$K(Mg,Fe)_3AlSi_3O_{10}(OH, F)_2$ Mca	-	-	-	-	-	-	-	11.9
Orthoclase	$K(AlSi_3O_8)$ Or	-	-	-	-	-	-	-	1.0
	Total	100.02	100.00	100.00	100.00	100.01	100.00	100.00	100.01
	% Crystallinity	7.43	88.1	11.6	89.1	90.5	56.3	78.7	89.2
	chi^2	18.8	2.12	13.6	3.27	4.11	1.56	3.97	1.41
	GOF	4.34	1.45	3.69	1.81	2.03	1.25	1.99	1.19

Additionally, other sulfate phases such as melanterite ($FeSO_4{\cdot}7H_2O$; 4%), gypsum (74%), or phosphates like brushite (5.9%) were linked to Fe and Ca reactivity in this section. The hydrochemistry of the water column in Ca3 and deeper strata suggested that Fe disulfate species were leading [$FeSO_4^+$ (73%) and Fe $(SO_4)_2^-$ (14%)], supporting the formation of melanterite, which would increase the proportion of hydrolyzed species, promoting the precipitation of saturated Fe phases like schwertmannite and jarosite [52].

On the other hand, in this profile position, a low 2θ crest seemed to indicate the presence of Al phases like boehmite (3.4%, Table 3). According to the hydrochemical model and in line with the data described by Sánchez-España et al. [41], at pH values greater than 4 (10 cm depth), the presence of Al oxyhydroxides (alunite and basaluminite, among others) could be expected. As previously outlined, the substitution of Fe by Al became more noticeable in this section of the profile (Figure 5c). In fact, the SEM data allowed the distinction of a common pattern of distribution for Al, S, and Mg (Figure 5d), which could imply the presence of other sulfate phases such as epsomite ($MgSO_4{\cdot}7H_2O$) or pickeringite ($MgAl_2(SO_4)_4{\cdot}22H_2O$). Although their presence could not be confirmed, the SEP data also supported this argument, as significant concentrations of Mg (280 mg/kg) associated with the labile fraction (F1) could be observed.

The samples at the depths of Ca4 and Ca5 did not show more Al-rich phases, except for the presence of diaspore, but this was in a low proportion (1.15% of the crystalline phases in sample Ca4). Both samples were characterized by the high presence of calcite from the reactive substrate, higher than 50% (Table 3, Figures S3 and S4). Additionally, the Ca4 sample showed representative amounts of gypsum (44%) and dolomite [$CaMg(CO_3)_2$; 3%], probably derived from the pH increase that could also promote the precipitation of these carbonate-type divalent metal phases in presence of high Mg concentrations [2]. Calcite and gypsum were the most frequent phases in the final section of DAS-Ca and in the DAS-Mg column (Table 3), due to the large amount of reactive calcium available (evidenced by the high proportion in sample Ca5, Figure 5e). In fact, the SEM study of the surface precipitates in the DAS-Mg column suggested that the white aggregates (Mg-SB) were essentially composed of gypsum (Figure 6a), while the green–blue ones (Mg-SV) also exhibited a nanocrystalline lattice composed of Fe, Cu, and C (Figure 6b). XRD suggested that this grid was composed of metal(II) hydroxide carbonate minerals (rosasite group, including malachite-Cu) and siderite [$Fe^{2+}(CO_3)$].

Under the conditions of the DAS-Mg column with a pH > 6, newly formed carbonate and bicarbonate phases could play an important role in removing divalent metals from the solution [2]. The presence of dolomite in this section was explained by the dissolution of the reactive magnesia in DAS-Mg. Thus, monohydrocalcite [$CaCO_3{\cdot}(H_2O)$], malachite, azurite, siderite, and rhodochrosite [$MnCO_3$] were identified (Table 3). The efficient removal of Cu and Mn within the initial centimeters of DAS-Mg (Mg1, Figure 4b) was evident, so subspherical Cu-C aggregates could be recognized by the SEM (Figure 6c).

Although iron chemistry is typically broad in more acidic environments (e.g., DAS-Ca), small amounts of available Fe were associated with the carbonate group as a divalent metal and precipitated as siderite (Figures 4b and S5; Table 3), representing 6.5% and 3.1% of the crystalline phases in samples Mg1 and Mg2, respectively. The SEP results clearly revealed high concentrations of Cu, Fe, Mn, and Zn in DAS-Mg (Figure 3g,h), associated with the carbonate system (F2). Additionally, based on the hydrochemical data of these environments (high Mn concentrations) and the mineralogical interpretations of the DAS-Mg samples, the presence of rhodochrosite could be confirmed (Figure 4b and Figure S5; 9% and 2% of Mg1 and Mg2, respectively).

According to Cortina et al. [32], metal hydroxides rapidly precipitate in these experiments, while sulfates precipitate more slowly and progressively. Although the literature suggests the presence of brucite [$Mg(OH)_2$], it only appeared in a negligible proportion (1.1% of the total crystalline phases of sample Mg2, Figure S5) in the deepest part of the Mg column. However, the SEM data demonstrated the presence of Na and Mg, associated with the formation of small-sized spherical aggregates (Figure 6d). This was also supported by the concentration of Mg and Na determined in F1 of the DAS-Mg samples (Figure 3g,h), which evidenced that carbonate-type divalent metal phases played an important role in the partition processes between AMD and solids in the final section of the treatment [2].

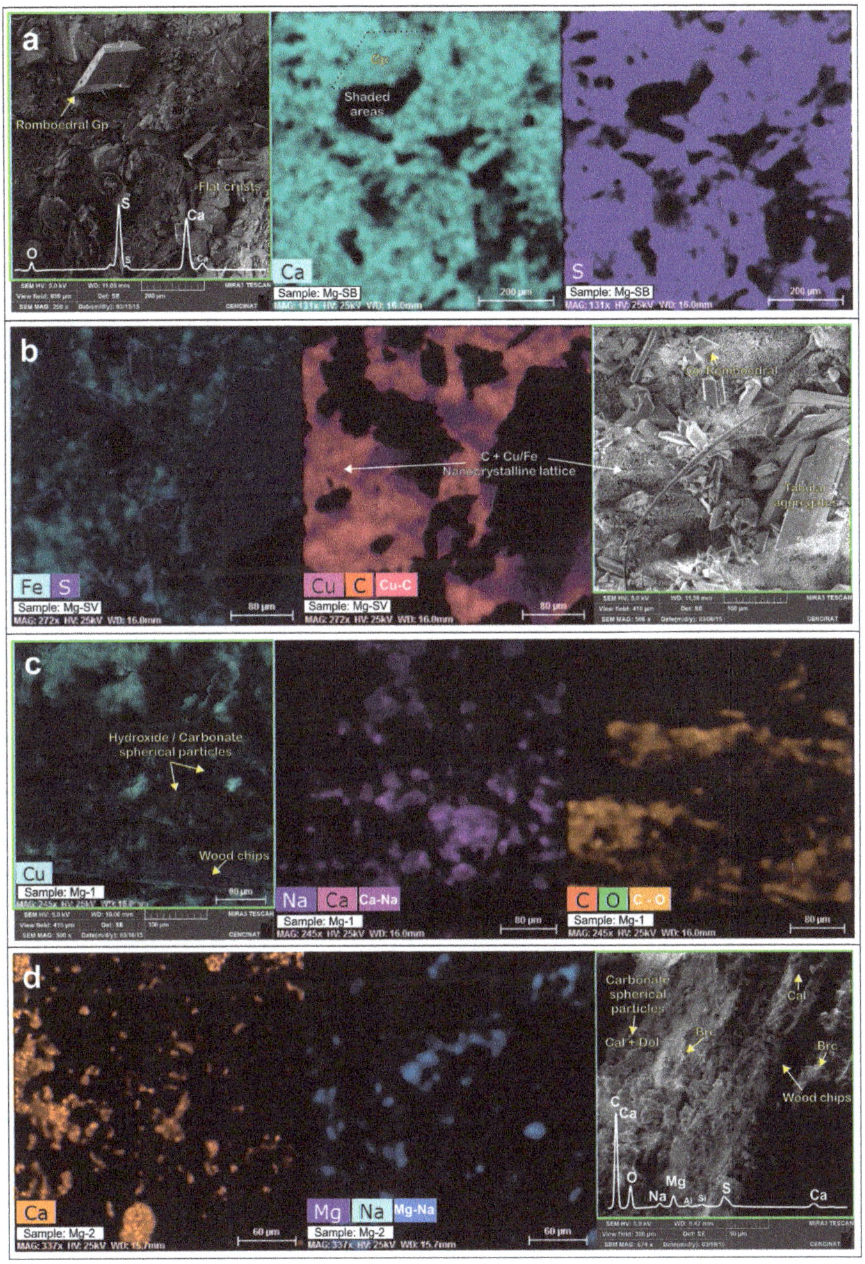

Figure 6. SEM images revealing the composition and morphology of mineralogical assemblages from DAS-Mg. The green frame highlights the correlation between the SE image and compositional mapping using Quantax. (**a**,**b**) Both precipitates collected on the surface of the residue from the DAS-Mg column, white (SB) and green (SV), respectively, showing diverse gypsum crystallization forms. Also, (**b**) presents a nanocrystalline lattice of Cu-Fe carbonate. (**c**) Sample Mg-1, where the relation of divalent metals and carbonate can be observed. (**d**) Sample Mg2, showing the presence subspherical aggregates of Ca-Mg (calcite and dolomite) and another spherical particulate with high Na-Mg.

Moreover, the alkaline conditions facilitated the saturation of carbonated phases, such as calcite or dolomite, which could be recognized as subspherical aggregates of Ca-Mg, displaying the same compositional pattern as C (Figure 6d). These phases could play a role in coprecipitation–sorption processes, influencing the solubility of Zn, Mn, and other trace elements [56]. In fact, higher Zn and Mn concentrations were associated with F2 of the Mg1 and Mg2 samples (516 and 465 g/kg of Zn and 13 and 75 g/kg of Mn, respectively). The data revealed that Zn was the divalent metal whose partition towards the solid phase occurred in the late stages of the treatment (DAS-Mg), but it was depleted by carbonate mineral phases such as aurichalcite $[(Zn, Cu^{+2})_5 (CO_3)_2{\cdot}(OH)_6]$ (7% in the Mg2 samples). The SEP data showed that 516 g/kg of Zn and 906 g/Kg of Cu were trapped in the carbonated phases (F2) of the Mg1 sample, supporting this assumption.

On the other hand, due to the high availability of sulfate hydroxides, newly formed phases linked to divalent metals such as brochantite $[Cu_4SO_4(OH)_6]$, antlerite $[(Cu^{+2})_3SO_4(OH)_4]$, goslarite $[ZnSO_4{\cdot}7H_2O]$, and anglesite $[PbSO_4]$, among others, are frequently described [47]. However, the mineralogical characterization of the samples was consistent with the hydrochemical model (Figure 4b in [2]) that showed the low depletion of aqueous sulphated phases in the final section of the treatment. While hydrated sulfates are frequently described as the main controlling factor in the partitioning process of divalent metals such as Cu, Zn, or even Fe [57], no sulfate phases involving these elements were identified in the samples of the present study.

Finally, the bottom sample of the treatment ("Arenas del Pindo"—sandy drainage; Table 3 and Figure S6) was characterized by the presence of quartz (72%), albite (8.7%), K-feldspar (microcline–orthoclase, 2.5%), and phyllosilicates (mainly biotite, 12%, and chlorite, 4% of the total). This mineralogy explains the presence of crystalline quartz in some samples such as Ca1 or Mg2, which probably accidentally transferred along the samples during the columns' opening process.

4.3. Possible Environmental Implication of Residues' Management

The SEP method is commonly used to evaluate the potential mobility of pollutants under various simulated environmental conditions. In this study, an improved SEP [35] was applied to assess the leaching behavior of PTEs from AMDp across different weathering scenarios (Figure 7), and the results were compared with the continuous concentration criterion (CCC) limits established by the European Community's recommended water quality standards [58]. This value represents a potential negative effect for aquatic life when uncontrolled leachates are generated from the inappropriate disposal of AMDp. The proposed scenarios were as follows. Scenario 1: the contact of the residue with rainwater (emulated by the F1 SEP fraction), which may occur in uncovered disposal facilities and is considered highly dangerous to the environment according to regulations [59]. Scenario 2: Interaction between the residue and weak acidic leachates under reducing conditions (F1 + F2 fraction). These conditions are relevant to scenarios such as underground disposal facilities. Scenario 3: The exposure of residues to weak acidic leachates in oxidizing environments (F1 + F3 + F4 fraction). These conditions aimed to simulate disposal scenarios in surface impoundments or waste piles, as has been previously described by [12].

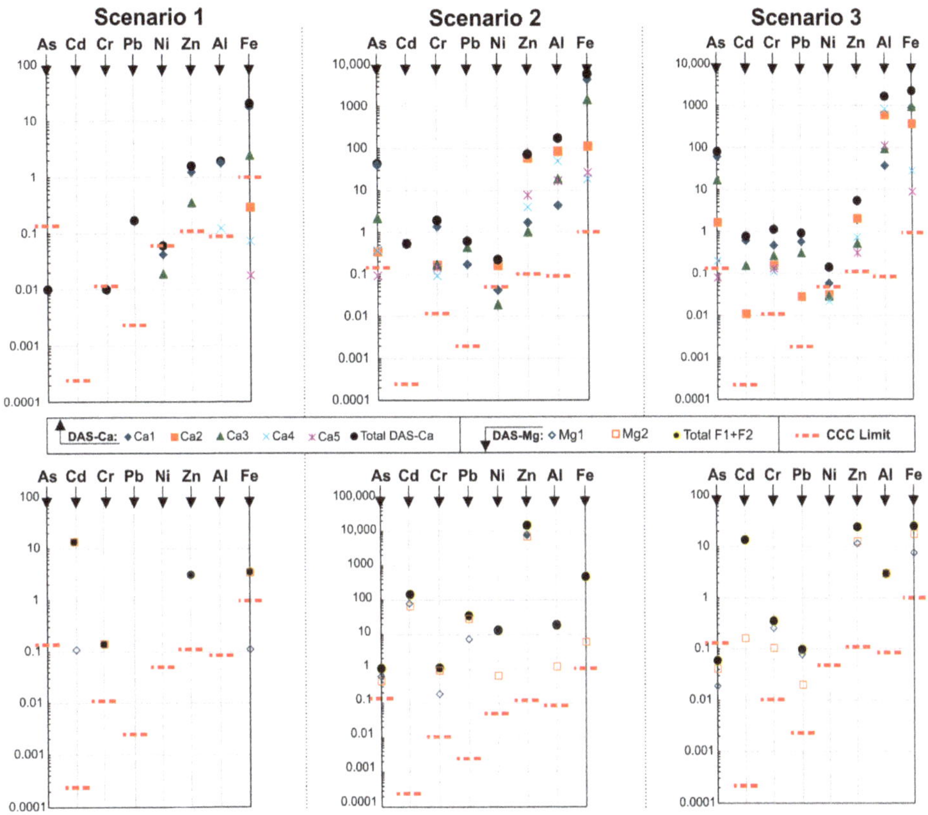

Figure 7. Charts showing the leached concentrations of As, Cd, Cr, Pb, Ni, Zn, Al, and Fe in the different weathering scenarios from the DAS-Ca (**upper**) and DAS-Mg (**down**) residues. The continuous concentration criterion (CCC limit) from the US EPA National Recommended Water Quality Criteria is also shown for comparison purposes. The data are in mg/L.

The data suggest that, although some samples may have occasionally fallen below the CCC limit, when the resulting whole-body residue from any DAS system is disposed of, it could pose danger to aquatic ecosystems and the life associated with them. In scenario 1 (exposure to rainfall), Pb, Zn, Al, and Fe exceeded the proposed values for DAS-Ca, while Cd, Cr, Zn, and Fe behaved similarly for DAS-Mg. For the rest of the more aggressive conditions, scenario 2 (underground disposal) and scenario 3 (surface impoundments or waste piles), the AMD residues generated leachates that surpassed the threshold for most of the elements studied, making them incompatible with aquatic life. In this regard, these metal-rich residues from passive treatments (both laboratory-scale and field-scale), when stored in an oxidizing environment, demonstrated the significant leaching of metals and sulfates [60–62]. Several studies have shown that residues from Fe-rich AMD are unstable and have a high risk of leaching. They should be stored underwater or in a neutral-pH environment to prevent contamination release and ensure safe disposal [5,6]. Meanwhile, specific studies on MgO-DAS also recommend storing them in a dry (not mixed with acidic or alkaline water) environment [15]. More recent studies suggest that, in addition to being stored under anoxic conditions in the presence of water to prevent the formation of new AMD, these residues should undergo stabilization/solidification before any storage [8].

In addition, Table 4 presents the regulatory limits for waste acceptance at landfills in the EU [59] in comparison to the concentrations of elements detected in AMDp (DAS-Ca and DAS-Mg) following the F1 leaching fraction. This fraction denotes the metals and metalloids capable of mobilization upon contact with water and those that are considered the leaching limit values to be applied to waste for storage in the three types of landfills proposed by the Royal Decree 646/2020 [63], under the auspices of EU regulation [59].

Table 4. A comparison between AMDp and the regulatory criteria for the acceptance of waste at different types of landfills and their classification of hazardousness (Council Decision, 2003/33/EC).

Landfill Type	As	Ba	Cd	Cr	Cu	Ni	Pb	Zn	Sulfate
DAS-Ca	<d.l	<d.l	<d.l	0.7	50	4.0	10	104	<5000
DAS-Mg	<d.l	<d.l	901	9.4	164	<d.l	<d.l	206	<5000
Inert wastes	0.5	20	0.04	0.5	2.0	0.4	0.5	4.0	1000
Non-Hazardous wastes	2.0	100	1	10	50	10	10	50	20,000
Hazardous wastes	25	300	5	70	100	50	50	200	50,000

<d.l. Below detection limit. Data in mg/kg.

According to the results obtained, the DAS-Ca residues would be classified as hazardous since the concentration of Zn (104 mg/kg) exceeded the reference value, while Cu (50 mg/kg) and Pb (10 mg/kg) were at the limit value. Meanwhile, the precipitates of the DAS-Mg would pose storage issues in landfills due to the excessive leached concentrations of Cd (901), Cu (164), and Zn (206 mg/kg).

Under these circumstances, the EU regulation [59] recommends additional treatment (such as further neutralization, encapsulation, etc.) before the storage of these materials in any type of regulated landfill. Also, recent directives have established and emphasized, in general terms, the necessity of subjecting waste destined for landfills to appropriate pretreatment [63], considered an essential requirement to ensure that landfill operations are conducted without endangering human health and the environment. Additionally, in accordance with the circularity principle, it is important to note that these treatments generate metal-rich solid residues considered generally hazardous wastes, but they can also present an opportunity for the potential recovery of metals and other critical elements [60,64,65]. In this regard, the European Regional Development Fund (FEDER, 2021–2027) aims to enhance these efforts by fostering the transition to a circular and resource-efficient economy (Raw Materials Initiative) through the generation of added value from waste. This initiative contributes to reducing the amount of waste sent to landfills and conserving natural resources by promoting reuse and recycling.

5. Conclusions

The mineralogy of solid residues from the DAS system experience with Ecuadorian AMD was consistent with the basic principles of metal retention from acidic solutions and helped the interpretation of the geochemical results and modelling. Elements like Fe and Al were retained in the DAS-Ca column, while most divalent metals fractionated towards the solid phase in DAS-Mg.

Among the minerals that retained Fe^{3+} and Al^{3+}, oxyhydroxysulfates such as schwertmannite and the jarosite–alunite group were primarily responsible for their fractionation towards the solid phase. In this process of mineral formation, other trace elements with diverse charges were retained through adsorption–coprecipitation mechanisms (As, Cr, Cu, Pb, and Zn). Additionally, sulfated efflorescent salts such as coquimbite and gypsum were responsible for removing sulfate from the solution.

The fractionation of divalent metals (Cu, Zn, and Mn) predominantly occurred in the DAS-Mg column, and it was controlled by the formation of carbonated phases (identified through XRD and SEM studies). In the initial stages of DAS-Mg, phases rich in Cu-Zn from the azurite–malachite group were abundantly formed, while other discrete phases such as rhodochrosite (Mn) appeared accompanying neoformed calcite, monohydrocalcite, and dolomite when the Mg supply increased in the column. Hydroxides of divalent metals have frequently been described as the main mechanism for controlling metal loads in passive treatment systems of the DAS type; however, although they were not absent, the results of this study suggest that newly formed carbonate and bicarbonate phases were the most evident neutralization mechanism trapping the metallic charge. This result confirms the initial hypothesis of the metal fractionation processes according to the hydrochemical model proposed by [2], highlighting the key role of carbonates in removing divalent metals from AMD solution.

The leachates from the SEP were analyzed against the limits set by the continuous concentration criterion (CCC) to assess the potential aquatic life exposure to contaminants in various disposal scenarios. In these scenarios (disposal in superficial environmental conditions, in underground facilities, and in surface impoundments or waste piles), the established values were exceeded for PTEs including As, Cd, Cr, Pb, Ni, and Zn (excluding the punctual samples of the profile). According to the regulatory normative, none of the DAS residues could be defined as inert wastes. The DAS-Ca solids could be classified as non-hazardous wastes for most elements (only Zn posed risk). Residues from DAS-Mg exceeded the hazard threshold for Cd, Cu, and Zn, indicating a significant environmental risk for all landfill options, emphasizing the need to prevent waste from directly contacting leaching solutions.

Combining different analytical and management approaches could improve waste management practices and mitigate potential environmental risks by providing a more comprehensive understanding of the composition, behavior, and potential environmental impacts of these residues. Efforts should focus on accurately characterizing waste to promote viable solutions aimed at reducing the volume of waste sent to landfills by using them as a new source of metals and other critical elements.

Supplementary Materials: The following supporting information can be downloaded at https://www.mdpi.com/article/10.3390/min15010015/s1. Figures S1–S6: The XRD pattern of samples in DAS-Ca (S1 (Ca2, 7.5 cm); S2 (Ca3, 8.5 cm); S3 (Ca4, 10 cm); S4 (Ca5, 15 cm)) and DAS-Mg (S5 (Mg2, 12 cm)), as well as S6 (the "Pindo", sand drain of the bottom columns), showing the main mineral phases identified* and their correlation with the patterns obtained from database by Match. The mineral symbols are according to the IMA–CNMNC. * Semiquantitative results were improved by Rietveld refinement (see also Table 3).

Author Contributions: Conceptualization, J.D.; methodology, O.L. and D.M.; validation, J.D., C.B.-B. and D.A.; investigation, J.D., C.B.-B. and D.M.; resources, D.A.; writing—original draft preparation, O.L.; writing—review and editing, J.D. and C.B.-B.; visualization, J.D. and C.B.-B.; supervision, J.D. and C.B.-B.; project administration, J.D. and D.A.; funding acquisition, J.D. All authors have read and agreed to the published version of the manuscript.

Funding: This work was supported by the PROMETEO Ecuadorian program (Secretary of Superior Education, Science, Technology and Innovation) in the framework of the project "Experiencia Piloto en la Remediación y Mitigación de la Oxidación de Sulfuros y la Generación de AMD en Relaveras del Distrito Minero de Zaruma-Portovelo" by the means of the postdoctoral contract nº 20140411 BP.

Data Availability Statement: Data is contained within the article or Supplementary Materials.

Acknowledgments: The authors acknowledge the material and human resources displayed from the INIGEMM (Ecuadorian National Research Institute of Geology, Mining and Metallurgy) and US (Seville University). Diana Ayala thanks Cumbal for the support in studying the samples using the SEM at the Center for Nanoscience and Nanotechnology (CENCINAT).

Conflicts of Interest: The authors declare that they have no known competing financial interests or personal relationships that could have appeared to influence the work reported in this paper.

References

1. Nordstrom, D.K.; Blowes, D.W.; Ptacek, C.J. Hydrogeochemistry and microbiology of mine drainage: An update. *Appl. Geochem.* **2015**, *57*, 3–16. [CrossRef]
2. Delgado, J.; Barba-Brioso, C.; Ayala, D.; Boski, T.; Torres, S.; Calderón, E.; López, F. Remediation experiment of Ecuadorian acid mine drainage: Geochemical models of dissolved species and secondary minerals saturation. *Environ. Sci. Pollut. Res.* **2019**, *26*, 34854–34872. [CrossRef] [PubMed]
3. Park, I.; Tabelin, C.B.; Magaribuchi, K.; Seno, K.; Ito, M.; Hiroyoshi, N. Suppression of the release of arsenic from arsenopyrite by carrier-microencapsulation using Ti-catechol complex. *J. Hazard. Mater.* **2018**, *344*, 322–332. [CrossRef]
4. Jouini, M.; Rakotonimaro, T.V.; Neculita, C.M.; Genty, T.; Benzaazoua, M. Stability of metal-rich residues from laboratory multi-step treatment system for ferriferous acid mine drainage. *Environ. Sci. Pollut. Res.* **2019**, *26*, 588–601. [CrossRef]
5. Genty, T.; Bussière, B.; Benzaazoua, M.; Zagury, G.J. Capacity of Wood Ash Filters to Remove Iron from Acid Mine Drainage: Assessment of Retention Mechanism. *Mine Water Environ.* **2012**, *31*, 273–286. [CrossRef]
6. Neculita, C.-M.; Zagury, G.J.; Bussière, B. Effectiveness of sulfate-reducing passive bioreactors for treating highly contaminated acid mine drainage: II. Metal removal mechanisms and potential mobility. *Appl. Geochem.* **2008**, *23*, 3545–3560. [CrossRef]
7. Rakotonimaro, T.V.; Neculita, C.M.; Bussière, B.; Genty, T.; Zagury, G.J. Performance assessment of laboratory and field-scale multi-step passive treatment of iron-rich acid mine drainage for design improvement. *Environ. Sci. Pollut. Res.* **2018**, *25*, 17575–17589. [CrossRef]
8. Hengen, T.J.; Squillace, M.K.; O'Sullivan, A.D.; Stone, J.J. Life cycle assessment analysis of active and passive acid mine drainage treatment technologies. *Resour. Conserv. Recy.* **2014**, *86*, 160–167. [CrossRef]
9. Skousen, J.; Zipper, C.E.; Rose, A.; Ziemkiewicz, P.F.; Nairn, R.; McDonald, L.M.; Kleinmann, R.L. Review of Passive Systems for Acid Mine Drainage Treatment. *Mine Water Environ.* **2017**, *36*, 133–153. [CrossRef]
10. Ayora, C.; Caraballo, M.A.; Macias, F.; Rötting, T.S.; Carrera, J.; Nieto, J.-M. Acid mine drainage in the Iberian Pyrite Belt: 2. Lessons learned from recent passive remediation experiences. *Environ. Sci. Pollut. Res.* **2013**, *20*, 7837–7853. [CrossRef]
11. Caraballo, M.A.; Macías, F.; Nieto, J.M.; Castillo, J.; Quispe, D.; Ayora, C. Hydrochemical performance and mineralogical evolution of a dispersed alkaline substrate (DAS) remediating the highly polluted acid mine drainage in the full-scale passive treatment of Mina Esperanza (SW Spain). *Am. Min.* **2011**, *96*, 1270–1277. [CrossRef]
12. Macías, F.; Pérez-López, R.; Caraballo, M.A.; Cánovas, C.R.; Nieto, J.M. Management strategies and valorization for waste sludge from active treatment of extremely metal-polluted acid mine drainage: A contribution for sustainable mining. *J. Clean. Prod.* **2017**, *141*, 1057–1066. [CrossRef]
13. Rakotonimaro, T.V.; Neculita, C.M.; Bussière, B.; Zagury, G.J. Effectiveness of various dispersed alkaline substrates for the pre-treatment of ferriferous acid mine drainage. *Appl. Geochem.* **2016**, *73*, 13–23. [CrossRef]
14. Rötting, T.S.; Thomas, R.C.; Ayora, C.; Carrera, J. Passive Treatment of Acid Mine Drainage with High Metal Concentrations Using Dispersed Alkaline Substrate. *J. Environ. Qual.* **2008**, *37*, 1741–1751. [CrossRef] [PubMed]
15. Macías, F.; Caraballo, M.A.; Nieto, J.M.; Rötting, T.S.; Ayora, C. Natural pretreatment and passive remediation of highly polluted acid mine drainage. *J. Environ. Manag.* **2012**, *104*, 93–100. [CrossRef]
16. Rakotonimaro, T.V.; Neculita, C.M.; Bussière, B.; Benzaazoua, M.; Zagury, G.J. Recovery and reuse of sludge from active and passive treatment of mine drainage-impacted waters: A review. *Environ. Sci. Pollut. Res.* **2017**, *24*, 73–91. [CrossRef] [PubMed]
17. Jouini, M.; Benzaazoua, M.; Neculita, C.M.; Genty, T. Performances of stabilization/solidification process of acid mine drainage passive treatment residues: Assessment of the environmental and mechanical behaviors. *J. Environ. Manag.* **2020**, *269*, 110764. [CrossRef] [PubMed]
18. Fanfani, L.; Zuddas, P.; Chessa, A. Heavy metals speciation analysis as a tool for studying mine tailings weathering. *J. Geochem. Explor.* **1997**, *58*, 241–248. [CrossRef]
19. Khorasanipour, M.; Tangestani, M.H.; Naseh, R.; Hajmohammadi, H. Hydrochemistry, mineralogy and chemical fractionation of mine and processing wastes associated with porphyry copper mines: A case study from the Sarcheshmeh mine, SE Iran. *Appl. Geochem.* **2011**, *26*, 714–730. [CrossRef]
20. Ribeta, I.; Ptacek, C.J.; Blowes, D.W.; Jambor, J.L. The potential for metal release by reductive dissolution of weathered mine tailings. *J. Contam. Hydrol.* **1995**, *17*, 239–273. [CrossRef]
21. Dold, B. Speciation of the most soluble phases in a sequential extraction procedure adapted for geochemical studies of copper sulfide mine waste. *J. Geochem. Explor.* **2003**, *80*, 55–68. [CrossRef]
22. Sahuquillo, A.; Rigol, A.; Rauret, G. Overview of the use of leaching/extraction tests for risk assessment of trace metals in contaminated soils and sediments. *TrAC Trends Anal. Chem.* **2003**, *22*, 152–159. [CrossRef]

23. USEPA (U.S. Environmental Protection Agency). *Guidelines for Human Exposure Assessment*; (EPA/100/B-19/001); Risk Assessment Forum, U.S. EPA: Washington, DC, USA, 2019.
24. Delgado, J.; Ayala, D.; Simón Páez, H. Treatment system to improve the drainage water quality of the Puyango River environmental liabilities. *Geogaceta* **2018**, *64*, 63–66.
25. Betancourt, O.; Narváez, A.; Roulet, M. Small-scale Gold Mining in the Puyango River Basin, Southern Ecuador: A Study of Environmental Impacts andHuman Exposures. *Ecohealth* **2005**, *2*, 323–332. [CrossRef]
26. Delgado, J.; Barba-Brioso, C.; Boski, T. Mine wastes characterisation and experimental design of a controlled tailing storage in the Zaruma-Portovelo mining district (SE Ecuador). *Geogaceta* **2018**, *64*, 135–138.
27. Guimaraes, J.R.D.; Betancourt, O.; Miranda, M.R.; Barriga, R.; Cueva, E.; Betancourt, S. Long-range effect of cyanide on mercury methylation in a gold mining area in southern Ecuador. *Sci. Total. Environ.* **2011**, *409*, 5026–5033. [CrossRef]
28. Tarras-Wahlberg, N.H.; Flachier, A.; Lane, S.N.; Sangfors, O. Environmental impacts and metal exposure of aquatic ecosystems in rivers contaminated by small scale gold mining: The Puyango River basin, southern Ecuador. *Sci. Total. Environ.* **2001**, *278*, 239–261. [CrossRef] [PubMed]
29. Tarras-Wahlberg, N.H.; Lane, S.N. Suspended sediment yield and metal contamination in a river catchment affected by El Niño events and gold mining activities: The Puyango river basin, southern Ecuador. *Hydrol. Process.* **2003**, *17*, 3101–3123. [CrossRef]
30. Rötting, T.S.; Ayora, C.; Carrera, J. Improved Passive Treatment of High Zn and Mn Concentrations Using Caustic Magnesia (MgO): Particle Size Effects. *Environ. Sci. Technol.* **2008**, *42*, 9370–9377. [CrossRef] [PubMed]
31. Cabrera, G.; Pérez, R.; Gómez, J.M.; Ábalos, A.; Cantero, D. Toxic effects of dissolved heavy metals on *Desulfovibrio vulgaris* and *Desulfovibrio* sp. strains. *J. Hazard. Mater.* **2006**, *135*, 40–46. [CrossRef] [PubMed]
32. Cortina, J.L.; Lagreca, I.H.; de Pablo, J.; Cama, J.; Ayora, C. Passive in situ remediation of metal-polluted water with caustic magnesia: Evidence from column experiments. *Environ. Sci. Technol.* **2003**, *37*, 1971–1977. [CrossRef] [PubMed]
33. Gibert, O.; de Pablo, J.; Cortina, J.L.; Ayora, C. Sorption studies of Zn(II) and Cu(II) onto vegetal compost used on reactive mixtures for in situ treatment of acid mine drainage. *Water Res.* **2005**, *39*, 2827–2838. [CrossRef]
34. Rötting, T.S.; Cama, J.; Ayora, C.; Cortina, J.L.; De Pablo, J. Use of caustic magnesia to remove cadmium, nickel, and cobalt from water in passive treatment systems: Column experiments. *Environ. Sci. Technol.* **2006**, *40*, 6438–6443. [CrossRef] [PubMed]
35. Caraballo, M.A.; Rötting, T.S.; Nieto, J.M.; Ayora, C. Sequential extraction and DXRD applicability to poorly crystalline Fe- and Al-phase characterization from an acid mine water passive remediation system. *Am. Min.* **2009**, *94*, 1029–1038. [CrossRef]
36. Delgado, J.; Barba-Brioso, C.; Nieto, J.M.; Boski, T. Speciation and ecological risk of toxic elements in estuarine sediments affected by multiple anthropogenic contributions (*Guadiana saltmarshes*, SW Iberian Peninsula): I. Surficial sediments. *Sci. Total. Environ.* **2011**, *409*, 3666–3679. [CrossRef] [PubMed]
37. Tessier, A.; Campbell, P.G.C.; Bisson, M. Sequential extraction procedure for the speciation of particulate trace metals. *Anal. Chem.* **1979**, *51*, 844–851. [CrossRef]
38. Rao, C.R.M.; Sahuquillo, A.; Lopez Sanchez, J.F. A Review of the Different Methods Applied in Environmental Geochemistry for Single and Sequential Extraction of Trace Elements in Soils and Related Materials. *Water Air Soil Pollut.* **2008**, *189*, 291–333. [CrossRef]
39. Consani, S.; Ianni, M.C.; Dinelli, E.; Capello, M.; Cutroneo, L.; Carbone, C. Assessment of metal distribution in different Fe precipitates related to Acid Mine Drainage through two sequential extraction procedures. *J. Geochem. Explor.* **2019**, *196*, 247–258. [CrossRef]
40. Hernández, J.; Meurer, E. Óxidos de hierro en los suelos: Sus propiedades y su caracterización con énfasis en los estudios de retención de fósforo. *Agrociencia* **1997**, *1*, 1–14. [CrossRef]
41. Sánchez-España, J.; Yusta, I.; Gray, J.; Burgos, W.D. Geochemistry of dissolved aluminum at low pH: Extent and significance of Al–Fe(III) coprecipitation below pH 4.0. *Geochim. Cosmochim. Acta* **2016**, *175*, 128–149. [CrossRef]
42. Hammarstrom, J.M.; Seal, R.R.; Meier, A.L.; Kornfeld, J.M. Secondary sulfate minerals associated with acid drainage in the eastern US: Recycling of metals and acidity in surficial environments. *Chem. Geol.* **2005**, *215*, 407–431. [CrossRef]
43. Marescotti, P.; Carbone, C.; Comodi, P.; Frondini, F.; Lucchetti, G. Mineralogical and chemical evolution of ochreous precipitates from the Libiola Fe–Cu-sulfide mine (Eastern Liguria, Italy). *Appl. Geochem.* **2012**, *27*, 577–589. [CrossRef]
44. Nordstrom, D.K.; Alpers, C.N. Geochemistry of acid mine waters. In *The Environmental Geochemistry of Mineral Deposits, Reviews in Economic Geology*; Plumlee, G.S., Logsdon, M.J., Eds.; Society of Economic Geologists: Littleton, CO, USA, 1999; Volume 6A, pp. 133–160.
45. Delgado, J.; Sarmiento, A.M.; Condesso De Melo, M.T.; Nieto, J.M. Environmental impact of mining activities in the southern sector of the Guadiana Basin (SW of the Iberian Peninsula). *Water Air Soil. Pollut.* **2009**, *199*, 323–341. [CrossRef]
46. Baes, C.F.; Mesmer, R.S. *The Hydrolysis of Cations*; John Wiley & Sons: New York, NY, USA; London, UK; Sydney, Australia; Toronto, ON, Canada, 1977; Volume 81, pp. 245–246. [CrossRef]
47. Carbone, C.; Dinelli, E.; Marescotti, P.; Gasparotto, G.; Lucchetti, G. The role of AMD secondary minerals in controlling environmental pollution: Indications from bulk leaching tests. *J. Geochem. Explor.* **2013**, *132*, 188–200. [CrossRef]

48. Zachara, J.M.; Cowan, C.E.; Resch, C.T. Sorption of divalent metals on calcite. *Geochim. Cosmochim. Acta* **1991**, *55*, 1549–1562. [CrossRef]
49. Papassiopi, N.; Zaharia, C.; Xenidis, A.; Adam, K.; Liakopoulos, A.; Romaidis, I. Assessment of contaminants transport in a watershed affected by acid mine drainage, by coupling hydrological and geochemical modeling tools. *Miner. Eng.* **2014**, *64*, 78–91. [CrossRef]
50. Hudson-Edwards, K.A.; Jamieson, H.E.; Lottermoser, B.G. Mine wastes: Past, present, future. *Elements* **2011**, *7*, 375–380. [CrossRef]
51. Behum, P.; Kim, K. *Workshop for the Application of Passive Technology to the Treatment of Acid Mine Drainage*; Office of Surface Mines: Alton, IL, USA, 1999; Presented at Jasper, Indiana.
52. Shum, M.; Lavkulich, L. Speciation and solubility relationships of Al, Cu and Fe in solutions associated with sulfuric acid leached mine waste rock. *Environ. Geol.* **1999**, *38*, 59–68. [CrossRef]
53. Barba-Brioso, C.; Martín, D.; Romero-Baena, A.; Campos, P.; Delgado, J. Revalorisation of Fine Recycled Concrete in Acid Mine Water Treatment: A Challenge to a Circular Economy. *Minerals* **2023**, *13*, 1028. [CrossRef]
54. Katsioti, M.; Tsakiridis, P.E.; Leonardou-Agatzini, S.; Oustadakis, P. Examination of the jarosite–alunite precipitate addition in the raw meal for the production of sulfoaluminate cement clinker. *J. Hazard. Mater.* **2006**, *131*, 187–194. [CrossRef] [PubMed]
55. Warr, L.N. IMA–CNMNC approved mineral symbols. *Mineral Mag.* **2021**, *85*, 291–320. [CrossRef]
56. Jensen, D.L.; Boddum, J.K.; Tjell, J.C.; Christensen, T.H. The solubility of rhodochrosite ($MnCO_3$) and siderite ($FeCO_3$) in anaerobic aquatic environments. *Appl. Geochem.* **2002**, *17*, 503–511. [CrossRef]
57. Jambor, J.L.; Nordstrom, D.K.; Alpers, C.N. Metal-sulfate Salts from Sulfide Mineral Oxidation. *Rev. Miner. Geochem.* **2000**, *40*, 303–350. [CrossRef]
58. USEPA. (U.S. Environmental Protection Agency). National Recommended Water Quality Criteria—Aquatic Life Criteria Table. 2016. Available online: https://www.epa.gov/wqc/national-recommended-water-quality-criteria-aquatic-life-criteria-table (accessed on 24 July 2024).
59. Decision 2003/33-2003/33/EC: Council Decision of 19 December 2002 Establishing Criteria and Procedures for the Acceptance of Waste at Landfills Pursuant to Article 16 of and Annex II to Directive 1999/31/EC-EU Monitor. Available online: https://www.eumonitor.eu/9353000/1/j9vvik7m1c3gyxp/vitgbgich8nt (accessed on 23 December 2024).
60. Mehdaoui, H.Y.; Guesmi, Y.; Jouini, M.; Neculita, C.M.; Pabst, T.; Benzaazoua, M. Passive treatment residues of mine drainage: Mineralogical and environmental assessment, and management avenues. *Miner. Eng.* **2023**, *204*, 108362. [CrossRef]
61. Vasquez, Y.; Neculita, C.M.; Caicedo, G.; Cubillos, J.; Franco, J.; Vásquez, M.; Hernández, A.; Roldan, F. Passive multi-unit field-pilot for acid mine drainage remediation: Performance and environmental assessment of post-treatment solid waste. *Chemosphere* **2022**, *291*, 133051. [CrossRef]
62. McCann, J.I.; Nairn, R.W. Characterization of residual solids from mine water passive treatment oxidation ponds at the tar creek 750 superfund site, Oklahoma, USA: Potential for reuse or disposal. *Clean. Waste Syst.* **2022**, *3*, 100031. [CrossRef]
63. Real Decreto 646/2020, de 7 de Julio, por el que se Regula la Eliminación de Residuos Mediante Depósito en Vertedero. Boletín Oficial del Estado, n. 187, de 8 de Julio de 2020. Available online: https://www.fao.org/faolex/results/details/es/c/LEX-FAOC203782/ (accessed on 25 July 2024).
64. Vass, C.R.; Noble, A.; Ziemkiewicz, P.F. The Occurrence and Concentration of Rare Earth Elements in Acid Mine Drainage and Treatment By-products: Part 1—Initial Survey of the Northern Appalachian Coal Basin. *Min. Metall. Explor.* **2019**, *36*, 903–916. [CrossRef]
65. Vaziri Hassas, B.; Shekarian, Y.; Rezaee, M.; Pisupati, S.V. Selective recovery of high-grade rare earth, Al, and Co-Mn from acid mine drainage treatment sludge material. *Miner. Eng.* **2022**, *187*, 107813. [CrossRef]

minerals

MDPI

Article

Mineralogical, Petrological and Geochemical Characterisation of Chrysotile, Amosite and Crocidolite Asbestos Mine Waste from Southern Africa in Context of Risk Assessment and Rehabilitation

Jessica Shaye Schapira [1,2,*], Robert Bolhar [1], Sharad Master [2] and Allan H. Wilson [1]

1 School of Geosciences, University of the Witwatersrand, Johannesburg 2050, South Africa; robert.bolhar@wits.ac.za (R.B.); allan.wilson@wits.ac.za (A.H.W.)
2 School of Molecular and Cell Biology, University of the Witwatersrand, Johannesburg 2000, South Africa; sharad.master@wits.ac.za
* Correspondence: jessschapira@gmail.com

Abstract: Derelict asbestos mine sites in South Africa pose a considerable risk to human, environmental and socio-economic health. Comprehensive mineralogical and geochemical datasets for the existing hazardous geological materials still exposed in Southern African derelict asbestos mines remain largely non-existent, as very little published and up-to-date literature is available. In this study, three representative types of asbestos mineral fibres from derelict asbestos mines in Southern Africa, namely chrysotile from Havelock mine, amosite from Penge mine and crocidolite from Prieska mine, are characterized mineralogically and geochemically to critically evaluate actual hazards in rural and asbestos-fibre-contaminated regions. The samples were examined using polarising light microscopy, X-ray fluorescence (major and trace elemental analysis), X-ray diffraction (including Rietveld refinement), specific surface area analysis and bio-durability testing. Data are discussed in view of their potential toxicities on both human health and the environment in the context of developing countries. Finally, information on the mineralogical and geochemical status of asbestos mine waste and its importance as baseline data for rehabilitation considerations is also evaluated.

Keywords: hazards; rehabilitation; dispersion; environmental; mine waste

1. Introduction

The mineral wealth of South Africa is impressive but matched perhaps only by the subsequent environmental and human health problems resulting from mining. There are ~6000 derelict mines throughout South Africa, which pose significant health and environmental concerns [1]. Of these, 249 are abandoned asbestos mines of which less than 40 have been rehabilitated [1]. A legacy of pollution remains in the wake of the asbestos mining industry, and considerable quantities of rock-waste generated by decades of asbestos mining continue to pose an insurmountable health and environmental risk to surrounding communities, especially those in developing countries [2]. Historically, poorly regulated asbestos mining operations in South Africa have resulted in the widespread contamination of the environment. Although occupational exposures diminished in the wake of the asbestos mining cessation in South Africa, the tenacious, contemporary, and vast asbestos contamination of the environment is indicative of an indeterminate and conceivably boundless epidemic of asbestos-related diseases (ARDs) [3].

The regulatory and commercial term, asbestos, defines a group of naturally occurring silicate minerals with a specific fibrous crystal habit and unique chemical, physical and technological properties [4]. There are six types of asbestos minerals that fall into one of two groups, namely serpentine and amphiboles [5]. Repeated experimentations have strongly proven the association between asbestos exposure and cancer types such as carcinoma [6–8].

The toxicity and carcinogenic effects of asbestos-mineral fibres are related to the physical and chemical properties, including morphology, durability and/or bio-persistence, high aspect ratio and chemical composition of these fibrous minerals [9–13]. The airborne dispersion of asbestos fibres through human activities and rock weathering increases the potential for inhalation of fibres, resulting in risks to human health [14–19]. Asbestos minerals also have a high capability to host a vast quantity of toxic species and elements, adding to the potential health risk problem [18,19]. Asbestos-related diseases (ARDs) acquired from environmental sources are undeniably a global concern [3,20]. Numerous types of asbestos-related diseases include both diseases of the pleura and diseases of the lung parenchyma [3]. Malignant mesothelioma, thickening and pleural plaques comprise the pleura diseases category, and asbestosis and lung cancer comprise the lung parenchyma diseases category [3,21]. The global scientific community, to date, agrees that, based on scientific evidence, there is no safe level or threshold of asbestos fibre exposure below which the risk of mesothelioma is negligible [22,23]. Different fibrous minerals display different toxicities due to their different physical and chemical properties [24,25].

Geochemical and mineralogical characterisation of asbestos-bearing mine rocks is important for both human and environmental health risk assessment and quantification, necessary for effective risk mitigation intervention [26]. Moreso, baseline geochemical and mineralogical data on asbestos fibre waste dumps are critical for directing rehabilitation interventions [26]. However, the costs involved in these assessments are a major issue for developing countries. The aim of this paper is to use traditional mineralogical, petrographic and geochemical methods conventionally applied to the characterisation of fibrous minerals, to demonstrate the importance of incorporating geologic-based delineation in derelict asbestos mine site rehabilitation programmes. The paper further highlights the necessity of basic geological knowledge in the context of asbestos mine reclamation and demonstrates the feasibility of relevant scientific methods that are also cost-effective, relatively rapid and easily accessible in fund-limited, developing countries.

2. Materials and Methods

2.1. Sampling Locations and Geological Background

Three different asbestos rock samples were investigated in this study (Table 1), namely (i) Chrysotile from Bulembu (Havelock Mine), Eswatini, Southern Africa (Figure 1A); (ii) amosite from Penge, Lydenberg district, South Africa (Figure 1B); and (iii) crocidolite from Prieska Division, Northern Cape, South Africa (Figure 1C).

Table 1. Location coordinates and description of sampling locations.

	Chrysotile	**Amosite**	**Crocidolite**
Sample number	Ch1	Am2	Cr3
Sampling locations	Havelock Mine, Bulembu, Swaziland (Eswatini)	Penge, Sekhukhune District, Limpopo, South Africa	Prieska Division, Northern Cape, South Africa
Location coordinates	25°57′21″ S 31°07′51″ E	24°25′07″ S 30°20′14″ E	28°19′01″ S 23°06′05″ E
Occurrence	Cross-vein fibres (growth of fibres at right angles to the walls of cracks)	Cross-fibre seams in banded ironstones	Cross-fibre seams in banded ironstones
Number of samples collected	Three	Two	Two

Figure 1. (**A**) chrysotile, (**B**) amosite and (**C**) crocidolite.

The chrysotile samples (Ch1) were collected from the Havelock mine, Eswatini. In southern Africa, the Havelock orebody is one of the largest chrysotile asbestos deposits [27]. The Havelock asbestos deposits occur within the Swartkoppie Formation of the Onverwacht Group in the south-eastern part of Precambrian layered ultramafic complexes of the Barberton greenstone belt [28,29]. Chrysotile asbestos occurs as serpentinite lenses or pods within the main lithological components of the Swartkoppie Formation [30]. The differentiated and serpentinised ultramafic bodies contain units of pyroxenite, metagabbro, dunite and peridotite, where relict and strongly altered olivine, clinopyroxene, orthopyroxene and chromite are the major rock forming minerals [31]. The asbestos deposits are structurally controlled and localised within the deformation zones of the host rocks in the form of cross-fibres in a stockwork of veins [27,30,32]. Hydration resulted in the partial or complete serpentinization of the ultramafic rock [31]. Chrysotile asbestos formation requires a specific combination of tectonic controls including folding, faulting, shearing, serpentinization and metamorphism [31]. The original material is provided by the host rocks to form veins and therefore fibre composition reflects vein composition [33].

The amosite samples (Am2) were collected from Penge, Limpopo Province where they occur as layered, extensive continuous seams within the banded iron formations (BIF) [34]. Situated in the metamorphic aureole of South Africa's Bushveld Complex, the Penge Iron Formation of the Transvaal Supergroup is a unique succession as it contains both amosite and crocidolite asbestos [35,36]. The distribution of amosite and crocidolite asbestos seams in the Penge Iron Formation are controlled by bulk-rock composition. However, some evidence suggests that with increasing metamorphic grades, amosite replaced crocidolite [36]. The fibrous amosite units in the BIF are in transitional contact with and underlain by a thick dolomitic sequence [31]. An angular disconformity marks the upper contact of the BIF, which is overlain by a quartzites and shale sequence. Alternating bands of dark-coloured magnetite, grunerite and graphite and light-coloured chert, quartz and siderite comprise the iron formation [36]. The bands are laterally extensive and range from several centimetres in thickness to microscopic [37]. The fibrous amosite asbestos is found as clearly defined lithological units [36], typically in several units of micro-banded magnetite-grunerite banded iron formation [36]. For the most part, amosite is found in lenses of fibrous masses with their long axes perpendicular to the country rock [38].

The crocidolite samples (Cr3) collected from the Prieska Mine in the Northern Cape occur as seams interbedded within the banded iron formations [34]. In the Northern Cape the crocidolite fields extend >450 km south of Prieska to the border of Botswana, where the blue fibres occur in the Asbestos Hills BIF as 1 to 50 mm cross-fibre seams [32,39].

Lengths of fibres are important when considering their removal from the lungs via macrophage cells that under normal circumstances eliminate foreign particles from the lungs [40]. Fibres with lengths greater than the diameter of macrophages cannot be removed, resulting in macrophage death and inflammatory cytokines release into surrounding tissues resulting in fibrosis or asbestosis if inhaled as collagen builds up [41]. Numerous studies have indicated that fibres with lengths less than 5 μm do not have any significant biological potency as they are cleared by macrophages [42]. Fibres with lengths > 10 to 15 μm have a greater probability of persisting in the lungs for extended time periods [42]. Thus, length characterisation of exposed asbestos-containing mine rock waste is important with regard to the relative health risks and effects [40].

Crocidolite is believed to have formed mainly from sodium-rich brines that moved through the iron formation [37]. Unlike the host rocks, crocidolite is extremely resistant to weathering and persists at the surface [31]. The fibrous crocidolite asbestos deposits demonstrate significant blue colour variation [31].

2.2. Polarised Light Microscopy (PLM)

Visible asbestos fibers extracted from their host rocks were cut with scissors and positioned on a glass slide. A small drop of eugenol (refractive oil index n = 1.54) was then deposited on the fibres after which they are covered with cover slips. The slide was optically scanned, and the asbestos mineral was identified using its optical properties (morphology, colour, pleochroism, birefringence, extinction characteristics and sign of elongation). Coatings on the fibres sometimes obscure the optical properties of the asbestos minerals and fibres finer than the microscopes resolving power (ca. 0.3 μm) and are not detectable. The asbestos mineral fibres colour and index of refraction may be altered or changed by acid and heat treatment.

2.3. Crushing

Simple crushing, using a mortar and pestle, enabled to explore whether the long, visibly elongated minerals in each of the rock samples are fibrous. Elongated minerals sometimes become matted together during crushing and form a ball and/or separate into needles or fibres; these are considered as fibrous and potentially asbestiform. Those that are easily crushed into a powder are not fibrous and therefore deduced not to be asbestiform. However, cleavage fragments and asbestiform fibres may occur in close association. Powdered samples were studied using X-ray diffraction (XRD), X-ray fluorescence (XRF) and Brunauer–Emmett–Teller (BET) surface area analysis. A rotary splitter was used to reduce the samples, which were then powdered in a mortar and pestle.

2.4. X-ray Diffraction (XRD)

The samples were analysed via X-ray diffraction (XRD) to determine the mineral type. The XRD analysis was performed by using the back-loading preparation method. Diffractograms were attained by employing a Malvern Panalytical Aeris diffractometer with PIXcel detector and fixed slits with Fe filtered Co-Kα radiation. Phases were determined by means of X'Pert Highscore plus software (version 2.1. PANalytical, Malvern, UK). The Rietveld method (quantitative analysis) [43] was used to estimate the relative phase amounts (weight %). The phases were identified using X'Pert Highscore plus software. The relative intensities and d-spacing were calculated from the diffractogram data.

2.5. X-ray Fluorescence (XRF)

Major and trace element concentrations of the four asbestos mineral fibre samples were determined using X-ray fluorescence (PANalytical PW2404 x-ray spectrometer). Major elements were determined using the Norrish Fusion 1 technique [44] using in-house correction procedures outlined in [45]. Sample weight used was 0.35 gm and flux weight of 2.5 gm. Samples were fused using Johnson Matthey Spectrolflux 105 at 1100 °C and raw data corrected. Standard calibrations were prepared using synthetic oxide mixtures, international standard as well as in-house controls. Calibration standards were from International Reference Materials USGS series (USA) and NIM series (South Africa). Pressed pellets were prepared for trace element analysis and the data corrected for matrix effects using Compton peak monitoring.

2.6. BET-N_2 Specific Surface Area Determination

Specific surface area is an important characteristic of mineral fibres [46] and a significant parameter related to the many attributes linked to its toxicity and carcinogenicity [47,48]. The specific surface area of the asbestos rock samples was determined via the BET method [49] using a Micromeritics TriStar 3000 V6.05 A surface area analyser with N_2 as absorbing gas. A mass of ~0.2 g of each sample was placed in BET sample tubes and degassed for 4 h. The samples were then loaded into the BET instrument and N_2 adsorption isotherms were obtained to determine the specific surface area.

2.7. Bio-Durability Tests

The ability of mineral fibres to resist chemical and/or biochemical alteration is referred to as biodurability [50–52]. The biodurability of the asbestos samples was determined through batch dissolution experiments at 37 °C and continuous agitation (90 rpm). Batch dissolution experiments (water–rock interaction study) allow for dissolution rates to be measured in a setting dominated by fluids [53,54]. The experiments were conducted at 37 °C to simulate body temperature [53]. Although the intricacy of the human body cannot be replicated, these experiments provide a basis to assess the biological disintegration of the different mineralogical types of asbestos [53]. Seven batch reactors were set up for each type of asbestos sample and the change in sample mass at different intervals of time was measured. The advancement of the dissolution reaction was determined at the following sampling times: 24 h, 48 h, 1 week, 2 weeks, 1 month, 2 months and 3 months. The batch experiments were conducted in 100 mL Erlenmeyer flasks containing water and HCl solution (50 mL) at a pH of 4 and 50 mg of sample. At each sampling time, the content of the flask was vacuum filtered using 0.22 μm ϕ (phi) cellulose Merck Millipore filters (ashless grade). The mass of the solid residue was determined by measuring the weight difference of the initial sample mass and the solid residues, after filtration and drying, known as the filter mass. The following equation was used to calculate the dissolved mass fraction (DMF) of the chrysotile, amosite and crocidolite (Equation (1)):

$$\text{DMF} = 1 - \frac{Mt}{Mo} \text{ or DMF} = \frac{Mo - Mt}{Mo} \tag{1}$$

where M_o is the initial mass of the solid at time = 0 and M_t is the mass of the solid at time t.

The dissolution efficiency of the chrysotile, amosite and crocidolite was also calculated using the raw data of total mass loss after 720 h. The dissolution efficiency (D%) of the asbestos rock sample in HCl-water solution was calculated according to Equation (2) [55,56]:

$$\text{D\%} = \frac{Mi - Mt}{Mi} * 100 \tag{2}$$

where M_i is the initial mass (grams), and M_t is the total mass loss (g).

3. Results

3.1. Bulk Material Description

Chrysotile (Ch1) collected from the Havelock Asbestos Mine (Bulembu) appears homogenously distributed within the rock (Figure S1A). These chrysotile fibres occur in veins in the serpentinite (Figure S1B). The fibres within the veins display a combination of straight, curved, and contorted forms. A partitioning parallel to the vein walls caused splitting of the cross-fibre vein and subsequent shortening of the fibre length in relation to the vein width (Figure S1D). Such disassociation affects the degree of fibre to sidewall cohesion enabling fibres to separate easily and become dispersed from the host rock. The highly fibrous components occupy ~80% of the rock sample and are easily separable. The bundles of chrysotile fibres are pale green in colour but separate to form a fluffy mass of white fibres (Figure S1C). The individual fibres are extremely fine, flexible and have a curved and wavy appearance. When observed individually, the chrysotile fibres are white in colour and have a silky lustre. The crystals comprising the fibres are short (~2.5 cm) and extremely hair-like. The mineral fibres retained their fibrous form and aspect ratio upon crushing in a mortar and pestle. The fibres can be bent and twisted without breaking indicating that the chrysotile fibres have high tensile strength.

The amosite (Am2) consists of long (~20 cm), thin, straight, brown coloured fibres that form in bundles and have a 'paintbrush-end' effect (Figure S2A). The fibres from the amosite sample are very brittle and 'shatter' easily when brushed with the dissecting forceps. The bundle of amosite fibres demonstrates a slight curvature because of their long length. The surface of the amosite fibre bundle shows individual matted and splintery fibres (Figure S2B).

The crocidolite sample (Cr3) consists of long (~10 cm), straight, greyish to pale blue coloured fibres in a bundle (Figure S3A). The poly-filamentous bundle is macroscopically curved and contorted (Figure S3B) indicating flexibility. The fibres are brittle and have a silky lustre. The bundle of fibres is easily parted, with a longitudinal fine structure, and is tufted at the ends. These fibres show partial flexibility when bent.

The elongated shape of the minerals visibly observed in hand sample allowed direct length measurements to be taken (Table 2).

Table 2. Fibre lengths for 21 counts for each mineral in hand sample.

	Chrysotile		**Amosite**		**Crocidolite**	
Counts	**Length (cm)**	**Width (μm)**	**Length (cm)**	**Width (μm)**	**Length (cm)**	**Width (μm)**
1	2.5	12.5	20	28.5	6	6.25
2	0.5	12	5	26.5	10	6.2
3	1.2	12.2	15	28	11	6.3
4	1.8	12.2	10	23.8	8	6
5	2.2	12.6	9	25.5	8	6
6	0.9	12.1	13	28	6	6
7	1.36	12.3	19	25	10	6.2
8	2.4	12.3	7	26.5	10	6.2
9	2.5	12.4	18	28	9	6.2
10	1.9	12.5	18	22.3	6	6.2
11	1.7	12	16	20.5	6	6.3
12	2.1	12.1	15	28.1	8	6
13	0.8	12.4	13	28.2	9	6.1
14	2	12.6	8	28	10	6
15	2	12.4	15	20.2	10	6.2
16	1.1	12.3	5	17	10	6.2
17	0.6	12.2	8	20	6	6.2
18	2.5	12.2	11	18.9	8	6.3
19	2.3	12.2	19	20	10	6.1
20	1.7	12.2	20	28.5	6	6.2
21	2	12.6	20	27.1	6	6.2
Average	1.72	12.3	13.52	24.7	8.24	6.16
Minimum	0.5	12	5	17	6	6
Maximum	2.5	12.6	20	28.57	11	6.3
Variance	0.40	0.30	25.11	13.87	3.13	0.01
Standard dev.	0.63	0.18	5.01	3.72	1.77	0.10

3.2. Polarised Light Microscopy (PLM)

The cross-polarizing light microscopy images of chrysotile, amosite and crocidolite are shown in Figure 2.

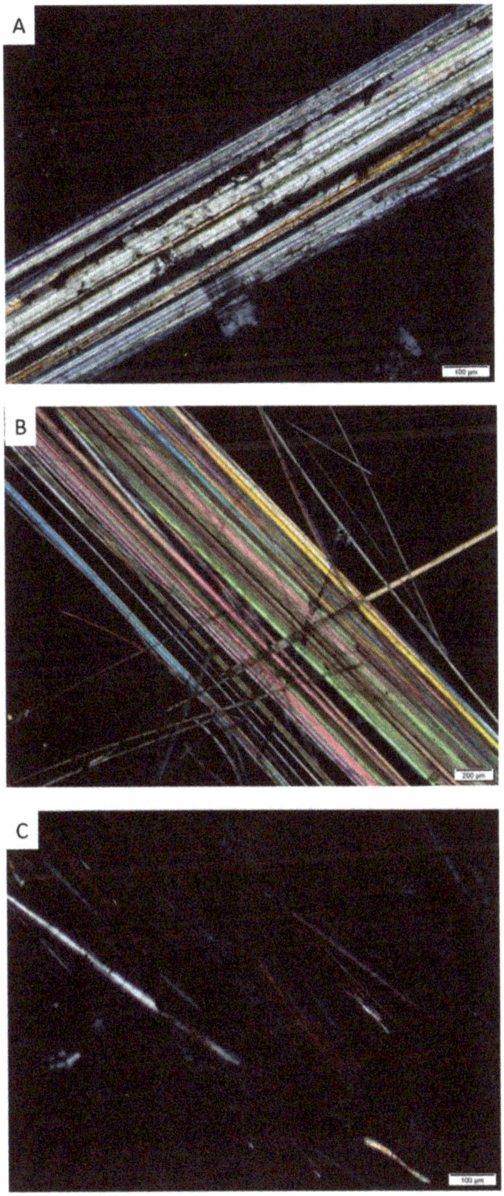

Figure 2. (**A**) Sample Ch1. Partially altered chrysotile fibres shown by amorphous, irregular material and cloudiness (XPL); (**B**) Sample Am2. Extremely fine amosite fibres showing parallel alignment and matting (XPL), and (**C**) Sample Cr3. Crocidolite fibres with splayed ends (XPL).

The morphology of the chrysotile fibres is asbestiform containing kinks and larger bundles with splayed ends. In plane polarised light (PPL), the chrysotile fibres have low relief and show weak pleochroism with purple, light, and dark brown and pale-yellow colours observed (Figure S4). In cross polarised light (XPL), the fibres give off pale interference colours, including white, yellow, orange, red and purple and show a parallel extinction angle (Figure S5). The individual fibres are distinct in both PPL and XPL by various colours and birefringence. The cloudy yellow-orange character of certain bands in XPL emphasises the structure of these veins. The margins of the fibres are outlined with serpentine or material shown as irregular, slightly anisotropic and potentially cryptocrystalline (Figures S6 and S7).

The morphology of the amosite fibres is asbestiform with straight fibres and fibre bundles. The ends of the fibre bundles show a broom-like or splayed appearance. Under plane polarised light (PPL), the amosite has a weak to medium relief and weak to moderate pleochroism displaying brown, purple, pale-green, and pale-yellow colours. In cross polarised light (XPL), pink, blue, purple, yellow, orange, and white interference colours are observed along with a parallel extinction angle. Extremely fine amosite fibres with parallel alignment and matting are shown in Figures S8 and S9. The maximum angle of pleochroism and birefringence is observed at 45° (Figures S10 and S11).

The morphology of the crocidolite fibres is asbestiform with straight fibres and fibre bundles. In plane polarised light (PPL), the crocidolite fibres have moderate relief and are weakly pleochroic with blue, black, and pale-yellow colours displayed (Figure S12). Under cross polarised light (XPL), blue and pale-yellow interference colours and a parallel extinction angle are observed (Figure S13). The fibre bundles have clearly observed splayed ends (Figures S14 and S15). Crocidolite shows the least number of interference colours amongst all three of the various asbestos mineral fibres studied.

3.3. *X-ray Diffraction (XRD)*

Although the fibres were individually extracted and clearly identifiable impurities removed, some mineralogical heterogeneity is unavoidable, and thus, XRD was employed to determine the bulk mineralogical compositions of the asbestos samples and the recognition and identification of any impurities in each sample. The XRD diffraction patterns (λ (Co$K\alpha$) = 1.78892) of the samples are shown in Figure 3 with the corresponding numerical 2θ position and intensity data given in Table S1.

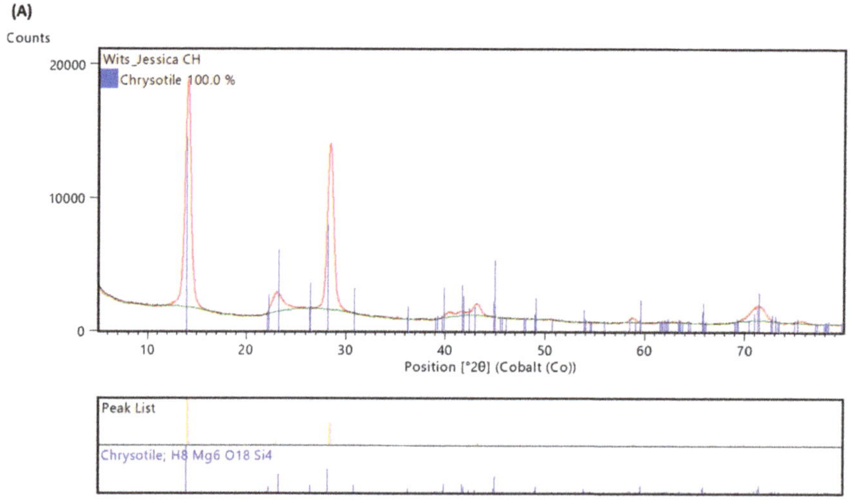

Figure 3. *Cont.*

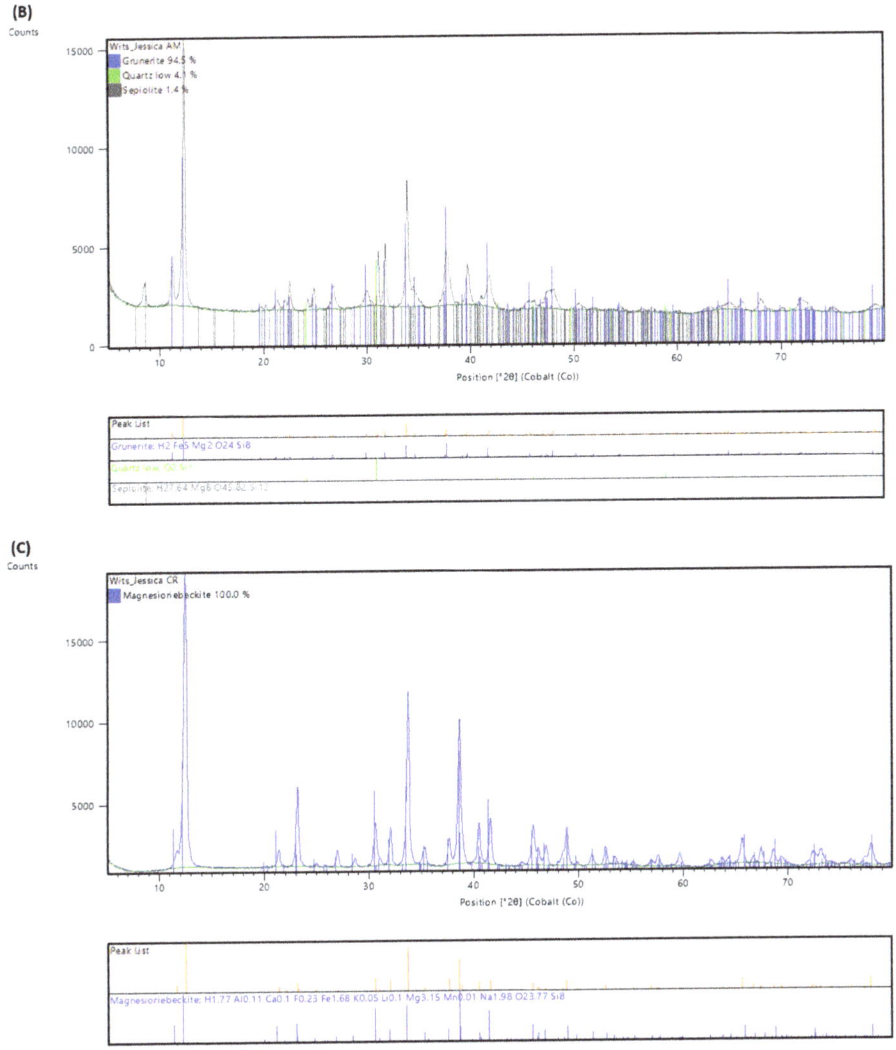

Figure 3. (**A**) Sample Ch1. Chrysotile asbestos XRD diffractogram and Rietveld refinement phases; (**B**) Sample Am2. Amosite asbestos XRD diffractogram and Rietveld refinement phases; and (**C**) Sample Cr3. Crocidolite asbestos XRD diffractogram and Rietveld refinement phases.

The relative intensity and *d*-spacing calculated for the chrysotile, amosite (grunerite) and crocidolite (magnesio-riebeckite) phases (Table S1) correspond to those of the known principal lattice spacings for each of the asbestiform minerals [57]. The Rietveld refinement data are shown in Table 3.

Table 3. Abundance (%) of mineral phases detected via XRD analysis.

Asbestos Rock Sample	Phases Detected (% Composition)
Chrysotile	Chrysotile (100%)
Amosite	Amosite (94.5%) >> Quarts low (4.1%) > Sepiolite (1.4%)
Crocidolite	Magnesio-riebeckite (100%)

% abundance represents the modal amounts of minerals (quantitative analysis) present in asbestos rock samples.

Chrysotile and crocidolite can be shown to be the only mineral phases indicating the purity of the fibres in each sample. In contrast, the amosite sample contains quartz and sepiolite as two other mineral phase.

3.4. *X-ray Fluorescence (XRF) Major and Trace Elemental Analysis*

X-Ray fluorescens spectroscopy (XRF) was used to quantify the major and trace elements in asbestos-containing mine waste rocks. The major and trace concentrations were determined to define the amount of harmful elements that can potentially be released into both the environment and possibly absorbed by the human body. Average major and trace elemental concentrations (n = 2) of each asbestos rock waste are presented below (Tables 4 and 5).

Table 4. Major element analysis (n = 2).

Oxides (wt%)	Chrysotile	Amosite	Crocidolite
SiO_2	42.08	48.93	51.52
Al_2O_3	0.59	0.36	0.07
Fe_2O_3	2.00	41.37	38.31
MnO	0.03	0.64	0.06
MgO	40.83	5.84	2.23
CaO	0.06	2.02	0.35
Na_2O	0.07	0.00	6.22
K_2O	0.01	0.24	0.10
TiO_2	0.03	0.03	0.03
P_2O_5	0.01	0.02	0.01
Cr_2O_3	0.01	0.02	0.01
NiO	0.19	0.01	0.01
LOI	12.18	0.48	1.24
Total	100.08	99.87	100.15

The samples display the following distinctive major element geochemical characteristics: Chrysotile samples contain the most aluminium and magnesium; amosite samples displayed the highest concentrations of iron and calcium, and crocidolite samples contained the greatest sodium.

Table 5. Trace element analysis (n = 2).

Element (ppm)	Chrysotile	Amosite	Crocidolite
Sc	6.4	4.64	D.L.*
V	16.42	3.51	3.91
Cr	83.4	4.61	D.L.*
Co	52.55	D.L.*	D.L.*
Ni	1518.54	51.24	11.86
Cu	21.99	36.77	35
Zn	16.19	41.62	12.66
Ga	D.L.*	D.L.*	1.63
Rb	D.L.*	20.77	1.02
Sr	0.74	26.63	0.86
Y	0.77	3.78	1.82
Zr	0.49	3.7	0.38
Nb	D.L.*	0.85	D.L.*
Mo	D.L.*	0.62	D.L.*
Ba	D.L.*	28.17	1.69
Pb	6.62	5.32	5.06
Th	D.L.*	D.L.*	D.L.*
U	D.L.*	D.L.*	D.L.*

* D.L.–detection limit.

3.5. BET-N_2 Specific Surface Aarea

The specific surface area (SSA), determined by the N_2 BET procedure, of fibrous minerals is defined as the surface area per unit volume. 'Specific' or 'reactive' are two ways in which the surface area may be defined [58]. Gas adsorption and the Brunauer–Emmett–Teller (BET) equation is used to determine the specific surface area, whereby the total surface area is divided by the mass of the sample [49]. The BET N_2 specific surface areas and associated parameters of the extracted fibres are given in Table S3.

3.6. Bio-Durability Tests

The solid asbestos residue was separated from the solution via filtration at the end of each time during the bio-durability test and the mass measured (Table 6).

Table 6. Measured mass (mg) of the solid asbestos sample residue after each time period during dissolution in acidic solution.

Time (Hours)	Chrysotile	Amosite	Crocidolite
0	50	50	50
24	25.2	39.1	37.7
48	24	38.6	33.6
168	21	36.3	32.3
334	18.6	35.5	31.7
720	16.6	35.4	31.1
Total mass loss (mg)	33.4	14.6	18.9

The weight loss for all three types of asbestos reached a plateau after 186 h indicating the completion of the chemical reaction. The mass loss of the experiments for each asbestos sample was summed up, although 100% solid mass loss is never achieved. The reason is not incomplete dissolution, but rather the precipitation of silica back out of the solution [23]. The dissolved mass fraction (DMF) calculated for the asbestos samples are given in Table S3 and graphically represented in Figure 4.

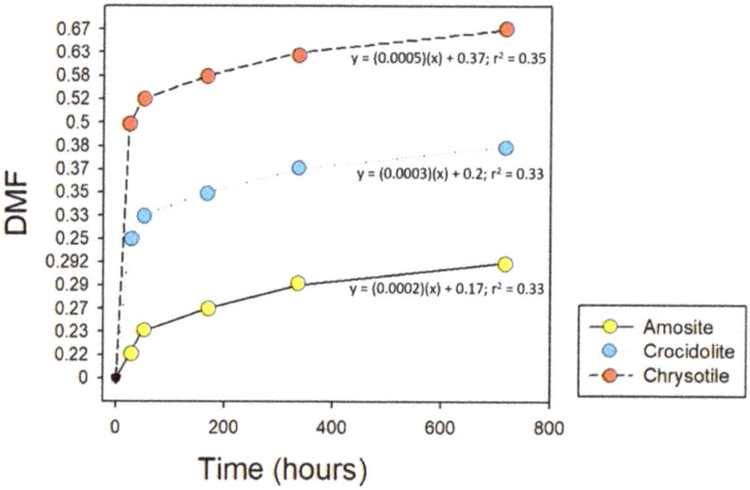

Figure 4. The dissolved mass fraction of the asbestos samples over time (hours).

The dissolution efficiency for chrysotile, amosite and crocidolite was calculated as 66.8%, 29.2% and 37.8%, respectively.

4. Discussion

4.1. Mineralogical Impact

Mineralogical and geochemical characterization of exposed asbestos-containing mine waste allows improved risk assessment with direct applications in predictive management. Mineral dust and fibres are heterogeneous substances with respiratory sensitizing properties. The complex nature and exposure to mineral fibres necessitate procedures that account for this complexity if the attenuation of the human health impacts is to be conducted reliably. Importantly, the selected methods are based on what is realistically practical for developing countries where funds and scientific engagement are limited. In hand sample, all three types of asbestos show asbestiform morphology, including fibre thickness, parallelism in arrangement, separability, and flexibility. The fibres retain their aspect ratio forming numerous finer fibres during breakage caused by crushing. All samples showed moderate cohesion as fibres were released from bulk material when vigorously disturbed by hand. The three types of asbestos minerals sampled are cross-fibres in veins for chrysotile and seams for the amphiboles.

The fibre geometry (Table 2), defined by its diameter (D) and length (L), is a key parameter in its pathogenicity, toxicity and inflammation [59,60]. Polarising light microscopy (PLM) is most used for semi-quantifying the percentage and identifying the type of fibre [61,62], but can also be used to accuratly measure fibre dimensions >1 μm in diameter [63,64]. The dimensions provide information on the respirability and biological activity of airborne fibres [65]. The 'Stanton Hypothesis' states that the optimum fibre morphology for generating intrapleural tumours is $D \leq 0.25$ μm and $L > 8$ μm, and was derived from experimental observations succeeding fibre implantation and injection into animals [66]. According to this model, 'frustrated phagocytosis' [67] results because phagocytic cells are unable to eliminate 'Stanton fibres', i.e., needle-shaped particles with $L > 8$ μm [66].

All samples consist of poly-filamentous fibre bundles with parallel-sided, long, thin fibres having straight extinction. Splayed ends and fibre curvature were characteristically demonstrated by the crocidolite samples. Under high magnification, amosite and crocidolite minerals do not appear compact having gaps and visible divisions between fibres further pointing to their asbestiform nature. Unlike amosite and crocidolite, individual chrysotile fibres are less clearly distinguishable under the polarising light microscope (PLM). Chrysotile appeared more compact; however, upon closer inspection, divisions between fibres were also identified, but less prominent than that of amosite and crocidolite. In PLM, the chrysotile fibres appear to be more tightly welded together with little porosity between them. Optically, the colours of the fibres are both homogenous and heterogenous under PLM.

Chemical heterogeneity suggested by colour variations in the fibre provides evidence of partial replacement of pre-existing fibres or generations of fibre growth. Homogenous colours were generally observed along the lengths of the amosite and crocidolite fibres (Figures S8, S10, S12 and S14). Heterogenous colours within individual fibres is only observed in chrysotile (Figures S4 and S6) and implies compositional variability. This is further documented under cross-polarising light (Figures S5 and S7). Both amphibole asbestos samples did not appear to show any modifications optically and texturally in PLM consistent with the homogenous colouring. In comparison with the amphibole asbestos samples, the chrysotile asbestos samples exhibit a more complex microstructure matrix under PLM, in which several coexisting textures are apparent.

The asbestos-containing rock materials are natural samples collected from mines, and thus, the presence of mineral impurities is expected [68,69]. Typically, natural asbestos-containing rock samples occurs with other non-asbestiform morphologies and minerals [70]. Although these associated phases are thought to be harmless, little information on their potential toxic effects exists [70]. Thus, the characterisation of natural assemblages should include all phases occurring with the suspected asbestos fibres [70]. The results obtained via XRD analysis facilitate the identification of co-occurring fine-grained minerals that are difficult to detect optically. Chrysotile (serpentine) and crocidolite (amphibole) samples are homogeneous containing no additional mineral phases. Amosite displays mineral phase heterogeneity and crystalline impurities of ~4.1% quartz (SiO_2) and ~1.4% sepiolite ($Mg_4Si_6O_{15}.6H_2O$). Fibrous and crystalline quartz (SiO_2) is known as a prolific cytotoxic particle that results in lung tumours upon inhalation [71]. The phyllosilicate mineral, sepiolite ($Mg_4Si_6O_{15}.6H_2O$), belongs to the mineral group hormite and is characterised by a fibrous habit [72]. Given limited studies and inadequate evidence for the carcinogenic effects of sepiolite in humans, only few animal studies have suggested that sepiolite is to be included in the Group 3 carcinogen category [73].

A survey of the published literature indicates very limited research and knowledge exists regarding the carcinogenic and pathogenic effects of asbestos associated mineral phases following chronic inhalation. However, asbestos-associated mineral phases should not be neglected when considering the combined factors encompassing the toxicity of asbestos-containing mine-wastes.

4.2. Geochemical Impact

Of further importance is the ability to quantify the amount of potentially harmful elements that can potentially be released both into the environment and ingested by the human body. The high concentrations of Al, Mg, Mn and Fe are not unexpected as they are major rock forming elements and are among the primary constituents in sediments and soils [74]. One of the most important factors for fibre-induced patho-biological activity is the total iron content of the asbestos minerals [60,75]. Siderosis is caused by the inhalation of iron-bearing compounds [76]. As iron acts as a catalyst for reactions involving release of reactive oxygen species and lipid, protein and DNA damage, it is a significant property in determining asbestos toxicity [77]. Iron becomes available at the reacting surface of fibres during dissolution where, through a Haber–Weiss chain reaction sequence, it promotes hydroxyl radical formation that damages DNA [78,79].

In addition to iron, other major chemical elements have been reported to participate in asbestos toxicity following inhalation [80,81]. Silicon (Si) is the second most abundant element, and when inhaled, this results in numerous pathologies such as silicosis [82]. Being the third most abundant element in the Earth's crust, the environmental toxicology of aluminium has been revealed in recent investigations to cause numerous diseases. It thus presents a major threat to plants, animals and humans [83,84]. The bulk chemical analysis of the chrysotile sample indicates only a slight deviation from the ideal composition of serpentine, containing very little Fe and Al. Substitution in chrysotile may occur in both the octahedral (O) and tetrahedral (T) sheets making up this layer silicate [85]. In the 1:1 T-and-O-sheet ratio, both Si^{4+} and Mg^{2+} can be replaced by Al^{3+}, respectively, with an average Al_2O_3 content of <0.9 wt.%, while the FeO content may be as much as 6 wt.% [86]. Mg^{2+} in the O sheet can also be replaced by Fe^{2+} and Fe^{3+} while Si^{4+} replacement in the T sheet is infrequent and minor. In the octahedral sheet, both Fe^{2+} and Fe^{3+} can replace Mg and the eventual replacement of Si^{4+} by Fe^{3+} may occur, although Al^{3+} is preferentially hosted in this position. The presence of both Fe^{2+} and Fe^{3+} exclusively in six-fold coordination has been suggested by [69].

The different samples show considerably variable Mn concentrations with the highest amount found in the amosite. These results concur with those reported in [87], which explained that Mg in all the M(1), M(2), M(3) and M(4) sites of magnesium-iron-manganese-lithium amphiboles may be substituted with Mn [11,12]. Manganese is an essential trace element for biological organisms. However, in excess, manganese poising ensues typically in the brain and lungs [88]. Thus, managing the environmental entrance and migration of manganese is a marked human health risk to humans [89].

Trace elements within mineral fibres may, in addition to the major elements, take part in the fibre toxicity [11,12,90–92]. The presence of trace metals in fibres and their effects on the carcinogenesis of asbestos has been documented by IARC, 2012 [93]. In our study, the highest content of Ni, Co, Cr and V was observed in chrysotile; amosite contained the greatest concentration of Zn, Rb, Sr, Zr, Nb and Ba; and crocidolite was the only sample in which Ga was measured. Overall, the amosite rock sample contained the greatest number of detected trace elements and crocidolite had the overall lowest concentration of trace elements. Mn and trace metals, such as Ni, Cr, Ni and Cu, in chrysotile almost exclusively represent isomorphous substitution of Mg [16,94,95]. Unlike that for antigorite and lizardite minerals, trace metal substitution in chrysotile is typically more restricted [96]. Interestingly, although characterised by different geological conditions of formation, the detection of Cu, Ni, Zn, Sr, Yt and Zr was shared by the three asbestos rock samples. Lead was detected in all samples, and is a considerably toxic metal [97]. Unlike other metals (e.g., copper, manganese, and zinc) lead serves no biological functions [97] and is highly toxic being listed as a hazardous heavy metal contaminant [98]. The toxicity of lead in living cells is caused by oxidative stress and ionic mechanisms [99,100]. Due to its high toxicity, lead is ranked among the 10 top priority substances of concern to the public (ATSDR, 2018). Several effects arise from the contamination of soil with lead including the reduction in soil fertility, microbial diversity and nutrients [98]. Nickel was detected in all samples with chrysotile exhibiting an exceptionally high concentration (1519 ppm). A variety of adverse human health effects, such as lung fibrosis, kidney diseases, contact dermatitis, cardiovascular diseases and cancer of the respiratory tract, are forms of nickel allergy that can result from contact with nickel compounds [101–103]. Bioavailable Ni^{2+} toxicity at the intracellular sites was postulated by [104]. In humans, CD4+T lymphocytes cause the greatest apoptosis and DNA damage, and caspase-9-positive T cells are induced by Ni^{2+} at a concentration of 0.05 mM [105].

Chromium was found in the chrysotile (83 ppm) and amosite (4.6 ppm) samples and represents a source of concern. Chromium results in the formation of hydroxyl and superoxide radicals described by the Fenton reaction [106]. Fenton reactions induced by Cr^{3+} damage proteins [107]. The direct binding of Cr^{3+} to numerous non-metalloproteins has been shown in Cr associated patients to result in the loss of their biological functions [108]. Chromium is also known to cause several health problems such as vomiting, kidney failure, mouth ulcers, lung cancer, stomach cancer, indigestion and acute tubular necrosis in humans following contact [109–111].

Vanadium was measured in chrysotile (16 ppm), crocidolite (3.9 ppm) and amosite (3.5 ppm) samples. Any of the three oxidation states of vanadium can produce genotoxic effects [112]. However, double-strand breaks are induced by V^{4+} causing lesions and creating aberrations in structural chromosomes [112,113]. Asthma, anaemia and rhinitis can be caused by excessive amounts of vanadium in the body and even increase the possibility of lung cancer and uraemia occurrence [114–117]. The release of vanadium from asbestos fibres into solution does not represent a concern as it is very low [112].

Molybdenum was measured in amosite (0.9 ppm). Biologically, molybdenum is an essential nutrient required by humans. However, inhalation and exposure to excess levels can decrease lung functioning, coughing and dyspnoea [118,119].

The substantial presence of potentially toxic trace elements at concentrations measured in the studied chrysotile and amphibole asbestos samples may be explained, primarily, because of isomorphic substitutions in particular crystallographic positions [94,120]. The variability in potentially toxic elements amongst the studied samples, on the other hand, is best explained by the shared chemical changeability exhibited by asbestos mineral particles [121] and the different petrological and geochemical processes occurring during their formation [122]. High levels of heavy metals in the wastes indicate the possibility of their release into the soil, water and atmospheric environments, presenting an interminable environmental hazard [123].

The heavy metals hosted in fibrous minerals accumulate in the lungs via dissolution following inhalation, altering the normal human lung baseline levels of these elements [124]. The surface area of asbestos has been proposed to play a role in fibre toxicity [125]. The surface area is a factor influencing the rate of dissolution and therefore clearance from the lungs [58,60]. Lung cancer, bronchogenic carcinoma, mesothelioma, etc., are caused when sufficient abundances of heavy metals are accumulated as the human lung tissue is damaged by metal-induced disease [11,12,76,92,124,126]. The concentration range of metals in normal human lungs are reported in Table 7; these ranges are greatly exceeded by their concentrations in the different asbestos types.

The solid mass loss of the bio-durability experiments for each asbestos sample is always less than 100% solid. The reason is not incomplete dissolution but rather the precipitation of silica out of the solution [23]. As demonstrated by the dissolution tests, chrysotile has the lowest bio-durability and amosite the highest. Based on the close link between bio-durability and bio-persistence, it is expected that amosite fibres will have a much longer retention time following inhalation when compared to both chrysotile and crocidolite. Therefore, amosite fibres have a greater toxicity than chrysotile and crocidolite following inhalation due to their greater persistence in the lungs [127].

More recently, in addition to the already stated mineralogical and geochemical properties influencing the toxicity to asbestos exposure, trace element concentrations hosted in asbestos mineral fibres and their role in fibre toxicity have come under the spotlight [122]. The obvious threat of exposure to asbestos is much publicised. Numerous rehabilitation strategies, focused solely on mitigating the dispersion of these mineral fibres, have been considered and implemented. As well as their role in determining fibre toxicity, the elevated concentrations of heavy metals hosted in asbestos minerals pose a profound influence on the quality of the environment. Many potentially toxic elements have been found to be hosted in all forms of asbestos minerals [128–130]. The fundamental factor surrounding these findings is that, in the natural setting, leaching, and weathering of asbestos-bearing rocks results in reduced heavy metal concentrations within the mineral particles themselves and the subsequent increase in concentrations in the surrounding soil and water ecosystems [130,131]. Compared to the maximum limits imposed by environmental governments and agencies, the concentrations of heavy metals in the proximity of asbestos-bearing geological sites are typically one order of magnitude greater [132], as documented, for instance, in the serpentine-derived soils of the Gimigliano—Mount Reventino Unit (GMRU), Calabria Region (S-Italy) [133]. In addition to soils, the interaction of water with asbestos-bearing rocks is also characterised by exceedingly high heavy metal concentrations due to the dissolution of these minerals [19]. The magnitudes of their concentrations and the fact that these toxic elements can be mobilised and dispersed into different terrestrial environments and subsequently absorbed by humans, makes their presence in asbestos-bearing mine waste a consequential public health and environmental threat.

Table 7. Comparison of geochemical data in this study and the concentration range of heavy metal in normal human lungs (ppm).

Metals (ppm)	Chrysotile	Amosite	Crocidolite	Concentration Range of Trace Elements in Normal Human Lungs (ppm) [122]
Al	11,147.8	6802.1	1322.6	
Fe	13,988	289,340	267,940	40–500
Mn	230	4987	465	0.01–3
Mg	246,200	35,220	13,449	
Cr	83.4	4.61	D.L.	0.002–0.50
Co	52.55	D.L.	D.L.	0.002–0.1
Ni	1518.54	51.24	11.86	0.01–1.00
Cu	21.99	36.77	35	1–5.00
Zn	16.19	41.62	12.66	1–30.00
Zr	0.49	3.7	0.38	
Ba	D.L.	28.17	1.69	>1.10
Pb	6.62	5.32	5.06	0.02–0.50

4.3. Geographic Impact and Rehabilitation

Substantial volumes of crushed rock-based wastes were produced during the mining of asbestos mineral resources in Southern Africa. The geographic impact of these derelict asbestos mine sites is vast, littering the landscape across most of rural South and Southern Africa. The accumulation of these historical asbestos-mining rock wastes resulted in large unmanaged mine dumps characterised by poly-mineral and rock assemblages, which tend to be unstable and display unfavourable physical and hydrological properties under prevalent physico-chemical conditions that increase their potentially toxicity [134–136]. Despite the increasing recognition of environmental, health and safety risks associated with derelict asbestos mine sites many of them are exposed, unconfined and still to be rehabilitated. In the context of risk assessment and exposure scenarios in Southern Africa numerous difficulties are faced as statistical information regarding the health response and number of asbestos related disease suffers has not been properly documented and/or acknowledged. The lack of background data makes fully comprehending and assessing the extent of risks from these asbestos sites challenging. The 'rehabilitation industry' in South Africa commonly practices vegetation establishment primarily aimed to generate surface stability with the aim to re-establish and return the site to a functional and sustainable ecosystem [137]. The main, governmental, and environmental agencies involved in derelict asbestos-mine land management have adopted the soil remediation approach to evaluate the rehabilitation requirements [137]. Asbestos mine dumps demonstrate huge variation in physical, geochemical and mineralogical characteristics [138]. Given the site-specific nature of mine rock wastes a thorough investigation and characterisation is the first critical step for the formulation of a rehabilitation plan [139]. Individually and interactively, the physical, mineralogical, and geochemical properties result in impediments and challenges to natural vegetation establishment during rehabilitation and its subsequent sustainability. The way asbestos mine dump rehabilitation is undertaken is based on the credence that ecological restoration and remediation on all derelict and post-mining terrains can be tackled and accomplished in short deadlines. This emphasises the lack of appreciation of the influence of geological conditions on the substrates, the community assembly and plant growth [140,141]. To predict the potential challenges to rehabilitation mineralogical and geochemical characterisation of asbestos mine waste should become a standard practice. Derelict asbestos mine sites in Southern Africa are a major issue due to the geographic expanse and their insufficient rehabilitation outcomes. To tackle this problem, greater attention and more research focus is required to understand how the edaphic conditions including geological factors hinder positive restorative outcomes and therefore impact progressive risk mitigation strategies. The potential exposure scenarios for communities surrounding derelict asbestos mine sites in Southern Africa are not as easily resolved for they include air, water and soil. As communities in these areas are rural and in large have livelihoods based on subsistence farming practices, their very existence is interconnected to the asbestos contaminated land in which they occupy. The complexities between the scientific and social aspects of derelict mine sites in Southern Africa make hazard and risk assessments more challenging compared to those of similar nature in developed countries. In such a context, specific scientific data and solutions to such environmental and human health issues require a paradigm shift for researchers developing solutions that have practical applicability to the location of interest.

5. Conclusions

Mineralogical and geochemical characterisations of asbestos mineral fibres left at derelict asbestos mine sites are important for two major reasons: firstly, to identify and assess their human health hazard and define the toxicity degree; secondly, to define the degree of potential environmental contamination in soils and (sub)surface waters in areas where these minerals occur. The results given in the study indicate that chrysotile, although being the least bio-durable, contains heavy metal concentrations that exceed those of the normal threshold concentrations in lungs beyond which result in functional respiratory problems. Amosite is of particular concern due to its high bio-durability and metal content above values of the safe lung-threshold. These high levels of heavy metals detected in both chrysotile and amosite are potentially harmful not only to human health, but also the environment in general, as they could contaminate the surrounding soil and water, which forms the basis of existence for rural communities in remote locations. To conclude, the cost-effective, reliable, and easily accessible analytical methods applied here substantiate that baseline values, pertinent to the geological material, require revision as very little data are currently reported in the literature and other official reports concerning South African regulations and guidelines.

Supplementary Materials: The following supporting information can be downloaded at: https://www.mdpi.com/article/10.3390/min13101352/s1, Figure S1: Sample Ch1 (A) Chrysotile rock sample; (B) length of fibres spanning the width of veins; (C) Individual masses of matted white fibres; and (D) parting at the centre of the vein width halving the length of the cross-vein fibres. Figure S2: Sample Am2 (A) amosite rock sample and (B) matted and splintery fibres. Figure S3: Sample Cr3 (A) crocidolite rock sample and (b) showing slight curvature of poly-filamentous bundles. Figure S4: Sample Ch1 chrysotile fibre bundle (PPL). Notice the break in the bundle in the top right. Figure S5: Sample Ch1 chrysotile fibre bundle (XPL). Figure S6: Sample Ch1. Partially altered chrysotile fibres shown by amorphous, irregular material and cloudiness (PPL). Figure S7: Sample Ch1. Partially altered chrysotile fibres shown by amorphous, irregular material and cloudiness (XPL). Figure S8: Sample Am2. Extremely fine amosite fibres showing parallel alignment and matting (PPL). Figure S9: Sample Am2. Extremely fine amosite fibres showing parallel alignment and matting (XPL). Figure S10: Sample Am2. Amosite fibres at maximum angle of pleochroism showing heterogenous colours (PPL). Figure S11: Sample Am2. Amosite fibres at maximum angle of birefringence showing heterogenous interference colours (XPL). Figure S12: Sample Cr3. Poly-filamentous crocidolite (PPL). Figure S13: Sample Cr3. Poly-filamentous crocidolite (XPL). Figure S14: Sample Cr3. Crocidolite fibres with spayed ends (PPL). Figure S15: Sample Cr3. Crocidolite fibres with spayed ends (XPL). Graph S1: Diffractogram and relative phases (weight %) of the chrysotile sample. Graph S2: Diffractogram and relative phases (weight %) of the amosite sample. Graph S3: Diffractogram and relative phases (weight %) of the crocidolite sample. Table S1: The values of 2θ and intensity (I) recorded for each the peaks of each phase from the X-ray diffraction record (λ (CoKα) = 1.78892). Table S2: BET surface area report. Table S3: The dissolved mass fraction (DMF) calculated for the asbestos samples.

Author Contributions: J.S.S., S.M. and R.B. conceptualized the research; R.B. and S.M. provided PhD supervision to J.S.S.; J.S.S. collected samples; J.S.S. and A.H.W. carried out analyses; J.S.S. prepared original draft of manuscript; R.B., S.M. and A.H.W. reviewed and corrected various versions of the manuscript; R.B. and S.M. acquired funding. All authors have read and agreed to the published version of the manuscript.

Funding: Financial support was provided by the DSI-NRF Centre of Excellence for Integrated Mineral and Energy Resource Analysis (DSI-NRF CIMERA) as a PhD scholarship to J. S. Schapira.

Data Availability Statement: All data used are presented as supplemental files. Additional data is available from the lead author on request.

Acknowledgments: Opinions expressed, and conclusions arrived at, are those of the authors and are not to be attributed to the DSI-NRF CIMERA. Paul Nex is thanked for providing some of the sample material. Constructive comments from four anonymous reviewers are acknowledged.

Conflicts of Interest: The authors declare no conflict of interest.

References

1. Cornelissen, H.; Watson, I.; Adam, E.; Malefetse, T. Challenges and strategies of abandoned mine rehabilitation in South Africa: The case of asbestos mine rehabilitation. *J. Geochem. Explor.* **2019**, *205*, 106354. [CrossRef]
2. Harris, L.V.; Kahwa, I.A. Asbestos: Old foe in 21st century developing countries. *Sci. Total Environ.* **2003**, *307*, 1–9. [CrossRef]
3. Ndlovu, N.; teWater Naude, J.; Murray, J. Compensation for environmental asbestos-related diseases in South Africa: A neglected issue. *Glob. Health Action* **2013**, *6*, 82–88. [CrossRef] [PubMed]
4. Williams, C.; Dell, L.; Adams, R.; Rose, T.; Van Orden, D.R. State-of-the-science assessment of non-asbestos amphibole exposure: Is there a cancer risk? *Environ. Geochem. Health* **2013**, *35*, 357–377. [CrossRef] [PubMed]
5. Hodgson, A.A. Nature and paragenesis of asbestos minerals. *Philos. Trans. R. Soc. Lond. A* **1977**, *286*, 611–624.
6. Okayasu, R.; Takahashi, S.; Yamada, S.; Hei, T.K.; Ullrich, R.L. Asbestos and DNA double strand breaks. *Cancer Res.* **1999**, *59*, 298–300. [PubMed]
7. Kilburn, K.H. Indoor air effects after building renovation and in manufactured homes. *Am. J. Med. Sci.* **2000**, *320*, 249–254. [CrossRef]
8. Unfried, K.; Schurkes, C.; Abel, J. Distinct spectrum of mutations induced by crocidolite asbestos clue for 8-hydroxydeoxyguanosine-dependent mutagenesis in vivo. *Cancer Res.* **2002**, *62*, 99–104.
9. Mossman, B.T.; Lippmann, M.; Hesterberg, T.W.; Kelsey, K.T.; Barchowsky, A.; Bonner, J.C. Pulmonary endpoints (lung carcinomas and asbestosis) following inhalation exposure to asbestos. *J. Toxicol. Environ. Health* **2011**, *14*, 76–121. [CrossRef]
10. Pugnaloni, A.; Giantomassi, F.; Lucarini, G.; Capella, S.; Bloise, A.; Di Primio, R.; Belluso, E. Cytotoxicity induced by exposure to natural and synthetic tremolite asbestos: An in vitro pilot study. *Acta Histochem.* **2013**, *115*, 100–112. [CrossRef]
11. Bloise, A.; Barca, D.; Gualtieri, A.F.; Pollastri, S.; Belluso, E. Trace elements in hazardous mineral fibres. *Environ. Pollut.* **2016**, *216*, 314–323. [CrossRef] [PubMed]
12. Bloise, A.; Punturo, R.; Catalano, M.; Miriello, D.; Cirrincione, R. Naturally occurring asbestos (NOA) in rock and soil and relation with human activities: The monitoring example of selected sites in Calabria (southern Italy). *Ital. J. Geosci.* **2016**, *135*, 268–279. [CrossRef]
13. Bloise, A.; Catalono, M.; Gualtieri, A.F. Effect of grinding on chrysotile, amosite and crocidolite and implications for thermal treatment. *Minerals* **2018**, *8*, 135. [CrossRef]
14. Harper, M. 10th Anniversary critical review: Naturally occurring asbestos. *J. Environ. Monit.* **2008**, *10*, 1394–1408. [CrossRef]
15. Culley, M.R.; Zorland, J.; Freire, K. Community responses to naturally occurring asbestos: Implications for public health practice. *Health Educ. Res.* **2010**, *25*, 877–891. [CrossRef]
16. Bloise, A.; Belluso, E.; Critelli, T.; Catalano, M.; Apollaro, C.; Miriello, D.; Barrese, E. Amphibole asbestos and other fibrous minerals in the meta-basalt of the gimigliano-mount reventino unit (Calabria, South-Italy). *Rend. Online Soc. Geol. Ital.* **2012**, *21*, 847–848.

17. Bloise, A.; Critelli, T.; Catalano, M.; Apollaro, C.; Miriello, D.; Croce, A.; Barrese, E.; Liberi, F.; Piluso, E.; Rinaudo, C.; et al. Asbestos and other fibrous minerals contained in the serpentinites of the Gimigliano-Mount Reventino Unit (Calabria, S-Italy). *Environ. Earth Sci.* **2014**, *71*, 3773–3786. [CrossRef]
18. Punturo, R.; Ricchiuti, C.; Mengel, K.; Apollaro, C.; De Rosa, R.; Bloise, A. Serpentinite-derived soils in southern Italy: Potential for hazardous exposure. *J. Mediterr. Earth Sci.* **2018**, *10*, 51–61.
19. Apollaro, C.; Fuoco, I.; Vespasiano, G.; De Rosa, R.; Cofone, F.; Miriello, D.; Bloise, A. Geochemical and mineralogical characterization of tremolite asbestos contained in the Gimigliano-Monte Reventino Unit (Calabria, south Italy). *J. Mediterr. Earth Sci.* **2018**, *10*, 5–15.
20. Braun, L.; Kisting, S. Asbestos-Related Disease in South Africa: The Social Production of an Invisible Epidemic. *Am. J. Public Health* **2006**, *96*, 1386–1396. [CrossRef] [PubMed]
21. Rees, D.; Myers, J.E.; Goodman, K.; Fourie, E.; Blignaut, C.; Chapman, R.; Bachmann, M.O. Case-Control Study of Mesothelioma in South Africa. *Am. J. Ind. Med.* **1999**, *35*, 213–222. [CrossRef]
22. Ramazzini, C. The global health dimensions of asbestos and asbestos related diseases. *J. Occup. Health* **2011**, *58*, 220–223. [CrossRef]
23. Gualtieri, A.F.; Viani, A.; Sgarbi, G.; Lusvardi, G. In vitro biodurability of the product of thermal transformation of cement–asbestos. *J. Hazard. Mater.* **2012**, *205–206*, 63–71. [CrossRef] [PubMed]
24. Fubini, B.; Otero-Arèan, C. Chemical aspects of the toxicity of inhaled mineral dusts. *Chem. Soc. Rev.* **1999**, *28*, 373–381. [CrossRef]
25. Fubini, B.; Bolis, V.; Cavenago, A.; Volante, M. Physicochemical properties of crystalline silica dusts and their possible implication in various biological responses. *Scand. J. Work Environ. Health* **1995**, *21*, 9–15. [PubMed]
26. Higueras, P.; Oyarzun, R.; Iraizoz, J.M.; Lorenzo, S.; Esbrí, J.M.; Martínez-Coronado, A. Low-cost geochemical surveys for environmental studies in developing countries: Testing a field portable XRF instrument under quasi-realistic conditions. *J. Geochem. Explor.* **2012**, *113*, 3–12. [CrossRef]
27. Anhaeusser, C.R. The nature of chrysotile asbestos occurrences in southern Africa: A review. *Econ. Geol.* **1976**, *71*, 96–116. [CrossRef]
28. Hunter, D.R.; Jones, D.H. *Geological Map Series (1:25,000)*; Sheet 2; Swaziland Geological Survey: Piggs Peak, Swaziland, 1969.
29. Viljoen, R.P.; Viljoen, M.J. The geology and geochemistry of the layered ultramafic bodies of the Kaapmuiden area, Barberton Mountain Land. *Geol. Soc. S. Afr. Spec. Publ.* **1969**, *1*, 661–688.
30. Biljon, V. The chrysotile deposits of the eastern Transvaal and Swaziland. In *The Geology of Some Ore Deposits in Southern Africa*; Haughton, S.H., Ed.; Geological Society of South Africa: Johannesburg, South Africa, 1964; pp. 625–669.
31. Abbott, P. A Review of Asbestos Resources. Master Thesis, Rhodes University, Grahamstown, South Africa, 1983.
32. Hart, H.P. Asbestos in South Africa. *J.S. Afr. Inst. Min. Metall.* **1988**, *88*, 185–198.
33. Riordon, P.H. The genesis of asbestos in ultrabasic rocks. *Econ. Geol.* **1955**, *50*, 67–81. [CrossRef]
34. Cairncross, B. *Field Guide to Rocks and Minerals of Southern Africa*; Struik Nature: Cape Town, South Africa, 2004.
35. Dreyer, C.J.B.; Sohnge, A.P.G. *The Crocidolite and Amosite Deposits of the Republic of South Africa and Bophuthatswana*; Handbook; Geological Survey of South Africa: Pretoria, South Africa, 1992; Volume 12, p. 126.
36. Miyano, T.; Beukes, N.J. Mineralogy and Petrology of the Contact Metamorphosed Amphibole Asbestos-bearing Penge Iron Formation, Eastern Transvaal, South Africa. *J. Petrol.* **1997**, *38*, 651–676. [CrossRef]
37. Genis, J.H. The Genesis of the Blue Amphibole Asbestos of the Union of South Africa. Ph.D. Thesis, University of Cape Town, Cape Town, South Africa, 1961.
38. Reinecke, L.; McClure, L. Variations in the quality of amosite asbestos at Penge, Transvaal. *Trans. Geol. Soc. S. Afr.* **1933**, *36*, 29–39.
39. Howlling, G.E. *Asbestos*; Imperial Institute: London, UK, 1937; p. 88.
40. Strohmeier, B.R.; Huntington, J.C.; Bunker, K.L.; Sanchez, M.S.; Allison, K.; Lee, R.J. What is asbestos and why is it important? Challenges of defining and characterizing asbestos. *Int. Geol. Rev.* **2010**, *52*, 801–872. [CrossRef]
41. Plumlee, G.S.; Morman, S.A.; Ziegler, T.L. The toxicological geochemistry of earth minerals: An overview of processes and the interdisciplinary methods used to understand them. *Rev. Mineral. Geochem.* **2006**, *64*, 5–57. [CrossRef]
42. Ilgren, E.B. The fiber length of coalinga chrysotile: Enhanced clearance due to its short nature in aqueous solution with a brief critique on 'Short Fiber Toxicity'. *Indoor Built Environ.* **2008**, *17*, 5–26. [CrossRef]
43. Rietveld, H.M. The Rietveld method. *Physica Scripta* **2014**, *89*, 098002. [CrossRef]
44. Norrish, K.; Hutton, J.T. An accurate X-ray spectrographic method for the analysis of geologic samples. *Geochem. Et Cosmochim. Acta* **1969**, *33*, 431–454. [CrossRef]
45. Wilson, A.H. A chill sequence to the Bushveld Complex: Insight into the first stage of emplacement and implications for the parental magmas. *J. Petrol.* **2012**, *53*, 1123–1168. [CrossRef]
46. Bismarck, A.; Aranberri-Askargorta, I.; Springer, J.; Lampke, T.; Wielage, B.; Stamboulis, A.; Shenderovich, I.; Limbach, H.-H. Surface Characterization of Flax, Hemp and Cellulose Fibers; Surface Properties and the Water Uptake Behavior. *Polym. Compos.* **2002**, *23*, 872–894. [CrossRef]
47. Boulanger, G.; Andujar, P.; Pairon, J.-C.; Billon-Galland, M.-A.; Dion, C.; Dumortier, P.; Brochard, P.; Sobaszek, A.; Bartsch, P.; Paris, C.; et al. Quantification of short and long asbestos fibers to assess asbestos exposure: A review of fiber size toxicity. *Environ. Health* **2014**, *13*, 59. [CrossRef] [PubMed]

48. Gualtieri, A.F. Bridging the gap between toxicity and carcinogenicity of mineral fibres by connecting the fibre crystal-chemical and physical parameters to the key characteristics of cancer. *Curr. Res. Toxicol.* **2021**, *2*, 42–52. [CrossRef]
49. Brunauer, S.; Emmett, P.H.; Teller, E. Adsorption of gases in multimolecular layers. *J. Am. Chem. Soc.* **1938**, *60*, 309–319. [CrossRef]
50. Aust, A.E.; Cook, P.M.; Dodson, R.F. Morphological and chemical mechanisms of elongated mineral particle toxicities. *J. Toxicol. Environ. Health Part B* **2011**, *14*, 40–75. [CrossRef]
51. Deng, Z.J.; Liang, M.L.; Tóth, I.; Monteiro, M.J.; Michin, R.F. Molecular interaction of poly (acrylic acid) gold nanoparticles with human fibrinogen. *ACS Nano* **2012**, *6*, 8962–8969. [CrossRef]
52. Di Giuseppe, D. Characterization of Fibrous Mordenite: A First Step for the Evaluation of Its Potential Toxicity. *Crystals* **2020**, *10*, 769. [CrossRef]
53. Kaszuba, J.; Yardley, B.; Andreani, M. Experimental perspectives of mineral dissolution and precipitation due to carbon dioxide-water-rock interactions. *Rev. Mineral. Geochem.* **2013**, *77*, 153–188. [CrossRef]
54. Oze, C.; Solt, K.-L. Biodurability of chrysotile and tremolite asbestos in simulated lung and gastric fluids. *Am. Mineral.* **2010**, *95*, 825–831. [CrossRef]
55. Rozalen, M.; Huertas, F.J. Comparative effect of chrysotile leaching in nitric, sulfuric and oxalic acid at room temperature. *Chem. Geol.* **2013**, *352*, 134–142. [CrossRef]
56. Rozalen, M.; Ramos, M.E.; Gervilla, F.; Kerestedjian, T.; Fiore, S.; Huertas, F.J. Dissolution study of tremolite and anthophyllite: pH effect on the reaction kinetics. *Appl. Geochem.* **2014**, *49*, 46–56. [CrossRef]
57. Karunaratne, P.C.T.; Fernando, G.W.A.R. Characterisation and radiation impact of corrugated asbestos roofing sheets in Sri Lanka. *J. Geol. Soc. Sri Lanka* **2015**, *17*, 31–40.
58. Fischer, C.; Kurganskaya, I.; Schäfer, T.; Lüttge, A. Variability of crystal surface reactivity: What do we know? *Appl. Geochem.* **2014**, *43*, 132–157. [CrossRef]
59. Donaldson, K.; Murphy, F.A.; Duffin, R.; Poland, C.A. Asbestos, carbon nanotubes and the pleural mesothelium: A review of the hypothesis regarding the role of long fibre retention in the parietal pleura, inflammation and mesothelioma. *Fibre Tox.* **2010**, *7*, 5. [CrossRef] [PubMed]
60. Gualtieri, A.F. Towards a quantitative model to predict the toxicity/pathogenicity potential of mineral fibers. *Toxicol. Appl. Pharmacol.* **2018**, *361*, 89–98. [CrossRef]
61. McCrone, W.; McCrone, L.; Delly, J. *Polarized Light Microscopy*; Ann Arbor Science Publishers, Inc.: Ann Arbor, MI, USA, 1978.
62. Middleton, A.P. The identification of asbestos in solid materials. In *Asbestos: Properties Applications and Hazards*; Michaels, L., Chissick, S.S., Eds.; John Wiley and Sons, Inc.: Chichester, UK, 1979.
63. Vaughan, N.P.; Rooker, S.J.; LeGuen, J.M. In situ identification of asbestos fibres collected on membrane filters for counting. *Ann. Occup. Hyg.* **1981**, *24*, 281–290.
64. Bernstein, D.; Dunnigan, J.; Hesterberg, T.; Brown, R.; Velasco, J.A.L.; Barrera, R.; Hoskins, J.; Gibbs, A. Health risk of chrysotile revisited. *Crit. Rev. Toxicol.* **2013**, *43*, 154–183. [CrossRef]
65. Walton, W.H. The nature, hazards, and assessment of occupational exposure to airborne asbestos dust: A review. *Ann. Occup. Hyg.* **1982**, *25*, 117–247.
66. Stanton, M.F.; Layard, M.; Tegeris, A.; Miller, E.; May, M.; Morgan, E.; Smith, A. Relation of particle dimension to carcinogenicity in amphibole asbestoses and other fibrous minerals. *J. Natl. Cancer Inst.* **1981**, *67*, 965–975. [PubMed]
67. Churg, A. Asbestos lung burden and disease patterns in man. In *Health Effects of Mineral Dust. Review in Mineralogy and Geochemistry*; Guthrie, G.D., Mossman, B.T., Eds.; Mineralogical Society of America: Chantilly, VA, USA, 1993; Volume 28, pp. 409–426.
68. Pollastri, S.; Gualtieri, A.F.; Gualtieri, M.L.; Hanuskova, M.; Cavallo, A.; Gaudino, G. The zeta potential of mineral fibers. *J. Hazard. Mater.* **2014**, *276*, 469–479. [CrossRef] [PubMed]
69. Pollastri, S.; D'Acapito, F.; Trapananti, A.; Colantoni, I.; Andreozzi, G.B.; Gualtieri, A.F. The chemical environment of iron in mineral fibers. A combined X-ray absorption and Mossbauer spectroscopic study. *J. Hazard. Mater.* **2015**, *298*, 282–293. [CrossRef] [PubMed]
70. Vigliaturo, R.; Ventura, G.D.; Choi, J.K.; Marengo, A.; Lucci, F.; O'Shea, M.J.; Pérez-Rodríguez, I.; Gieré, R. Mineralogical Characterization and Dissolution Experiments in Gamble's Solution of Tremolitic Amphibole from Passo di Caldenno (Sondrio, Italy). *Minerals* **2018**, *8*, 557. [CrossRef] [PubMed]
71. Oberdörster, G. Toxicokinetics and effects of fibrous and nonfibrous particles. *Inhal. Toxicol.* **2002**, *14*, 29–56. [CrossRef]
72. Singer, A.; Kirsten, W.F.A.; Buhmann, C. Occurance of sepiolite in the northern Transvaal, South Africa. *S. Afr. J. Geol.* **1992**, *95*, 165–170.
73. Turci, T.; Tomatis, M.; Pacella, A. Surface and bulk properties of mineral fibres relevant to toxicity. *Eur. Mineral. Union Notes Mineral.* **2017**, *18*, 171–214.
74. Iqbal, J.; Shah, M.H. Occurrence, risk assessment, and source apportionment of heavy metals in surface sediments from Khanpur Lake, Pakistan. *J. Anal. Sci. Technol.* **2014**, *5*, 28–32. [CrossRef]
75. Gualtieri, A.F.; Pollastri, S.; Gandolfi, N.B.; Gualtieri, M.L. In vitro acellular dissolution of mineral fibres: A comparative study. *Sci. Rep.* **2018**, *8*, 7071. [CrossRef] [PubMed]
76. Nemery, B. Metal toxicity and the respiratory tract. *Eur. Respir. J.* **1990**, *3*, 202–219. [CrossRef]
77. Fubini, B.; Mollo, L. Role of iron in the reactivity of mineral fibers. *Toxicol. Lett.* **1995**, *82/83*, 951–960. [CrossRef]

78. Gazzano, E.; Foresti, E.; Lesci, I.G.; Tomatis, M.; Riganti, C.; Fubini, B.; Roveri, N.; Ghigo, D. Different cellular responses evoked by natural and stoichiometric synthetic chrysotile asbestos. *Toxicol. Appl. Pharmacol.* **2005**, *206*, 356–364. [CrossRef] [PubMed]
79. Gazzano, E.; Turci, F.; Foresti, E.; Putzu, M.G.; Aldieri, E.; Silvagno, F.; Lesci, I.G.; Tomatis, M.; Riganti, C.; Romano, C.; et al. Iron loaded synthetic chrysotile: A new model solid for studying the role of iron in asbestos toxicity. *Chem. Res. Toxicol.* **2007**, *20*, 380–387. [CrossRef] [PubMed]
80. Shimizu, Y.; Dobashi, K.; Kusakbe, T.; Nagamine, T.; Oikawa, M.; Satoh, T.; Haga, J.; Ishii, Y.; Ohkubo, T.; Kamiya, T.; et al. In-air micro-particle induced X-ray emission analysis of asbestos and metals in lung tissue. *Int. J. Immunopathol. Pharmacol.* **2008**, *21*, 567–576. [CrossRef]
81. Nakamura, E.; Makishima, A.; Hagino, K.; Okabe, K. Accumulation of radium in ferruginous protein bodies formed in lung tissue: Association of resulting radiation hotspots with malignant mesothelioma and other malignancies. *Proc. Jpn. Acad. Ser. B Phys. Biol. Sci.* **2009**, *85*, 229–239. [CrossRef]
82. Hamilton, R.F., Jr.; Thakur, S.A.; Holian, A. Silica binding and toxicity in alveolar macrophages. *Free Radic. Biol. Med.* **2008**, *44*, 1246–1258. [CrossRef] [PubMed]
83. Barabasz, W.; Albinska, D.; Jaskowska, M.; Lipiec, J. Ecotoxicology of Aluminium. *Pol. J. Environ. Stud.* **2002**, *11*, 199–203.
84. Gupta, N.; Gaurav, S.S.; Kumar, A. Molecular Basis of Aluminium Toxicity in Plants: A Review. *Am. J. Plant Sci.* **2013**, *4*, 21–37. [CrossRef]
85. Deer, W.A.; Howie, R.A.; Zussman, J. Rock Forming Minerals. 3B. In *Layered Silicates Excluding Micas and Clay Minerals*, 2nd ed.; The Geological Society: London, UK, 2009.
86. Wicks, F.J.; Plant, A.G. Electron-Microprobe and X-Ray-Microbeam studies of serpentine textures. *Can. Mineral.* **1979**, *17*, 785–830.
87. Oberti, R.; Hawthorne, F.C.; Cannillo, E.; Camara, F. Longrange order in amphiboles. In *Amphiboles: Crystal Chemistry, Occurrence, and Health Issues*; Hawthorne, F.C., Oberti, R., Della Ventura, G., Mottana, A., Eds.; Mineralogical Society of America: Chantilly, VA, USA, 2007; pp. 125–172.
88. Hudnell, H.K. Effects from environmental Mn exposures: A review of the evidence from non-occupational exposure studies. *Neurotoxicology* **1999**, *20*, 379–398. [PubMed]
89. Aschner, J.L.; Aschner, M. Nutritional aspects of manganese homeostasis. *Mol. Asp. Med.* **2005**, *26*, 353–362. [CrossRef]
90. Fukuda, H.; Ebara, M.; Yamada, H.; Arimoto, M.; Okabe, S.; Obu, M.; Yoshikawa, M.; Sugiura, N.; Saisho, H. Trace elements and cancer. *Jpn. Med. Assoc. J.* **2004**, *47*, 391–395.
91. Bargagli, E.; Monaci, F.; Bianchi, N.; Bucci, C.; Rottoli, P. Analysis of trace elements in bronchoalveolar lavage of patients with diffuse lung diseases. *Biol. Trace Elem. Res.* **2008**, *124*, 225–235. [CrossRef]
92. Wei, B.; Yang, L.; Zhu, O.; Yu, J.; Jia, X. Multivariate analysis of trace elements distribution in hair of pleural plaques patients and health group in a rural area from China. *Hair Ther. Transplant. J.* **2014**, *4*, 2167–3118. [CrossRef]
93. IARC. *A Review of Human Carcinogens. Part C: Arsenic, Metals, Fibers, and Dusts*; International Agency for Research on Cancer: Lyon, France, 2012.
94. Bloise, A.; Barrese, E.; Apollaro, C. Hydrothermal alteration of Ti-doped forsterite to chrysotile and characterization of the resulting chrysotile fibers. *Neuses Jahrb. Fur Mineral. Abh.* **2009**, *185*, 297–304. [CrossRef]
95. Bloise, A.; Belluso, E.; Fornero, E.; Rinaudo, C.; Barrese, E.; Capella, S. Influence of synthesis conditions on growth of Ni-doped chrysotile. *Microporous Mesoporous Mater.* **2010**, *132*, 239–245. [CrossRef]
96. Wicks, F.J.; O'Hanley, D.S. Serpentine minerals; structures and petrology. *Rev. Mineral. Geochem.* **1988**, *19*, 91–167.
97. Jaishankar, M.; Tseten, T.; Anbalagan, N.; Mathew, B.B.; Beeregowda, K.N. Toxicity, mechanism and health effects of some heavy metals. *Interdiscip. Toxicol.* **2014**, *7*, 60–72. [CrossRef]
98. Alengebawy, A.; Abdelkhalek, S.T.; Qureshi, S.R.; Wang, M.Q. Heavy Metals and Pesticides Toxicity in Agricultural Soil and Plants: Ecological Risks and Human Health Implications. *Toxics* **2021**, *9*, 42. [CrossRef]
99. Mathew, B.B.; Tiwari, A.; Jatawa, S.K. Free radicals and antioxidants: A review. *J. Pharm. Res.* **2011**, *4*, 4340–4343.
100. Wadhwa, N.; Mathew, B.B.; Jatawa, S.; Tiwari, A. Lipid peroxidation: Mechanism, models and significance. *Int. J. Curr. Sci.* **2012**, *3*, 29–38.
101. Oller, A.R.; Costa, M.; Oberdörster, G. Carcinogenicity assessment of selected nickel compounds. *Toxicol. Appl. Pharmacol.* **1997**, *143*, 152–166. [CrossRef] [PubMed]
102. McGregor, D.B.; Baan, R.A.; Partensky, C.; Rice, J.M.; Wilbourn, J.D. Evaluation of the carcinogenic risks to humans associated with surgical implants and other foreign bodies—A report of an IARC Monographs Programme Meeting. *Eur. J. Cancer* **2000**, *36*, 307–313. [CrossRef]
103. Seilkop, S.K.; Oller, A.R. Respiratory cancer risks associated with low-level nickel exposure: An integrated assessment based on animal, epidemiological, and mechanistic data. *Regul. Toxicol. Pharmacol.* **2003**, *37*, 173–190. [CrossRef]
104. Horie, M.; Nishio, K.; Fujita, K.; Kato, H.; Nakamura, A.; Kinugasa, S.; Endoh, S.; Miyauchi, A.; Yamamoto, K.; Murayama, H.; et al. Ultrafine NiO particles induce cytotoxicity in vitro by cellular uptake and subsequent Ni(II) release. *Chem. Res. Toxicol.* **2009**, *22*, 1415–1426. [CrossRef] [PubMed]
105. Caicedo, M.; Jacobs, J.J.; Reddy, A.; Hallab, N.J. Analysis of metal ion-induced DNA damage, apoptosis, and necrosis in human (Jurkat) T-cells demonstrates Ni^{2+} and V^{3+} are more toxic than other metals: Al^{3+}, Be^{2+}, Co^{2+}, Cr^{3+}, Cu^{2+}, Fe^{3+}, Mo^{5+}, Nb^{5+}, Zr^{2+}. *J. Biomed. Mater. Res.* **2007**, *84*, 905–913. [CrossRef] [PubMed]

106. Briffa, J.; Sinagra, E.; Blundell, R. Heavy metal pollution in the environment and their toxicological effects on humans. *Heliyon* **2020**, *6*, e04691. [CrossRef]
107. Scharf, B.; Clement, C.C.; Zolla, V.; Perino, G.; Yan, B.; Elci, G.; Purdue, E.; Goldring, S.; Macaluso, F.; Cobelli, N.; et al. Molecular analysis of chromium and cobalt-related toxicity. *Sci. Rep.* **2014**, *4*, 5729. [CrossRef]
108. 106 Gualtieri, A.F.; Andreozzi, G.B.; Tomatis, M.; Turci, F. Iron from a geochemical viewpoint. Understanding toxicity/pathogenicity mechanisms in iron-bearing minerals with a special attention to mineral fibers. *Free Radic. Biol. Med.* **2019**, *133*, 21–37. [CrossRef]
109. O'Flaherty, E.J. A physiologically-based model of chromium kinetics in the rat. *Toxicol. Appl. Pharmacol.* **1996**, *138*, 54–64. [CrossRef]
110. Beaumont, J.J.; Sedman, R.M.; Reynolds, S.D.; Sherman, C.D.; Li, L.H.; Howd, R.A.; Sandy, M.S.; Zeise, L.; Alexeeff, G.V. Cancer mortality in a Chinese population exposed to hexavalent chromium in drinking water. *Epidemiology* **2008**, *19*, 12–23. [CrossRef] [PubMed]
111. Shekhawat, K.; Chatterjee, S.; Joshi, B. Chromium Toxicity and its Health Hazards. *Int. J. Adv. Res.* **2015**, *3*, 167–172.
112. Gualtieri, A.F.; Lusvardi, G.; Zoboli, A.; Di Giuseppe, D.; Gualtieri, M.L. Biodurability and release of metals during the dissolution of chrysotile, crocidolite and fibrous erionite. *Environ. Res.* **2019**, *171*, 550–557. [CrossRef]
113. Rodríguez-Mercado, J.J.; Mateos-Nava, R.A.; Altamirano-Lozano, M.A. DNA damage induction in human cells exposed to vanadium oxides in vitro. *Toxicol. In Vitro* **2011**, *25*, 1996–2002. [CrossRef] [PubMed]
114. WHO. *Environmental Health Criteria 81: Vanadium*; WHO: Geneva, Switzerland, 1990; pp. 1–35.
115. WHO. *Vanadium Pentoxide and other Inorganic Vanadium Compounds*; Concise International Chemical Assessment Document: 29; WHO: Geneva, Switzerland, 2001.
116. ATSDR (Agency for Toxic substances and Disease Registry). *Draft Toxicological Profile for Several Trace Elements*; US Dept. Health Human Services, Agency for Toxic Substances and Disease Registry: Atlanta, GA, USA, 2002.
117. Crans, D.C.; Smee, J.J.; Gaidamauskas, E.; Yang, L. The chemistry and biochemistry of vanadium and the biological activities exerted by vanadium compounds. *Chem. Rev.* **2004**, *104*, 849–902. [CrossRef] [PubMed]
118. Chan, P.C.; Herbert, R.A.; Roycroft, J.H.; Haseman, J.K.; Grumbein, S.L.; Miller, R.A.; Chouf, B.J. Lung Tumor Induction by Inhalation Exposure to Molybdenum Trioxide in Rats and Mice. *Toxicol. Sci.* **1998**, *45*, 58–65. [CrossRef] [PubMed]
119. Ott, H.C.; Prior, C.; Herold, M.; Riha, M.; Laufer, G.; Ott, G. Respiratory symptoms and bronchoalveolar lavage abnormalities in molybdenum exposed workers. *Wien. Klin. Wochenschr.* **2004**, *116*, 25–30. [PubMed]
120. Ballirano, P.; Bloise, A.; Gualtieri, A.F.; Lezzerini, M.; Pacella, A.; Perchiazzi, N.; Dogan, M.; Dogan, A.U. The crystal structure of mineral fibres. In *Mineral Fibres: Crystal Chemistry, Chemical Physical Properties, Biological Interaction and Toxicity*; Gualtieri, A.F., Ed.; European Mineralogical Union and the Mineralogical Society of Great Britain & Ireland: London, UK, 2017; Volume 18, pp. 17–64.
121. Andreozzi, G.B.; Ballirano, P.; Gianfagna, A.; Mazziotti-Tagliani, S.; Pacella, A. Structural and spectroscopic characterization of a suite of fibrous amphiboles with high environmental and health relevance from Biancavilla (Sicily, Italy). *Am. Mineral.* **2009**, *94*, 1333–1340. [CrossRef]
122. Bloise, A.; Ricchiuti, C.; Punturo, R.; Pereira, D. Potentially toxic elements (PTEs) associated with asbestos chrysotile, tremolite and actinolite in the Calabria region (Italy). *Chem. Geol.* **2020**, *558*, 119896. [CrossRef]
123. Lu, S.; Wang, Y.; Teng, Y.; Yu, X. Heavy metal pollution and ecological risk assessment of the paddy soils near a zinc-lead mining area in Hunan. *Env. Monit Assess* **2015**, *187*, 627. [CrossRef]
124. Vanoeteren, C.; Cornelis, R.; Verbeeck, P. Evaluation of trace elements in human lung tissue III. *Corresp. Analysis. Sci. Total Environ.* **1986**, *54*, 237–245. [CrossRef] [PubMed]
125. Lippmann, M. Asbestos Exposure Indices. *Environ. Res.* **1988**, *48*, 86–106. [CrossRef] [PubMed]
126. Dixon, J.R.; Lowe, D.B.; Richards, D.E.; Cralley, L.J.; Stokinger, H.E. The role of trace metals in chemical carcinogenesis: Asbestos cancers. *Cancer Res.* **1970**, *30*, 1068–1074. [PubMed]
127. Visonà, S.D.; Capella, S.; Bodini, S.; Borrelli, P.; Villani, S.; Crespi, E.; Colosio, C.; Previderè, C.; Belluso, E. Evaluation of Deposition and Clearance of Asbestos (Detected by SEM-EDS) in Lungs of Deceased Subjects Environmentally and/or Occupationally Exposed in Broni (Pavia, Northern Italy). *Front. Public Health* **2021**, *9*, 678040. [CrossRef]
128. Scambelluri, M.; Piccardo, G.B.; Philippot, P.; Robbiano, A.; Negretti, L. High salinity fluid inclusions formed from recycled seawater in deeply subducted alpine serpentinite. *Earth Planet. Sci. Lett.* **1997**, *148*, 485–499. [CrossRef]
129. Tiepolo, M.; Oberti, R.; Zanetti, A.; Vannucci, R.; Foley, S.F. Trace-element partitioning between amphibole and silicate melt. In *Amphiboles: Crystal Chemistry, Occurrence, and Health Issues*; Hawthorne, F.C., Oberti, R., Della Ventura, G., Mottana, A., Eds.; Mineralogical Society of America: Chantilly, VA, USA, 2007; pp. 417–452.
130. Kumar, A.; Maiti, S.K. Assessment of potentially toxic heavy metal contamination in agricultural fields, sediment, and water from an abandoned chromite-asbestos mine waste of Roro hill, Chaibasa, India. *Environ. Earth Sci.* **2015**, *74*, 2617–2633. [CrossRef]
131. Holmes, A.; Morgan, A.; Sandalls, F.J. Determination of iron, chromium, cobalt, nickel, and scandium in asbestos by neutron activation analysis. *Am. Ind. Hyg. Assoc. J.* **1971**, *32*, 281–286. [CrossRef] [PubMed]
132. Tashakor, M.; Yaacob, W.Z.W.; Mohamad, H.; Ghani, A.A. Geochemical characteristics of serpentinite soils from Malaysia. *Malays. J. Soil Sci.* **2014**, *18*, 35–49.

133. Punturo, R.; Bloise, A.; Critelli, T.; Catalano, M.; Fazio, E.; Apollaro, C. Environmental implications related to natural asbestos occurrences in the ophiolites of the Gimigliano-Mount Reventino Unit (Calabria, Southern Italy). *Int. J. Environ. Res.* **2015**, *9*, 405–418.
134. Mendez, M.O.; Maier, R.M. Phytoremediation of mine tailings in temperate and arid environments. *Rev. Environ. Sci. Biotechnol.* **2008**, *7*, 47–59. [CrossRef]
135. Huang, L.; Baumgartl, T.; Mulligan, D. Is rhizosphere remediation sufficient for sustainable revegetation of mine tailings. *Annu. Bot.* **2012**, *110*, 223–238. [CrossRef]
136. Kumaresan, D.; Cross, A.T.; Moreira-Grez, B.; Kariman, K.; Nevill, P.; Stevens, J.; Allcock, R.J.N.; O'Donnell, A.G.; Dixon, K.W.; Whiteley, A.S. Microbial functional capacity is preserved within engineered soil formulations used in mine site restoration. *Sci. Rep.* **2017**, *7*, 564. [CrossRef]
137. Schimmer, C. Synergistic Use of Soil Microbes and Plants to Facilitate Rehabilitation on Gold Tailings Materials. Master's Thesis, North-West University, Potchefstroom, South Africa, 2017.
138. Northwatch, Mining Watch Canada and Associates. *The Boreal Below: Mining Issues and Activities in Canada's Boreal Forest*; Mining Watch Canada: Ottawa, ON, Canada, 2008; p. 205.
139. Pentreath, R.J. The Discharge of Waters from Active and Abandoned Mines. In *Mining and its Environmental Impact*; Hester, R.E., Harrison, R.M., Eds.; Royal Society of Chemistry: London, UK, 1994; pp. 121–131.
140. Lamb, D.; Erskine, P.D.; Fletcher, A. Widening gap between expectations and practice in Australian mine site rehabilitation. *Ecol. Manag. Restor.* **2015**, *16*, 186–195. [CrossRef]
141. Cross, A.T.; Stevens, J.C.; Sadler, R.; Moreira-Grez, B.; Ivanov, D.; Zhong, H.; Dixon, K.W.; Lambers, H. Compromised root development constrains the establishment potential of native plants in unamended alkaline post-mining substrates. *Plant Soil* **2019**, *461*, 163–179. [CrossRef]

minerals

MDPI

Article

Geochemical Characterization of Rock Samples from Selected Fiji Mine Sites to Evaluate On-Site Environmental Vulnerabilities

Apete Tuiyaro Soro [1,*], Walubita Mufalo [2], Takahiko Arima [2], Carlito Baltazar Tabelin [3] and Toshifumi Igarashi [2]

[1] Division of Sustainable Resources Engineering, Graduate School of Engineering, Hokkaido University, Sapporo 060-8628, Japan
[2] Division of Sustainable Resources Engineering, Faculty of Engineering, Hokkaido University, Sapporo 060-8628, Japan
[3] Department of Materials and Resources Engineering Technology, Mindanao State University—Iligan Institute of Technology, Iligan City 9200, Philippines
* Correspondence: apetesoro@gmail.com

Abstract: Although mining contributes to about 1.4% of Fiji's gross domestic product (GDP), the excavated rocks from mining may have detrimental effects on the environment. In this study, rock samples from five Fiji mine sites were selected to assess their geochemical characteristics from an environmental point of view. The mineralogical and chemical constituents, release and retention mechanisms of hazardous elements, and acid/neutralization potential of the rock samples were investigated to understand their environmental impacts on-site. The results showed that sulfide minerals typically found in the rock samples, such as pyrite, chalcopyrite, and sphalerite were responsible for the release of hazardous elements such as Cu, Pb, and Zn via oxidation. Leachates of rock samples from Mt Kasi, Nukudamu, and Wainivesi exceeded the World Health Organization (WHO) regulatory limit for Cu (2 mg/L), Pb (0.01 mg/L), and Zn (3 mg/L) in drinking water. In contrast, no hazardous elements were leached from the Tuvatu and Vatukoula rock samples, which could be attributed to the dissolution of calcite and dolomite that buffered the pH and limited heavy metal mobility. The acid–base accounting (*ABA*) and accelerated weathering test by hydrogen peroxide indicated that most of the rock samples containing sulfide minerals were likely to generate acidity. Furthermore, the results highlighted that once carbonate minerals are depleted in the rock samples, acid mine drainage (AMD) generation is inevitable. These findings reaffirm the need for committed effort in environmental management of the mine sites to prevent environmental issues associated with AMD.

Keywords: geochemical characterization; sulfide and carbonate minerals; hazardous elements; leaching; acid mine drainage

1. Introduction

1.1. Background

Mining is generally considered an accepted form of development contributing to global growth. Despite this, it remains inherently disruptive to the environment and affected communities wherever it is undertaken. Sub-standard mining practices, insufficient monitoring and control, and lack of prudent rehabilitation programs often lead to long-term environmental degradation. Due to its extractive nature of accessing and extracting target minerals, mining operations often require the removal of surface vegetation and soils, including alteration of habitats and landforms, which changes surface water hydrology, hydrogeology, and soil profiles, and finally affects human health [1]. Mining operations are also notorious for inducing erosion and land subsidence as well as generating blasting noise, vibration, dust, hazardous wastes, rocks and tailings, and polluted effluents such as acid mine drainage (AMD). The negative impacts of mining are often exacerbated in developing countries because of the large knowledge gaps in leading practices and limited applications of advanced technologies for ore extraction, mineral processing, and mine waste management [2–5]. For instance, local communities' concerns only focus on physically observable changes such as river water discoloration, odor, taste, or feel rather than chemical quantification and understanding of underlying scientific processes unknown to them [2].

Sustainable mine waste management remains an ongoing challenge for the mining industry. Waste rocks and tailings, for example, pose special concerns because of their complex geochemical characteristics and the huge volume generated. In fact, for gold as well as copper (Cu) mining, about 99% of the extracted ores are deposited back into the environment as tailings [6,7]. When geochemical properties of mine wastes are not thoroughly understood, it leads to inappropriate treatments with limited functionalities that can worsen their long-term impacts. Waste rocks may contain sulfide minerals such as pyrite (FeS_2), which not only generate AMD due to their oxidation in the presence of water and atmospheric oxygen but also release hazardous elements into the environment [7–9]. AMD is a notorious discharge from mining areas, and its generation is considered one of the most pressing and serious environmental problems facing the mining industry. Once generated, AMD's strongly acidic pH has the potential to mobilize other hazardous elements present in coexisting minerals that have detrimental effects on nearby ecosystems [9,10].

Despite Fiji's long mining history and growing interest in mineral development, there remain limited available narratives about mining, which concentrate mostly on the benefits of mineral extraction with little literature explaining the orebody's geochemical nature. The Vatukoula Gold Mine (VGML), because of its long existence, is the most studied mine in Fiji [11–17]. However, only a few studies have been conducted on the other mining areas of the country. Moreover, the majority of these previous works focused on the chemical properties of river water and its sediments, including heavy metals and metalloids in Fiji's mine sites. Kumar et al. (2021), for example, highlighted the influence of the discharge from the Toko tailings storage facility (TSF) of Vatukoula on the nearby river sediments [11]. These authors conducted a similar study on potentially toxic elements across the Wainivesi River, which also receives discharge from the Wainivesi mine [13]. In both studies, they assumed that mining discharges contributed to the contamination of the rivers since the mine sites are located within the study area but failed to directly relate the contamination with certainty to the respective mines.

This means that the properties of excavated rocks from mines and how they relate to their environmental impacts remains poorly understood in Fiji. This knowledge gap can be addressed by the present study by investigating the geochemical characteristics of Fiji mine sites. In this study, sixteen fresh orebodies' rock samples were evaluated from five representative mine sites. The study rationale focused on the general geochemical characterization of the rock types of the orebodies of each mine to provide an understanding of how such rock types influence the vulnerability of each site due to their geochemical properties and make-up when disturbed by mining activities. Therefore, the specific objectives of this work are to (1) characterize the mineralogy and geochemical compositions of the rock samples, (2) analyze the leachate characteristics, and (3) investigate the acid-generating potential of the rocks using acid–base accounting and pH tests with hydrogen peroxide. The current study is important for the sustainable development of Fiji's growing mining sector and is equally significant for better and more comprehensive evaluation of mining's environmental impacts.

1.2. Study Sites

Fiji is an archipelago of about 330 islands with a combined landmass of 18,376 km^2 and lies southwest of the Pacific Ocean as shown in Figure 1. As a developing nation, Fiji is largely isolated along the edge of the Pacific Ring of Fire and is richly endowed with minerals, which the country exploits to support its economy. Fiji's climate is mainly tropical, characterized by a high rainfall season (2000–4000 mm) from November to April and dry season from May to October with temperatures of 20–30 °C throughout the year [18,19].

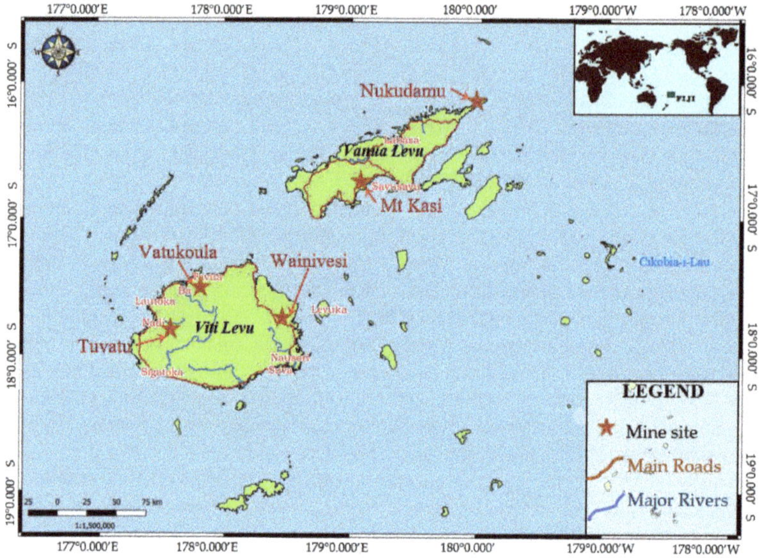

Figure 1. The map of Fiji and locations of Mt Kasi, Nukudamu, Tuvatu, Vatukoula, and Wainivesi mining sites where rock samples were collected.

The rocks collected for this study are fresh samples and are assumed to be representative of the orebodies of the studied mines. A summary of the geology of each mine site is presented herein based primarily on the work of Colley and Flint (1995). At Mt Kasi, the underlying rock was described as basaltic andesite flows, breccias, and tuffs, cut by an altered feldspar porphyry of dacitic composition. The primary ore-bearing minerals were native gold, pyrite, enargite (luzonite), tennantite, goldfieldite, chalcopyrite, tellurides, and cassiterite, whereas secondary minerals included covellite, chalcocite, and neodigenite [20,21]. The Nukudamu mine site sits on the rocks of the Udu volcanic group, which is composed of massive flows, breccias, tuffs, and associated sediments, mostly of submarine origin, generally of fine grained or glassy and containing phenocrysts of andesine, quartz, pyroxene, classified as rhyodacite due to the high content of silica. The Nukudamu deposit is described as Kuroko type, and the mineralization consists primarily of massive Cu, zinc (Zn), and iron (Fe) sulfides (pyrite) with coexisting minerals such as kaolinite, smectite, chalcedony, and quartz [20,22,23].

The Tuvatu mine site is surrounded by the Nadele breccia of the Wainimala Group and comprises massive, coarse, polymict, multicolored breccia, and sedimentary units of thin layered siltstone, sandstone, and mudstone [24]. The Nadele breccia is unconformably overlain by the Koroimavua volcanic group, which has shoshonitic affinities and includes andesitic and biotite-bearing dacitic lithic and crystal tuffs, grits, and agglomerates. Intrusive monzonite is locally brecciated with quartz–tourmaline replacement and veins infilled with potassium feldspar (K-feldspar) and biotite [20,25]. The Vatukoula mine site rocks are derived from a potassium-rich shoshonitic magma with olivine basalt (absarokite) parent magma to shoshonite, trachyandesite (banakite), and monzonite by-products [20]. Its mineralization is described as a low-grade, porphyry Cu type associated with steep caldera faults spatially related to epithermal gold mineralization [26–29].

As for Wainivesi, its host rocks are highly silicified andesites and dacites displaying brecciation with rare basic rocks, suggested to be originally calcareous [30], and contain other rock units including tuffs, shaly volcaniclastic beds, and breccias. The origin of the Wainivesi deposit is still ambiguous, but its mineralization is described as massive sulfide occurring within the hydrothermal breccia, banded, and brecciated with a variety of mineral assemblages [20].

2. Materials and Methods

2.1. Sample Collection, Preparation, and Characterization

The rock samples (0.5–1 kg) were collected from five selected mine sites (Mt Kasi, Nukudamu, Tuvatu, Vatukoula, and Wainivesi) as shown in Figure 1, with the respective details provided in Table S1. The Tuvatu and Vatukoula mines are located on the western side of Viti Levu island, with active mining leases, and Wainivesi on the east with a special exploration license. Mt Kasi and Nukudamu mines are both located on Vanua Levu island. Mt Kasi is currently under special prospecting license to further explore its mineral potential after abandonment of more than 30 years while Nukudamu was abandoned since it ceased operation in 1968.

The bulk rock samples were crushed to <2 mm sizes and stored in air-tight containers. The samples were further crushed to <50 μm, powder pressed into two separate pellets, and then analyzed by XRD (MultiFlex, Rigaku Corporation, Tokyo, Japan) and X-ray fluorescence (XRF) (SpectroXepos, Rigaku Corpoation, Tokyo, Japan) for their mineralogical and chemical compositions, respectively. All the X-ray diffraction (XRD) patterns were identified using Match!® software (Crystal Impact, version 3.3.8, Bonn, Germany) and quantified into relative percentage abundance. Furthermore, validation of mineralogy of selected samples (M2, N1, T3, V1, V4, and W1) was carried out using an optical microscope. The total carbon (TC) and inorganic carbon (IC) of the rock samples (<2 mm) were measured using the total carbon analyzer (TOC-VCSH, Shimadzu Corporation, Kyoto, Japan) with a solid sample combustion unit (SSM-5000A, Shimadzu Corporation, Kyoto, Japan). Finally, loss on ignition (LOI) test was conducted to assess analytical accuracy of total volatile measurements, mainly added to other oxides totaling 100 ± 1.0%. Loss on ignition (LOI) was determined through gravimetry by heating the sample (<2 mm) inside a furnace at 750 °C for 1 h after drying in an oven at 110 °C for 24 h [31,32].

2.2. Batch Leaching Tests

The leaching experiment was based on the Environmental Agency of Japan Notification No. 46 [33], which was applied to evaluate the leaching concentrations of Al, Cu, Ca, Fe, K, Mg, Na, SO_4^{2-}, Si, Pb, and Zn from the rock samples [34]. This method was considered relevant as it closely simulates the effects of rainfall on mined rocks considering that Fiji has a predominantly tropical climate and still does not have a national standard leaching test protocol. For the experiment, 15 g of crushed rock sample (<2 mm) was added to 150 mL of deionized (DI) water in a 250 mL Erlenmeyer flask and placed in the shaker for 24 h at 120 rpm. This method is widely used in excavated and hydrothermally altered rocks that simulate on-site environmental vulnerabilities via leaching [35–37]. The pH (Laqua pH meter F-71, SS112, Horiba Ltd., Kyoto, Japan), electrical conductivity (EC) (CM-31P, CM-31P-W, DKK-TOA Corporation, Tokyo, Japan), oxidation–reduction potential (ORP) (RM 30P, DKK-TOA Corporation, Tokyo, Japan) and temperatures were measured after the leaching tests. The leachates were then filtered through 0.45 μm Millex® membrane filters. All filtrates were acidified with 1% nitric acid (HNO_3) before chemical analysis. Alkalinity was measured in leachates with pH > 4.8 by titrating a known volume of the sample's leachate to pH 4.8 with 0.01 M of sulfuric acid (H_2SO_4). Hazardous elements and major ions were determined by inductively coupled plasma atomic emission spectroscopy (ICP-AES ICPE 9800, Shimadzu Corporation, Kyoto, Japan).

2.3. Quality Assurance and Quality Control

The analysis was conducted in triplicate together with certified reference materials (Table S2) traceable to NIST SRM® (Merck KGaA, Darmstadt, Germany) for accurate leachate concentrations. All the chemicals used were reagent grade. Blank solutions

(Millipore Milli-Rx 12α system, Merck Millipore, California, CA, USA) were analyzed 3 times for each analytical session and calibrations were conducted on concentrations ranging from 0.1 to 20 mg/L to obtain the mean and relative standard deviation for each element using ICP-AES (margin of error ± 2%–3%, detection limit 0.01–0.001 mg/L) [38].

2.4. Statistical Analysis and Thermodynamic Modeling

Principal component analysis (PCA) was performed using Origin Pro® software (Version 9.8, OriginLab Corporation, Northampton, MA, USA) on the batch leaching results. This multivariate statistical method is especially useful in finding correlations in complex systems with high-dimensional data where several components must be considered at the same time. The PCA was applied to understand the correlations between the different variables and identify dominant factors of the batch leaching results. The thermodynamic calculations of the leachates were determined using Visual MINTEQ 3.1 (USEPA, Stockholm, Sweden) [39]. The input parameters include the pH, temperature, pe, alkalinity as HCO^{3-}, and components of the leachates (Al, Cu, Ca, Fe, K, Mg, Na, SO_4^{2-}, Si, Pb, and Zn).

2.5. Acid–Base Accounting and pH Test with Hydrogen Peroxide

The acid generation potential of the rock samples was investigated to determine their likelihood of producing acidity using acid–base accounting (*ABA*) and a pH test with hydrogen peroxide (pH (H_2O_2)). The *ABA* test was used to evaluate the acid-neutralizing potential (*NP*) and acid-generating potential (*AP*) of rock samples [40,41]. In this experiment, *NP* was determined by weighing 2 g of crushed rock samples (<2 mm) into a 250 mL Erlenmeyer flask with 90 mL of deionized (DI) water, which was then thoroughly mixed. The mixture was then treated with 1 mL of 1M hydrochloric acid (HCl), which was added again after 2 h. The suspension was allowed to react at room temperature for 24 h and then titrated to pH 8.3 with 1M sodium hydroxide (NaOH). The *NP* (kg $CaCO_3$/t) was calculated using Equation (1). Meanwhile, the *AP* was determined using 1 g of crushed rock sample (<2 mm) mixed with 100 mL of 15% H_2O_2 (pH = 7), which was kept at room temperature for 48 h before being heated to remove residual H_2O_2. After cooling, its acidity was determined by titration with 1 M NaOH to pH 8.3. The *AP* (kg $CaCO_3$/t) was calculated using Equation (2) from the content of sulfide sulfur ($S_{sulfide}$%), which was determined by leaching with H_2O_2 and calculated according to Equation (3). The difference between the values of *NP* and *AP* is the net acid-neutralizing potential (*NNP* = *NP* − *AP*):

$$NP = (50 \times (Xa - Yb))/c \quad (1)$$

where *X* is volume of HCl (mL), *Y* is volume of NaOH (mL), *a*: normality of HCl (mol/L), *b* refers to normality of NaOH (mol/L), and *c* represents the mass of sample (g):

$$AP = 31.25 \times S_{sulfide} \quad (2)$$

$$S_{sulfide} = 1.6 \times V \quad (3)$$

where, *V* is the volume of NaOH (mL).

If the *NNP* value is below −20 kg $CaCO_3$/t, acid generation is likely to occur; if the *NNP* value is between −20 and 20 kg $CaCO_3$/t, the acid generation is uncertain; and if the *NNP* value is above 20 kg $CaCO_3$/t, the rock sample is unlikely to generate any acidity.

The pH (H_2O_2) test was considered alongside the *ABA* technique to accurately predict the likelihood of acid generation of the rock samples. The method follows the Japanese Geotechnical Standard—JGS 0271-2016, [42] a streamlined technique that simulates the long-term natural process of mineral breakdown through weathering in cases of mineral extraction caused by mining. The pH (H_2O_2) acidification test uses the oxidation principle, in which H_2O_2, a strong oxidant, oxidizes sulfide minerals exposed during the test. In the experiment, H_2O_2 solution was first prepared by neutralizing a pre-determined volume of 30% H_2O_2 to pH 6 using 10 mmol/L of NaOH. Rock samples weighing 2 g (<2 mm) each were transferred into 100 mL beakers (mt = wt. of beaker + wt. of sample), and then 20 mL of H_2O_2 was carefully added. The beakers were covered in plastic wrap with holes to release pressure during oxidation. The solution was then placed in a sand bath heated to 60 °C in a fume hood. Any violent reaction was controlled by stirring the solution or temporarily removing it from the sand bath to cool. The reaction was completed when fizzing or bubble formation became negligible. The solution was then allowed to cool down to room temperature. The beaker and the remaining solution were reweighed, and DI water was added until the new weight of *mt* + 20 g was reached. The resultant solution was thoroughly mixed and the pH was measured. The pH threshold acidification potential is 3.5, and samples with pH < 3.5 have AMD-generating potential [43,44].

3. Results and Discussion

3.1. Mineralogical and Chemical Properties of the Rock Samples

The chemical and mineralogical composition of Mt Kasi (M1–M3), Nukudamu (N1–N4), Tuvatu (T1–T4), Vatukoula (V1–V4), and Wainivesi (W1), including their TC and IC contents, are summarized in Table 1 and Figure 2. The Mt Kasi samples (M1, M2, and M3) are mainly composed of SiO_2(79.3 to 90.4 wt%), Al_2O_3 (2.0 to 17.0 wt%), and sulfur (S; 6.28 to 6.76 wt%), with nacrite ($Al_2Si_2O_5(OH)_4$), pyrite (Fe_2S), and arsenopyrite (FeAsS) as identified minerals (Figure 2a). Pyritic minerals (Figure S1) could not be confirmed in M2 with XRD; however, an optical microscope confirmed the presence of both quartz and pyritic minerals (Figure S1a). In the Nukudamu samples, N1 and N3 had high contents of SiO_2 at 79.1 wt% and 68.5 wt%, respectively, while N4 and N2 contain 40.0 to 96.5 wt% Fe_2O_3. These results are consistent with the mineralogical compositions of the samples that identified cristobalite (SiO_2), albite ($Na(AlSi_3O_8)$), goethite (FeOOH), and hematite (Fe_2O_3) as major mineral constituents (Table S3). Sulfide-bearing minerals such as arsenopyrite, enargite (Cu_3AsS_4), and pyrite (Figure 2b) were also detected in N2 and N3 with substantial amounts of Pb, Zn, Cu, and As ranging from 208 to 986 mg/kg. The presence of pyritic minerals, plagioclase, and magnetite (Fe_3O_4) was also confirmed with an optical microscope in N1 (Figure S1b).

At Tuvatu, anorthite ($CaAl_2Si_2O_8$), augite $(Ca,Na)(Mg,Fe,Al,Ti)(Si,Al)_2O_6$, and orthoclase ($KAlSi_3O_8$) were the major minerals, and minor minerals included quartz, calcite ($CaCO_3$), diopside ($MgCaSi_2O_6$), and dolomite ($CaMg(CO_3)_2$) (Table S3). Surprisingly, pyritic minerals were detected with an optical microscope in T3 (Figure S1c), but were not detected by XRD (Table S3). The Tuvatu samples were predominantly of silicate origin but unlike the Mt Kasi and Nukudamu samples, carbonate minerals such as calcite and dolomite were detected in minor quantities (Figure 2c), which is consistent with (Clark et al., 2022, 2023), who found plagioclase and pyroxene as major minerals and rare disseminated pyritic minerals with minor carbonates [45,46]. The average chemical composition of the samples indicates relatively high contents of SiO_2 (59.1 wt%), with MgO (6.2 wt%) and CaO (5.8 wt%), which correlated with the silicate and carbonate minerals detected by XRD. The TC and IC were generally low, with contents ranging from <0.01 to 0.76 wt%. The XRD results of the Vatukoula rock samples (V1, V2, V3, and V4) revealed augite, orthoclase, and diopside as their major mineral constituents, while calcite, quartz, and augite are present as minor minerals (Figure 2d). Trace amounts of dolomite (Table S3) including pyritic minerals (V1 and V4) were detected (Figure S1d,e). These identified minerals are well supported by the XRF average chemical composition of SiO_2 (51.7 wt%), MgO (8.2 wt%), CaO (9.5 wt%), TC (up to 1.26 wt%), and IC contents (up to 1.69 wt%). Meanwhile, the Wainivesi (W1) (Figure 2e) sample is mainly composed of sphalerite (ZnS) and quartz with nacrite, and chalcopyrite as a minor mineral component. This is consistent with the high chemical contents of Zn (358,000 mg/kg), Cu (28,700 mg/kg), and S (70.7 wt%) in the sample. The oxidation of sulfide minerals from rock samples could pose a threat to the surrounding soil and surface water, as highlighted by Kumar et al. (2021) [13], which linked the presence of toxic elements in the sediments and surface water of the Wainivesi River to the weathering of exposed rocks around the mine.

Table 1. Chemical compositions of the rock samples from the XRF analysis and the TC/IC test.

Sample	SiO_2 (wt%)	TiO_2 (wt%)	Al_2O_3 (wt%)	Fe_2O_3 (wt%)	MnO (wt%)	MgO (wt%)	CaO (wt%)	Na_2O (wt%)	K_2O (wt%)	P_2O_5 (wt%)	As (mg/kg)	Cd (mg/kg)	Cu (mg/kg)	Pb (mg/kg)	Zn (mg/kg)	S (wt%)	TC (wt%)	IC (wt%)	LOI %
M1	79.3	0.64	17.0	4.90	<0.01	<0.01	0.04	0.02	0.02	0.11	92	2	61	15	10	6.28	0.10	<0.01	7.06
M2	90.4	0.34	2.00	2.76	<0.01	<0.01	0.03	0.14	0.13	0.08	137	1	29	28	3	6.31	0.01	<0.01	1.62
M3	84.1	0.39	12.9	3.79	<0.01	0.03	0.05	0.00	0.03	0.17	38	<2	67	11	4	6.76	<0.01	<0.01	5.46
N1	79.1	0.47	13.8	2.35	<0.01	0.31	1.93	6.51	0.50	<0.01	50	1	50	5	140	0.37	0.02	<0.01	0.99
N2	2.88	0.36	1.47	96.5	0.03	<0.01	0.11	<0.01	0.03	<0.01	619	<2	986	208	535	1.30	0.13	<0.01	7.01
N3	68.5	0.01	0.54	26.4	0.09	<0.01	0.05	<0.01	0.05	0.36	590	<1	327	519	460	0.81	0.01	<0.01	5.91
N4	1.84	0.11	6.50	40.0	0.01	<0.01	0.08	0.63	0.02	0.06	7	<1	83	21	59	52.3	0.67	<0.01	7.60
T1	62.7	0.52	21.2	3.89	0.06	8.27	3.35	5.25	0.12	0.27	<1	<1	6	<1	59	0.05	0.76	0.61	1.68
T2	53.7	0.73	16.3	11.4	0.22	8.21	10.6	2.71	2.93	0.67	6	<1	152	3	87	0.19	0.01	<0.01	0.52
T3	58.0	0.57	18.6	8.05	0.23	5.49	7.03	3.06	4.64	0.47	4	<1	190	8	112	0.83	0.04	0.06	1.17
T4	62.4	0.45	20.2	2.69	0.07	3.18	2.40	3.92	7.63	0.35	66	1	30	21	180	0.76	0.01	0.46	1.11
V1	43.2	0.44	11.9	9.93	0.24	9.16	13.7	1.81	2.59	0.33	4	<1	144	4	76	2.30	1.26	1.69	4.87
V2	57.9	0.46	20.3	6.47	0.15	4.89	5.04	3.28	4.64	0.64	1	1	82	9	78	0.05	0.65	0.56	3.51
V3	46.5	0.49	12.4	10.2	0.20	14.0	13.0	1.23	0.83	0.30	3	<1	126	5	81	0.51	0.98	0.72	3.97
V4	59.1	0.50	19.5	6.82	0.16	4.93	6.24	4.52	4.26	0.67	2	<1	161	9	82	0.21	0.31	0.37	1.36
W1	23.6	0.04	3.51	10.7	<0.01	0.02	<0.01	<0.01	0.42	0.03	80	1600	28,700	507	358,000	70.7	<0.01	<0.01	4.41

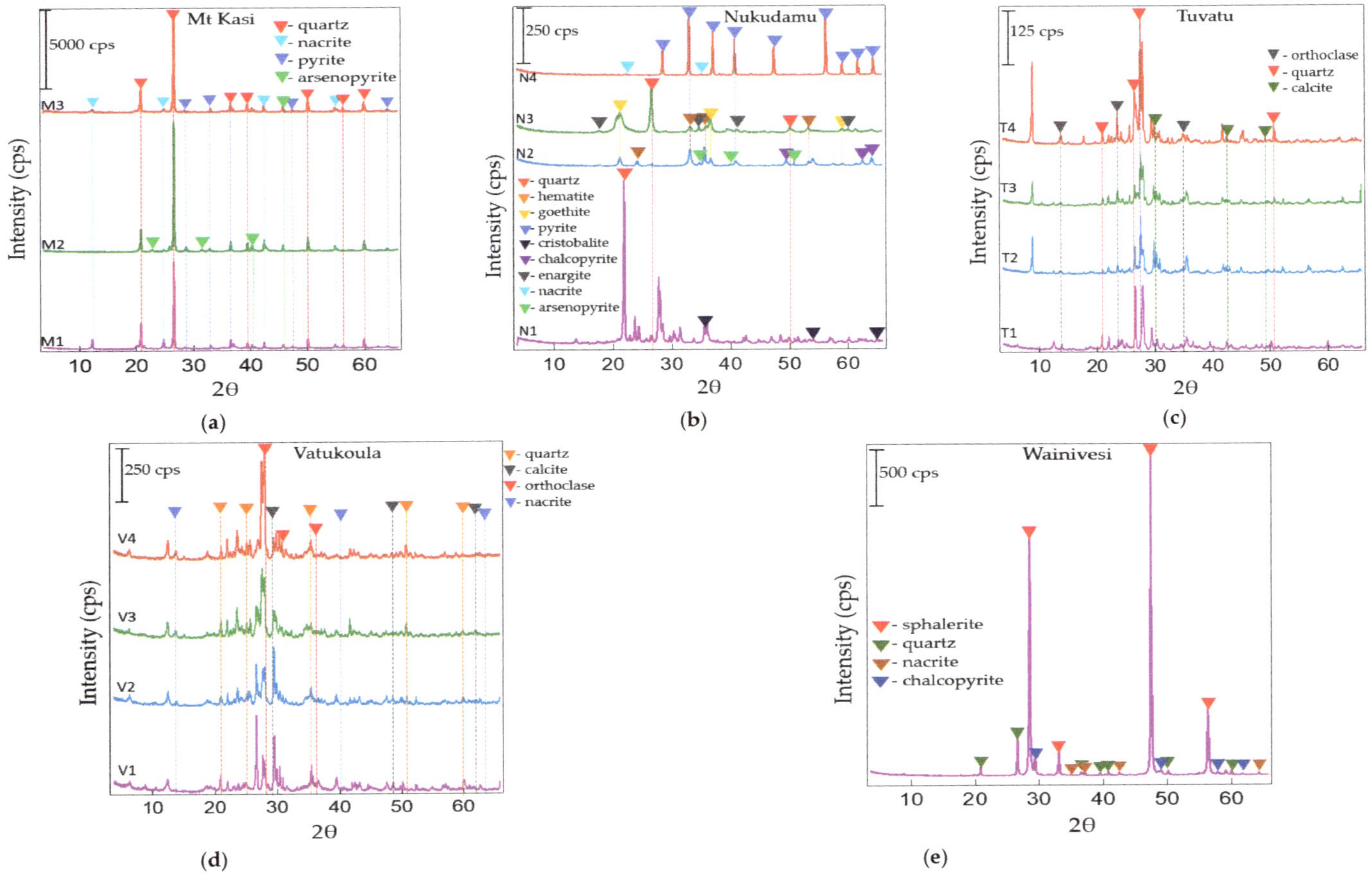

Figure 2. XRD patterns for the mineralogical composition of (**a**) Mt Kasi, (**b**) Nukudamu, (**c**) Tuvatu, (**d**) Vatukoula, and (**e**) Wainivesi.

3.2. Leachate Characteristics of the Rock Samples—Release and Retention Mechanisms

Figures 3 and 4 illustrate the pH and concentrations of Cu, Fe, Pb, Zn and coexisting ions (Ca, K, Mg, and SO_4^{2-}) after batch leaching tests, which were compared with the WHO guideline values for drinking water [47] and is the same standard adopted for drinking water quality in Fiji. Lower pH values were obtained for Nukudamu (2.6 to 5.1) and Mt Kasi (3.7 to 4.6). The Wainivesi rock sample indicated a weakly acidic pH of 6.2. Based on the WHO drinking water guidelines for pH, Mt Kasi, Nukudamu, and Wainivesi leachates were all below the regulatory range (pH 6.5 to 8.5). The low pH might be attributed to the oxidation of sulfide minerals, i.e., pyrite, chalcopyrite, arsenopyrite, and sphalerite found in the rocks [48,49]. Pyrite is typically ubiquitous, associated with most of the base metals, and usually problematic due to AMD generation when exposed to oxidizing conditions. Oxidation products, such as Fe(III) hydroxysulfate [50] (pH < 3), sulfate (SO_4^{2-}), ferrous iron (Fe^{2+}), and hydrogen (H^+) ions are released into the solution, and the ferrous iron is further oxidized to ferric iron (Fe^{3+}) generating more acidity [49]. Fe^{3+} in the acidic solution also tends to oxidize other metal sulfide minerals in the rock samples, consequently releasing more metal, H^+, and SO_4^{2-}. The mobility of the dissolved metal ions in the rocks, soils, or sediments is governed by the balance between the release and retention mechanisms under a given geochemical condition [37]. Sulfide oxidation is likely the most dominant release mechanism of Cu, Pb, Zn, and SO_4^{2-} (Figure 4) in the Mt Kasi, Nukudamu, and Wainivesi rock samples. This is also reflected in the higher Eh values of leachates from these three rock samples (Table S4). Kelderman and Osman (2006) and Popenda (2013) highlighted that the increase in solubilization of ions including hazardous metals such as Cu, Pb, and Zn due to higher redox potential could be attributed to the oxidation of heavy metal sulfide bindings [51,52].

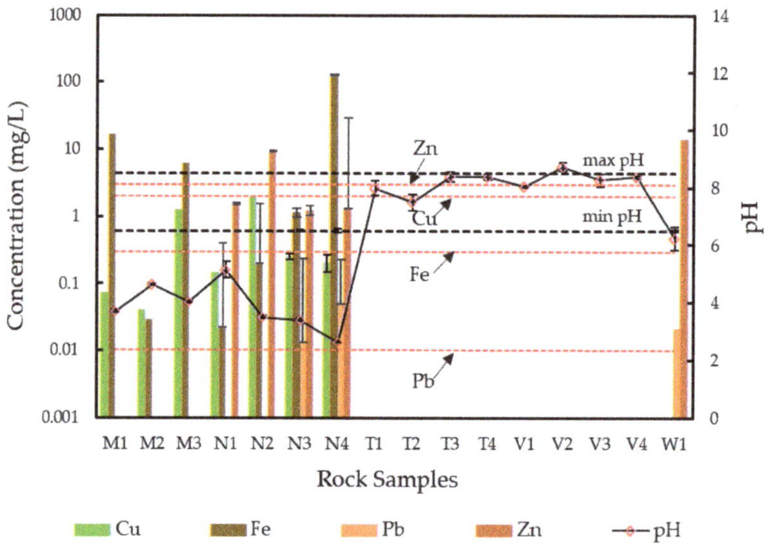

Figure 3. Leachate concentrations of Cu, Fe, Pb, and Zn and the pH of Mt Kasi (M1–M3), Nukudamu (N1–N4), Tuvatu (T1–T4), Vatukoula (V1–V4), and Wainivesi (W1). The dotted lines indicate the WHO drinking water guideline for Cu, Fe, Pb, Zn, and pH (minimum and maximum values; 6.5–8.5).

Although Cu, Fe, Pb, and Zn were detected in the leachates, they were not consistently leached from all the rock samples (Figure 3; Table S4). Higher concentrations of Fe at 16.8 ± 3.72 mg/L, 6.08 ± 1.45 mg/L, 1.15 ± 0.20 mg/L, and 129 ± 0.20 mg/L were observed at Mt Kasi (M1, M3) and Nukudamu (N3, N4), respectively, which exceeded the limit for Fe (<0.3 mg/L) in drinking water. Copper leached only from the Mt Kasi and Nukudamu samples at lower concentrations of up to 1.94 ± 0.41 mg/L (N2) and was around the WHO drinking water standard at 2 mg/L. High Zn concentration was observed in the leachates of samples from Wainivesi (14.2 ± 3.22 mg/L) and Nukudamu (N2; 9.54 ± 1.32 mg/L). Meanwhile, Pb leached only from two of the four Nukudamu rock samples (N3 and N4) with concentrations of 0.02 ± 0.02 and 0.05 ± 0.06 mg/L, respectively, and concentrations of 0.02 ± 0.04 mg/L for the Wainivesi sample. Compared with the WHO guidelines for dissolved Pb and Zn in drinking water, the Nukudamu (N3 and N4) and Wainivesi rock samples exceeded the threshold limit of 0.01 mg/L for Pb, while N2 exceeded the limit for Zn 3 mg/L. The high Pb leaching concentration from the Wainivesi rock samples could be the possible reason for the elevated levels of Pb found in the Wainivesi River [13].

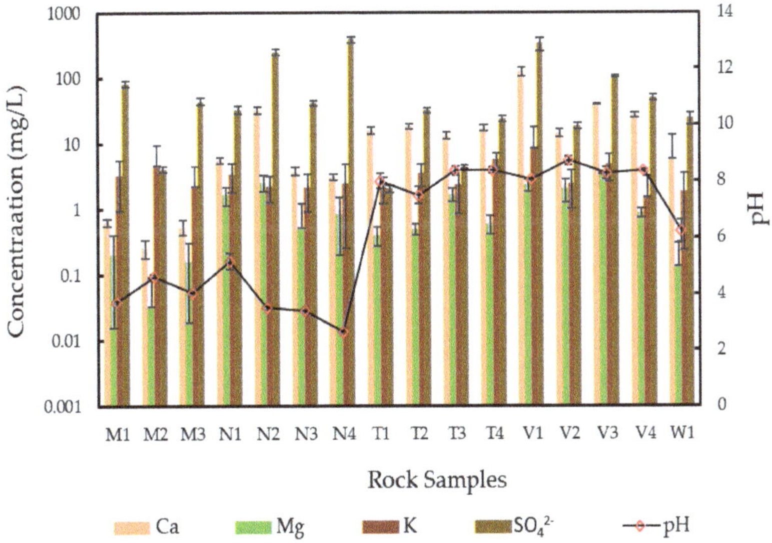

Figure 4. Leachate concentrations of Ca, Mg, K, and SO_4^{2-} and the pH of Mt kasi (M1–M3), Nukudamu (N1–N4), Tuvatu (T1–T4), Vatukoula (V1–V4), and Wainivesi (W1) samples.

In contrast, the Tuvatu and Vatukoula rock samples showed higher pH ranging from 7.5 to 8.7. The alkaline pH could be attributed to the dissolution of carbonate minerals (i.e., calcite and dolomite) in the rock samples (Figure 2c,d and Table S3), which releases divalent cations (Ca, Mg, Fe) and HCO_3^- that raise the leachate pH to near neutral. This is supported by the higher Ca concentrations in T1–T4 and V1–V4 samples (Figure 4), and no hazardous elements were released from these rock samples. To gain more insights into the retention mechanism of the hazardous elements, thermodynamic calculations using Visual MINTEQ were conducted. The results indicate that calcite and dolomite were undersaturated (Table S5), suggesting the dissolution of these minerals. Equilibrium dissolution of calcite and dolomite consumes H^+ and generates aqueous carbonate species, e.g., hydrozincite, smithsonite, hydrocerussite, and cerussite [53,54], which are later reflected in Nukudamu (N1) and Wainivesi (W1) leachates. Although the saturation indices (*SI*s) for N1 and W1 indicated the dissolution of calcite and dolomite, Pb and Zn were released probably due to the lack of carbonate minerals to consume H^+. This is supported by the undersaturation of hydrozincite, smithsonite, hydrocerussite, and cerussite (Table S5). The *SI*s showed that the mobility of Pb and Zn in the two samples could be controlled by the dissolution of calcite and dolomite in the system because it promotes Pb/Zn carbonate surface precipitations. The gradual co-precipitations of Pb, Zn, and Cu with carbonates could enhance the stability of the hazardous elements in the system [48].

However, no such speciation (Pb/Zn-carbonates) was observed in the Tuvatu and Vatukoula rock samples. The high buffering effect of Tuvatu and Vatukoula samples may have been enhanced to a lesser extent by clay minerals such as kaolinite ($Al_2Si_2O_5(OH)_4$) [55] as indicated by positive *SI*s for this mineral. Although the pH buffering reaction of Al-bearing phases between pH 4 to 4.5 [56] could occur through gibbsite dissolution (*SI*s < 0) in Mt Kasi and Nukudamu and may lead to the sequestering of hazardous elements, this contribution is not as pronounced as those observed with carbonate minerals in the Tuvatu and Vatukoula rock samples. Once the carbonate minerals are depleted, however, environmental vulnerability might be inevitable via the release of hazardous elements from the rock samples. It is also relevant to note that arsenic (As), even though identified by the chemical analysis in rock samples of Mt Kasi and Nukudamu, was outside the focus of the leaching analysis and should be considered for future investigations.

3.3. Relationship between the Rock Samples and Their Vulnerability in Generating Acid Mine Drainage

To identify the factors controlling the leachate chemistry when the rock samples are in contact with water, PCA was used to extract important and inter-correlated dependent variables of the samples from the leachates using 15 variables (pH, EC, ORP, Al, Cu, Fe, Mg, Mn, Pb, Zn, Ca, K, Na, SO_4^{2-}, and Si) (Table 2, Figure S2). The results showed that the 4 principal components accounted for 77.5% of leachate concentration with eigenvalues > 1. The PC1 (31.6%) had positive loadings of Pb, Fe, ORP, and EC, and to some extent also Cu and SO_4^{2-}, though with lower loadings, which could be related to chalcopyrite and pyrite. The loadings of these variables were related to the high loadings of negative pH. This component indicates that the release of Pb, Cu, and SO_4^{2-} were associated with the higher ORP and more acidic pH of the rock samples, which was attributed to the enhanced oxidation of sulfide minerals [48]. The PC2 (21.1%) indicated high loadings of EC, Mg, Ca, K, and SO_4^{2-}. The highest contribution of Ca and SO_4^{2-} was related to the minerals controlling the EC of the leachate pH such as calcite, gypsum, and dolomite. This means that these minerals influenced the buffering mechanism of the rock samples. Meanwhile, the PC3 and PC4 accounted for only 16.6 and 8.6%, respectively. Contributions of Al and Si could infer the effects of phyllosilicate and clay minerals on the leachate chemistry (e.g., kaolinite, orthoclase, and anorthite); however, the contributions of Al and Si on the leachate chemistry from PC3 and PC4 were not very pronounced.

The *ABA* test results in Figure 5 show that rock samples from Mt Kasi, Nukudamu, Wainivesi, Tuvatu (T1), and V4 from Vatukoula had negative *NNP* values (<−20 kg $CaCO_3$/t). This means that rock samples from these sites were likely to generate acidity when exposed to oxidizing conditions. The negative *NNP* supports the presence of sulfide minerals in these samples. Although T1 contained some carbonate minerals, the *ABA* test suggests that once its carbonate neutralization capacity is depleted, acid generation might occur. Meanwhile, sample V1 from Vatukoula had *NNP* (20.5) (>20 kg $CaCO_3$/t), indicating that it is a non-acid-producing rock and unlikely to generate AMD. The remaining samples, T3, T4, V2, and V3, had *NNP* values between −20 to 20 kg $CaCO_3$/t, suggesting uncertainty to generate acidity.

The pH (H_2O_2) acidification test results are illustrated in Figure 6. The rock samples M1, M3, N2, N3, N4, T3, and W1 had pH < 3.5, which means that they are likely to generate acidity. Although the Tuvatu (T3) rock sample was characterized as having an alkaline leachate pH with a mineralogy dominated by silicates and carbonates, the *ABA* test, indicated uncertainty while the pH (H_2O_2) test showed a pH of 3.2. This means that the sample remains likely to generate acidity if the neutralization capacity of the carbonate minerals in the sample are exhausted. These results could be extended to the other rock samples that were in the *ABA* uncertainty range. On the other hand, the Mt Kasi sample (M2) and Nukudamu (N1), which initially recorded acidic leaching pH, had pH > 3.5, indicating non-acid forming rocks based on the pH (H_2O_2) test. This implies that a single acid test could not completely provide the certainty of acid production of rock from these mine sites. In comparing the acid-forming potentials of the samples using the two techniques, the *ABA* method showed that 68.7% of the samples were acid forming while pH (H_2O_2) captured 43.8%. This discrepancy could be attributed to the disparity in approaches applied in these techniques. On the one hand, *ABA* is based on *AP* and *NP* by measurement of sulfur and carbonate, respectively, for the *NNP* calculation [57,58]. However, *ABA* is a widely applied technique that is found to be more appropriate for short- to long-term acid generation evaluation, but its testing procedure can be exhaustive and give three overlapping outcomes. The pH (H_2O_2) tests, on the other hand, use oxidation via a strong oxidant to completely oxidize sulfides exposed during the experiment but occasionally tend to overestimate the acid-producing potential of samples. In terms of its ease of use, this test is a more streamlined approach to determine acid-forming potential made popular in Japan with an immediate and definite outcome [43,44]. The significant differences in the tests could be attributed to the reaction rate and the nature of the reactant used (i.e., 1M HCl and 15% H_2O_2 for the *ABA* test and 30% H_2O_2 for the pH(H_2O_2) test) with the availability of the minerals (i.e., sulfide and carbonates) to react also influencing the different outcomes of the 2 methods. The pH(H_2O_2) method uses strong H_2O_2 which might accelerate and react intensely with sulfides and, in the process, cause the dissolution of carbonates neutralizing its effect compared with *ABA*, which reacts individually with sulfide and carbonate minerals with less intensity. In addition, the contrast may be attributed to *ABA* potential to assess the capacity of rocks to neutralize acid, which is essential for effective mine waste management and rehabilitation as well as regulatory compliance. In contrast, the pH(H_2O_2) test, which originated in Japan, is a simpler method that provides immediate results based on the oxidation of sulfide minerals.

Table 2. Principal component analysis (PCA) including total variance, eigenvalue, and cumulative frequency for the leachates of rock samples.

Variables	PC1	PC2	PC3	PC4
pH	**−0.381**	0.141	−0.078	−0.276
EC	**0.308**	**0.393**	−0.082	0.103
ORP	**0.362**	−0.171	0.181	**0.310**
Al	0.253	−0.049	**0.471**	−0.081
Cu	0.238	−0.097	**0.469**	−0.077
Fe	**0.317**	0.220	−0.299	0.166
Mg	−0.076	**0.325**	**0.345**	−0.062
Mn	0.121	−0.186	0.049	**0.316**
Pb	**0.331**	0.129	−0.358	−0.058
Zn	0.178	−0.111	0.139	−0.618
Ca	−0.123	**0.452**	0.209	−0.125
K	−0.163	**0.395**	0.072	0.143
Na	−0.297	0.052	0.047	0.290
SO_4^{2-}	0.262	**0.451**	0.097	0.021
Si	−0.216	−0.008	**0.306**	**0.410**
Eigenvalue	5.00	3.03	2.42	1.30
Variance%	31.6	21.1	16.6	8.6
Cumulative%	31.6	52.7	69.2	77.5

Note: Bold text indicates variables that accounted significantly under each PC.

Incorporating the two methods coupled with geochemical characterization and thermodynamic calculation provided insights into the acid production and neutralization potentials of rock samples. Thus, it is still safe to assume that the samples (M2, N1, and T3) have the potential to generate acidity as confirmed by their mineralogy and leaching behavior. The tests have established that Mt Kasi, Nukudamu, and Wainivesi mine sites are the most vulnerable because of sulfide minerals that have the potential to generate AMD. Furthermore, the various geochemical characterization suggests the susceptibility of those samples in uncertainty range, which discloses important information about their vulnerability once the carbonate minerals are depleted. The results demonstrate that the degree and scope of environmental vulnerability are often dictated by the geochemical properties of extracted ores. The vulnerability worsens if mining and waste management best practices are lacking; therefore, AMD generation would likely persist for a long time even after mining has ceased [9,59–61]. Thus, close attention should be paid to the abandoned sites and proper environmental management practices should be seriously applied to currently operating mine sites. The above knowledge adds valuable contributions to a more prudent mining permitting process and strengthens mining policies and frameworks to ensure improved post-mining rehabilitation programs.

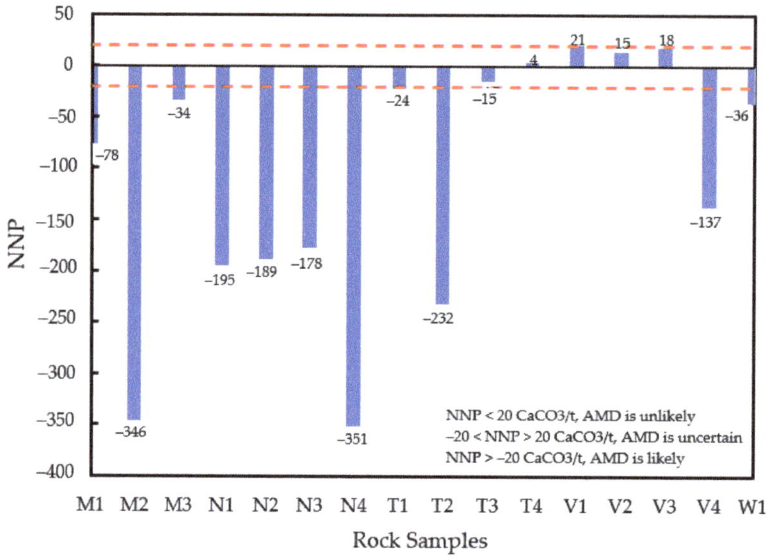

Figure 5. Acid–base accounting (*ABA*) for Mt Kasi (M1–M3), Nukudamu (N1–N4), Tuvatu (T1–T4), Vatukoula (V1–V4), and Wainivesi (W1) rock samples.

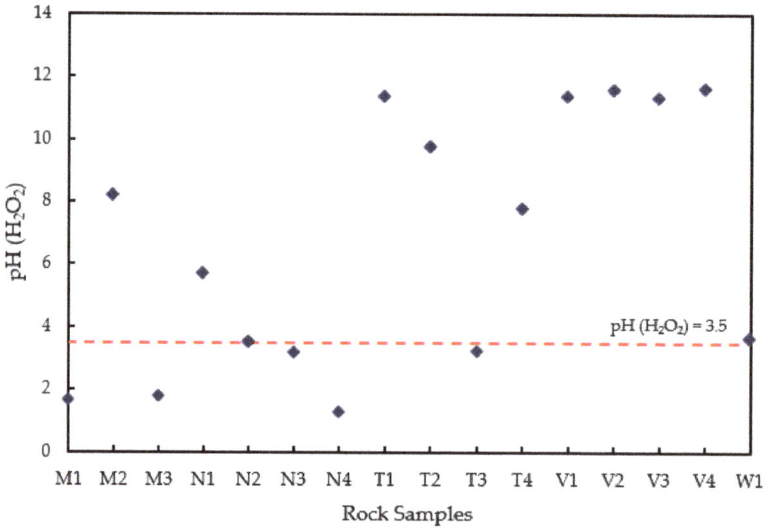

Figure 6. Hydrogen peroxide pH test (pH (H_2O_2)). The dotted line indicates the threshold for acidification potential (pH = 3.5).

4. Conclusions

The geochemical characterization of rock samples from Fiji provides valuable insights into the potential environmental impacts of mining activities. The leachate concentrations of Cu, Pb, and Zn in the rock samples from Mt Kasi, Nukudamu, and Wainivesi exceeded the drinking water regulatory thresholds established by WHO. This is primarily attributed to the oxidation of sulfide minerals, which has been conclusively elucidated via a leaching experiment. In contrast, the retention of hazardous elements in the Tuvatu and Vatukoula rock samples was linked to the strong buffering effect of calcite and dolomite. Acidic pH and high redox potential of the rock samples were found to be key factors contributing to the release of hazardous elements. The *ABA* test, coupled with leaching tests and geochemical characterization, provides a comprehensive assessment of the acid potential of the rock samples from Fiji. The *ABA* test revealed that the rock samples from Mt Kasi, Nukudamu, and Wainivesi have a high potential to generate AMD, which was consistent with the results of the leaching test and are likely to persist compared with the Tuvatu and Vatukoula rock samples with minor carbonate minerals such as calcite and dolomite. Overall, the study highlights the importance of conducting thorough geochemical studies before approving mining activities to identify potential risks and mitigate environmental impacts. Characterization methods used in this study should be considered a mandatory requirement for mining approval to strengthen and improve mining development, waste management, and post-mining rehabilitation in developing countries such as Fiji.

Supplementary Materials: The following supporting information can be downloaded at: https://www.mdpi.com/article/10.3390/min13050661/s1, Figure S1. Microscopic observation of rock samples M2, N1,T3, V1, V4, and W1. Figure S2. Graphical presentation of the PCA showing the first 3 components with significant inter-correlated dependent variables. Table S1. Rock sample details. Table S2. Quality assurance (QA) and quality control (QC). Table S3. XRD diffraction percentage mineralogical abundances (%) of the rock samples. Table S4. The EC, Eh, ORP, pe, and mean concentrations of co-existing elements, Al, Na, and Si, of the rock samples. Table S5. Saturation indices (SIs) of selected mineral phases in the leachates of the rock samples.

Author Contributions: Conceptualization, A.T.S. and T.I.; sample collection and methodology, A.T.S., T.A. and T.I.; validation, W.M., T.A., C.B.T. and T.I.; formal analysis, A.T.S., W.M., C.B.T. and T.A.; writing—original draft preparation, A.T.S.; writing—review and editing, W.M., T.A., C.B.T. and T.I.; visualization, A.T.S. and M.W; supervision, T.A. and T.I.; project administration, A.T.S.; funding acquisition, A.T.S., T.A. and T.I. All authors have read and agreed to the published version of the manuscript.

Funding: This research received no external funding.

Data Availability Statement: Data for this study are presented in the paper.

Acknowledgments: The authors would like to thank the Fiji Mineral Resources Department for providing samples for this study, the Horonobe Research Institute for the Subsurface Environment for XRD and XRF analysis, and the guidance and assistance of the supervisors.

Conflicts of Interest: The authors declare no conflict of interest.

References

1. Xu, Q.; Wang, G.; Liang, X.; Qu, S.; Shi, Z.; Wang, X. Determination of mining-induced changes in hydrogeological parameters of overburden aquifer in a coalfield, northwest China: Approaches using the water level responses to earth tides. *Geofluids* **2021**, *2021*, 5516997. [CrossRef]
2. McKinnon, E. The environmental effects of mining waste disposal at Lihir Gold Mine, Papua New Guinea. *Rural Remote Health* **2002**, *1*, 40–50.
3. Banks, G.; Paull, D.; Mockler, S. *The Social and Environmental Impact of Mining in Asia-Pacific: The Potential Contribution of a Remote-Sensing Approach*; Resource Management in Asia-Pacific Working paper No. 60; The Australian National University: Canberra, Australia, 2005.

4. Mudd, G.M.; Roche, C.; Northey, S.A.; Jowitt, S.M.; Gamato, G. Mining in Papua New Guinea: A complex story of trends, impacts and governance. *Sci. Total Environ.* **2020**, *741*, 140375. [CrossRef]
5. Worlanyo, A.S.; Jiangfeng, L. Evaluating the environmental and economic impact of mining for post-mined land restoration and land-use: A review. *J. Environ. Manag.* **2021**, *279*, 111623. [CrossRef] [PubMed]
6. Fashola, M.O.; Ngole-Jeme, V.M.; Babalola, O.O. Heavy metal pollution from gold mines: Environmental effects and bacterial strategies for resistance. *Int. J Environ. Res. Public Health* **2016**, *13*, 1047. [CrossRef]
7. Park, I.; Tabelin, C.B.; Jeon, S.; Li, X.; Seno, K.; Ito, M.; Hiroyoshi, N. A review of recent strategies for acid mine drainage prevention and mine tailings recycling. *Chemosphere* **2019**, *219*, 588–606. [CrossRef] [PubMed]
8. Saharan, M.R.; Gupta, K.K.; Jamal, A.; Sheoran, A.S. Management of acidic effluents from tailing dams in metalliferous mines. *Mine Water Environ.* **1995**, *14*, 85–93. [CrossRef]
9. Akcil, A.; Soner, K. Acid Mine Drainage (AMD): Causes, treatment and case studies. *J. Clean. Prod.* **2006**, *14*, 1139–1145. [CrossRef]
10. Foli, G.; Apeah, O.B.; Amedjoe, C.G. Pre-mining water quality prediction from non-weathered sulphide ores along the Ashanti metallogenic belt in Ghana using Acid-Base accounting procedure. *Am. J. Sci. Ind. Res.* **2011**, *2*, 827–833. [CrossRef]
11. Kumar, S.; Islam, A.R.M.T.; Islam, H.T.; Hasanuzzaman, M.; Ongoma, V.; Khan, R.; Mallick, J. Water resources pollution associated with risks of heavy metals from Vatukoula Goldmine region, Fiji. *J. Environ. Manag.* **2021**, *293*, 112868. [CrossRef] [PubMed]
12. Singh, S.; Mosley, L. Trace metal levels in drinking water on Viti Levu, Fiji Islands. *South Pac. J. Nat. Sci.* **2003**, *21*, 31–34. [CrossRef]
13. Kumar, S.; Islam, A.R.M.T.; Hasanuzzaman, M.; Salam, R.; Islam, M.S.; Khan, R.; Rahman, M.S.; Pal, S.C.; Ali, M.M.; Idris, A.M.; et al. Potentially toxic elemental contamination in Wainivesi River, Fiji impacted by gold-mining activities using chemometric tools and SOM analysis. *Environ. Sci. Pollut. Res.* **2022**, *29*, 42742–42767. [CrossRef]
14. Ko, B.G.; Anderson, C.W.; Bolan, N.S.; Huh, K.Y.; Vogeler, I. Potential for the phytoremediation of arsenic-contaminated mine tailings in Fiji. *Soil Res.* **2008**, *46*, 493–501. [CrossRef]
15. Matanitobua, V.; Ng, J.; Chiswell, B.; Harris, H.; Aalbersberg, W.; Noller, B. Arsenic adsorption and desorption processes downstream of a Fijian gold mine. In *Arsenic in Geosphere and Human Diseases, Proceedings of the Arsenic 2010: Third International Congress on Arsenic in the Environment (As-2010), Tainan, Taiwan, 17–21 May 2010*; CRC Press: London, UK, 2010; pp. 176–178.
16. Matakarawa, S. Gold Mining and Acute Respiratory Infection in Children: A Retrospective Cohort Study in Vatukoula, Fiji. Master's Thesis, University of Canterbury, Christchurch, New Zealand, 2018.
17. Rabuku, A.T.W.; Abdul, Q.M. Natural radioactivity measurement of gold mine tailings in Vatukoula, Fiji Islands. *Renew. Energy Environ. Sustain.* **2020**, *5*, 10. [CrossRef]
18. Mataki, M.; Koshy, K.C.; Lal, M. Baseline Climatology of Viti Levu (Fiji) and Current Climatic Trends. *Pac. Sci.* **2006**, *60*, 49–68. [CrossRef]
19. Ongoma, V.; Rahman, M.A.; Ayugi, B.; Nisha, F.; Galvin, S.; Shilenje, Z.W.; Ogwang, B.A. Variability of diurnal temperature range over Pacific Island countries, a case study of Fiji. *Meteorol. Atmos. Phys.* **2021**, *133*, 85–95. [CrossRef]
20. Colley, H.; Flint, D.J. *Metallic Mineral Deposits of Fiji, No.4, Mineral Resources Department Memoir*; Government of Fiji: Suva, Fiji, 1995; 196p.
21. Turner, S.J. Fluid Inclusion, Alteration, and Ore Mineral Studies of an Epithermal, Mineralized Vein System: Mount Kasi, Vanua Levu, Fiji. Master's Thesis, University of Auckland, Auckland, New Zealand, 1986.
22. Colley, H.; Rice, C.M. A kuroko type ore deposit in Fiji. *Econ. Geol.* **1975**, *70*, 1373–1386. [CrossRef]
23. Colley, H.; Rice, C.M. Kuroko-type deposits in Vanua Levu, Fiji. *N. Z. J. Geol. Geophys.* **1978**, *21*, 277–285. [CrossRef]
24. Ashley, P.M.; Andrew, A.S. Petrographic, fluid inclusion and stable isotope investigation of a suite of samples from the Tuvatu prospect, Fiji. *Commonwealth Sci. Indust. Res. Org.* **1989**, *50*, 36.
25. Feudigmann, P.; Taylor, I.; Woodward, A.; Lee, A.F.; Morgan, D.M. *Lion One Metals Tuvatu Gold Project Preliminary Economic Assessment*; National Instrument 43-101 Technical Report; Lion One: Nadi, Fiji, 2015.
26. Ahmad, M.; Solomon, M.; Walsh, J.L. Mineralogical and geochemical studies of the Emperor gold telluride deposit, Fiji. *Econ. Geol.* **1987**, *82*, 345–370. [CrossRef]
27. Anderson, W.B.; Eaton, P.C. Gold mineralisation at the Emperor mine, Vatukoula, Fiji. *J. Geochem. Explor.* **1990**, *36*, 267–296. [CrossRef]
28. Eaton, P.C.; Setterfield, T.N. The relationship between epithermal and porphyry systems within the Tavua Caldera, Fiji. *Econ. Geol.* **1993**, *88*, 1053–1083. [CrossRef]
29. White, N.C.; Leake, M.J.; McCaughey, S.N.; Parris, B.W. Epithermal gold deposits of the southwest Pacific. *J. Geochem. Explor.* **1995**, *54*, 87–136. [CrossRef]
30. Houtz, R.E. *Geology of North Tailevu, Viti Levu*; Bulletin 1; Fiji Geological Survey: Viti Levu, Fiji, 1958; p. 19.
31. *JIS A 1226*; Test Method for Ignition Loss of Soils. Japanese Standards Association: Tokyo, Japan, 2020; pp. 1–10.
32. Marove, C.A.; Tangviroon, P.; Tabelin, C.B.; Igarashi, T. Leaching of hazardous elements from Mozambican coal and coal ash. *J. Afr. Earth Sci.* **2020**, *168*, 103861. [CrossRef]
33. *JLT-13*; Departmental Notification No. 13 on Leaching Test Method for Landfill Wastes. Japanese Environmental Agency: Tokyo, Japan, 1973.
34. Tabelin, C.B.; Hashimoto, A.; Igarashi, T.; Yoneda, T. Leaching of boron, arsenic, and selenium from sedimentary rocks: II. pH dependence, speciation, and mechanisms of release. *Sci. Total Environ.* **2014**, *473*, 244–253. [CrossRef] [PubMed]

35. Huyen, D.T.; Tabelin, C.B.; Thuan, H.M.; Dang, D.H.; Truong, P.T.; Vongphuthone, B.; Kobayashi, M.; Igarashi, T. The solid-phase partitioning of arsenic in unconsolidated sediments of the Mekong Delta, Vietnam and its modes of release under various conditions. *Chemosphere* **2019**, *233*, 512–523. [CrossRef]
36. Khoeurn, K.; Sasaki, A.; Tomiyama, S.; Igarashi, T. Distribution of zinc, copper, and iron in the tailings dam of an abandoned mine in Shimokawa, Hokkaido, Japan. *Mine Water Environ.* **2019**, *30*, 119–129. [CrossRef]
37. Tabelin, C.B.; Silwamba, M.; Paglinawan, F.C.; Mondejar, A.J.S.; Duc, H.G.; Resabal, V.J.; Opiso, E.M.; Igarashi, T.; Tomiyama, S.; Ito, M.; et al. Solid-phase partitioning and release-retention mechanisms of copper, lead, zinc, and arsenic in soils impacted by artisanal and small-scale gold mining (ASGM) activities. *Chemosphere* **2020**, *260*, 127574. [CrossRef]
38. Mufalo, W.; Tangviroon, P.; Igarashi, T.; Ito, M.; Sato, T.; Chirwa, M.; Nyambe, I.; Nakata, H.; Nakayama, S.; Ishizuka, M. Solid-phase partitioning and leaching behavior of Pb and Zn from playground soils in Kabwe, Zambia. *Toxics* **2021**, *9*, 248. [CrossRef]
39. Gustafsson, J.P. *Visual Minteq 3.1 User Guide*; KTH, Department of land and Water Resources: Stockholm, Sweden, 2014; p. 550.
40. Lawrence, K.A.; Wang, Y. Determination of neutralization potential in the prediction of acid rock drainage. In Proceedings of the 4th International Conference of Acid Rock Drainage, Vancouver, BC, Canada, 31 May–6 June 1997; pp. 451–464.
41. Wang, L.; Li, Y.; Cui, X.; Wang, X.; Lu, A.; Wang, X.; Wang, C.; Gan, D. Weathering behavior and metal mobility of tailings under an extremely arid climate at Jinchuan Cu-Ni sulfide deposit, Western China. *J. Geochem. Explor.* **2017**, *173*, 1–12. [CrossRef]
42. *JGS 0271-2016*; Test Method for Acidification Potential of Soils and Rock by Using Hydrogen Peroxide Solution. Japanese Geotechnical Society: Tokyo, Japan, 2016.
43. Igarashi, T.; Oyama, T.; Saito, N. Experimental study on acidification potential of leachate from sedimentary rocks containing pyrite. *J. Jpn. Soc. Eng. Geol.* **2001**, *42*, 214–221. [CrossRef]
44. Inui, T.; Katayama, M.; Katsumi, T.; Takai, A.; Kamon, M. Evaluating the long-term leaching characteristics of heavy metals in excavated rocks. *J. Soc. Mater. Sci. Jpn.* **2014**, *63*, 73–78. [CrossRef]
45. Clarke, R.; Smith, D.; Naden, J. Source controls on: Regional geology and magmatic evolution of Fiji. *Lithos* **2022**, *432–433*, 106897. [CrossRef]
46. Clarke, R.; Smith, D.J.; Naden, J.; Holwell, D.; Mann, S. Mineralogical constraints on the genesis of an alkalic-type epithermal Au-Te deposit: Tuvatu, Fiji. *Ore Geol. Rev.* **2023**, *154*, 105279. [CrossRef]
47. World Health Organization. *Guidelines for Drinking-Water Quality*, 4th ed.; World Health Organization: Geneva, Switzerland, 2011.
48. Tabelin, C.B.; Igarashi, T.; Villacorte-Tabelin, M.; Park, I.; Opiso, E.M.; Ito, M.; Hiroyoshi, N. Arsenic, selenium, boron, lead, cadmium, copper, and zinc in naturally contaminated rocks: A review of their sources, modes of enrichment, mechanisms of release, and mitigation strategies. *Sci. Total Environ.* **2018**, *645*, 1522–1553. [CrossRef]
49. Igarashi, T.; Herrera, P.S.; Uchiyama, H.; Miyamae, H.; Iyatomi, N.; Hashimoto, K.; Tabelin, C.B. The two-step neutralization ferrite-formation process for sustainable acid mine drainage treatment: Removal of copper, zinc, and arsenic and the influence of coexisting ions on ferritization. *Sci. Total Environ.* **2020**, *715*, 136877. [CrossRef]
50. Bao, Y.; Guo, C.; Lu, G.; Yi, X.; Wang, H.; Dang, Z. Role of microbial activity in Fe (III) hydroxysulfate mineral transformations in an acid mine drainage-impacted site from the Dabaoshan Mine. *Sci. Total Environ.* **2018**, *616*, 647–657. [CrossRef]
51. Kelderman, P.; Osman, A.A. Effect of redox potential on heavy metal binding forms in polluted canal sediments in Delft (The Netherlands). *Water Res.* **2007**, *41*, 4251–4261. [CrossRef]
52. Popenda, A. Effect of redox potential on heavy metals and As behavior in dredged sediments. *Desalination Water Treat.* **2014**, *52*, 3918–3927. [CrossRef]
53. Tangviroon, P.; Noto, K.; Igarashi, T.; Kawashima, T.; Ito, M.; Sato, T.; Mufalo, W.; Chirwa, M.; Nyambe, I.; Nakata, H.; et al. Immobilization of lead and zinc leached from mining residual materials in Kabwe, Zambia: Possibility of Chemical Immobilization by Dolomite, Calcined Dolomite, and Magnesium Oxide. *Minerals* **2020**, *10*, 763. [CrossRef]
54. Khoshraftar, Z.; Hadiseh, M.; Ahad, G. An insight into the potential of dolomite powder as a sorbent in the elimination of heavy metals: A review. *Case Stud. Chem. Environ. Eng.* **2022**, *7*, 100276. [CrossRef]
55. Mufalo, W.; Tangviroon, P.; Arima, T.; Igarashi, T.; Ito, M.; Noto, K.; Kawashima, T.; Nyambe, I.; Nakata, H.; Nakayama, S. Immobilization of Pb and Zn leached from mining residue materials in Kabwe, Zambia: Performance of calcined dolomite in column experiments. *J. Geochem. Explor.* **2023**, *249*, 107209. [CrossRef]
56. Gu, X.; Evans, L.J. Surface complexation modelling of Cd (II), Cu (II), Ni (II), Pb (II), and Zn (II) adsorption onto kaolinite. *Geochim. Cosmochim. Acta* **2008**, *72*, 267–276. [CrossRef]
57. Gunsinger, M.R.; Ptacek, C.J.; Blowes, D.W.; Jambor, J.L.; Moncur, M.C. Mechanisms controlling acid neutralization and metal mobility within a Ni-rich tailings impoundment. *J. Appl. Geochem.* **2006**, *21*, 1301–1321. [CrossRef]
58. Brady, K.B.; Cravotta, C.A. Acid base accounting: An improved method of interpreting overburden chemistry to predict quality of coal mine drainage. In Proceedings of the 13th Annual West Virginia Surface Mine Drainage Task Force Symposium, Morgantown, WV, USA, 8–9 April 1992.
59. Stewart, W.A.; Miller, S.D.; Smart, R. Advances in acid rock drainage (ARD) characterization of mine wastes. In Proceedings of the 7th International Conference on Acid Rock Drainage (ICARD), American Society of Mining and Reclamation (ASMR) 2006, St. Louis, MI, USA, 26–30 March 2006.

60. Simate, G.S.; Ndlovu, S. Acid mine drainage: Challenges and opportunities. *J. Environ. Chem. Eng.* **2014**, *2*, 1785–1803. [CrossRef]
61. Assawincharoenkij, T.; Hauzenberger, C.; Ettinger, K.; Sutthirat, C. Mineralogical and geochemical characterization of waste rocks from a gold mine in northeastern Thailand: Application for environmental impact protection. *Environ. Sci. Pollut. Res.* **2018**, *25*, 3488–3500. [CrossRef] [PubMed]

minerals

MDPI

Article

Controls on the Generation and Geochemistry of Neutral Mine Drainage: Evidence from Force Crag Mine, Cumbria, UK

Adam P. Jarvis [1,*], Catherine J. Gandy [1] and John A. Webb [2]

[1] School of Engineering, Newcastle University, Newcastle upon Tyne NE1 7RU, UK; catherine.gandy@newcastle.ac.uk
[2] Discipline of Ecology and Environment, La Trobe University, Melbourne 3086, Australia; john.webb@latrobe.edu.au
* Correspondence: adam.jarvis@newcastle.ac.uk

Abstract: Neutral mine drainage (NMD) at Force Crag mine in north-west England has a circumneutral pH and high levels of Zn contamination. A long-term geochemical and hydrological dataset from this site was analysed using a novel molar mass balance approach, which demonstrated that the water chemistry is dominated by species released by the oxidation of sulphides: sphalerite (Zn, Cd, Ni), galena (Pb, mostly removed by adsorption to ferrihydrite) and pyrite (Fe, mostly precipitates as ferrihydrite). The calculations show that the sphalerite:galena:pyrite oxidation ratio is ~1:2:1, but the mine water chemistry is dominated by Zn due to the removal of Pb and Fe by adsorption/precipitation. The acidity released by pyrite oxidation is neutralised by the dissolution of calcite and, to a lesser extent, chlorite. The presence of pyrite is responsible both for the release of acidity and the removal of some contaminant metals by adsorption on ferrihydrite. The concentrations of sulphate, Zn, Cd and Ni in the mine water decrease with increasing flow due to dilution; modest increases in metal flux with flow probably reflect increased oxidation due to greater amounts of oxygenated water flowing through the mine. In contrast, Al, Pb and Cu concentrations are positively correlated with flow due to the flushing of these metals adsorbed to ferrihydrite particles. The influence of temperature is relatively subtle; metal fluxes are a balance between abiotic oxidation (which increases at higher temperatures and flows) and bacterially mediated oxidation (which is depressed at high flow rates when temperatures decrease below 10 °C). These conclusions apply to NMD mine water throughout the UK and elsewhere in the world, including mines hosted in both limestone and silicate rocks. The molar mass balance approach, together with synchronous flow and geochemistry data, provides crucial information for effective mine-water-treatment system design by elucidating the critical roles of flow rate and temperature in determining contaminant concentrations and loads.

Keywords: Force Crag; neutral mine drainage; mass balance; hydrology; water quality; metals; sphalerite; galena

1. Introduction

Contaminated drainage from metalliferous mines is commonly acidic (acid mine drainage, AMD), but at some mines it has a circumneutral pH (neutral mine drainage, NMD). This is generally due to neutralisation by carbonate (mostly limestone) dissolution [1]. NMD is less common than AMD globally, and because the environmental impacts are not as severe, it has received relatively little attention. Nevertheless, NMD often contains high levels of dissolved metals, frequently Zn. NMD is a widespread problem in many regions around the world [2–7], and consequently, it has received international attention with respect to both its evolution and treatment (e.g., [8–10]).

NMD is a particular problem in the UK. A total of 260 abandoned mine sites across England and Wales have been confirmed as causing contamination of freshwaters, and a large majority of these have near-neutral pH drainage, with elevated Zn concentrations in particular [11–13]. The combined flux of Zn from abandoned base-metal mines across England and Wales is more than 250 t/annum [14], and although the levels of Zn are typically low compared to many examples of AMD, they still have serious environmental impacts [15,16]. Consequently, there is an ongoing programme of UK government investment in designing and building treatment systems to remediate these sites wherever possible. Estimates of the cost for remediation of all abandoned metal mine discharges in England and Wales are substantial: over £372 million for system construction and operation for a period of 10 years [11].

One of the first metal-mine water discharges to be remediated in the UK was the main NMD drainage at the Force Crag mine in the Lake District National Park, in north-west England, which was the last working metal mine in the region (finally closed in 1991). The UK Coal Authority built a passive treatment system at Force Crag that was commissioned in March 2014 [13,17]. As part of the remedial programme, detailed monitoring of water quality and hydrology across the site has been carried out, and these data have provided an ideal opportunity to understand in detail the controls on the composition of the NMD issuing from the mine.

The aim of the current work was to elucidate key processes controlling the water quality of the main contaminated discharge emerging from the mine using the 57 months of detailed water quality and hydrology data available. To do so, a novel geochemical mass balance approach was developed, both to quantify subsurface processes at the mine and to assess any evolution of the mine water quality over time. The results of the study can then be applied to understanding the variability in discharge flow and water quality at the Force Crag mine site, as this is crucial for long-term prognoses of the performance of the operational treatment system. This approach can also be used to understand the variability in NMD composition at mines across the UK and elsewhere in the world and to plan remediation at these sites.

2. Materials and Methods

2.1. Study Site

Force Crag mine (Figure 1) is in the Lake District of Cumbria, north-west England, at 54.5835° N 3.2397° W, towards the head of the Coledale Beck valley. Force Crag is the cliff from which the mine takes its name and lies 300 m to the west-south-west of the mine. The Coledale Beck valley is narrow and steep-sided, and the beck (stream) is therefore fast-flowing. A tributary, Pudding Beck, descends the cliffs adjacent to the workings. The valley floor consists of a glacial deposit of yellow boulder clay overlain by peat.

The mineralised vein mined at Force Crag occupies a roughly E-W-trending, steeply dipping fault cutting through slates of the Ordovician Skiddaw Group [18]. The slates are regionally metamorphosed siltstones of turbidite origin and comprise predominantly quartz, K- and Na-rich white micas and chlorite, with relatively little compositional variation [19–21].

In the upper part of the Force Crag vein, barite is the main constituent, accompanied by abundant manganese and iron oxides, cerussite and rare malachite [18]. Below this, the vein is composed of massive white quartz, underlain in turn by a zone with abundant sulphides, predominantly sphalerite; galena is less common, and minor barite is present. Pyrite and chalcopyrite occur locally as small crystals in the lower parts of the vein, and siderite is common; dolomite, calcite, ankerite and rare fluorite are also present [18,22,23]. Near the top of the sphalerite and galena-rich part of the vein, supergene Pb, Cu, Ag, Co and Zn minerals have been recorded: erythrite, acanthite, anglesite, brochantite, cerussite (most common), cuprite, langite, lautenthalite, linarite, pyromorphite, serpierite and native silver [24]. Gypsum, melanterite and jarosite occur as post-mining encrustations on the mine walls [24]. Ferric hydroxide (ochre) coatings up to 1.5 m thick line the walls, roof and floor in places [25]. Although this material has not been identified mineralogically, it was probably deposited as ferrihydrite, which is the amorphous ferric hydroxide mineral stable at the pH of the mine water (pH > 5 [26]).

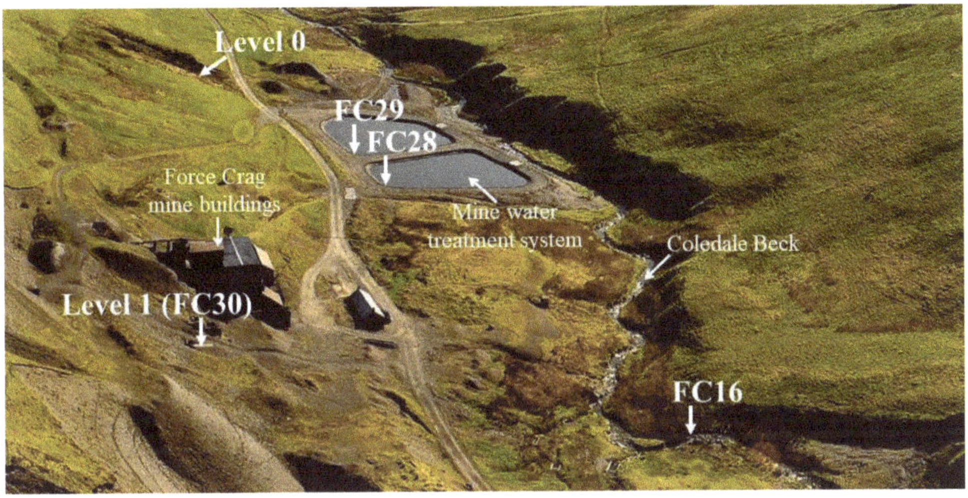

Figure 1. Force Crag mine buildings and treatment ponds, showing sampling sites: main mine water discharge (Level 1, FC30), inlets to mine water treatment ponds (FC28/29) and Coledale Beck immediately upstream of the mine (FC16). Photo by John Malley, National Trust (taken in April 2015).

At Force Crag mine, the mineralisation was accessed from a series of levels (adits) numbered from 0 to 7 (from lowest to highest) [27]. Mining extended over a horizontal distance of $\approx$1.2 km and a vertical extent of $\approx$350 m. Various operators mined for lead at the site between 1839 and 1865, and then for barite and zinc periodically between 1867 and 1991, when the mine was finally abandoned [27]. More than 60,000 tons of barite, 1248 tons of sphalerite, 624 tons of galena and 20,000 ounces of silver were extracted [28].

Level 1 (FC30) forms the main drain of the mine; some drainage also occurs from Level 0, which is 200 m to the east of Level 1, at an elevation of 20 m below it (Figure 1). Discharge water from the mine contains high levels of zinc, cadmium and lead; consequently, concentrations of all three metals in the Coledale Beck exceed relevant regulatory standards most of the time [13]. Before the passive treatment system commenced operation in 2014, the mine discharge polluted the Coledale Beck for at least 10 km downstream of the mine.

2.2. Water Quality and Hydrology Monitoring

Water samples for the Level 1 mine water discharge were collected at location FC29 (Figure 1), one of the two inlet points to the treatment system (FC28 is the other; Figure 1), rather than at the discharge point (FC30; Figure 1), because evaluation of the treatment system was a key part of the wider programme of work (not reported here). However, the flow rate of the discharge from Level 1 was measured directly at FC30 because some of the water from Level 1 bypasses the treatment system before FC28/29. Flow at FC30 was measured immediately after the collection of water samples at FC29 using a sharp-crested V-notch weir just below the level entrance. There is no difference in water quality between FC30 and FC29 because water is rapidly transported from one location to the other via a 0.315 m diameter buried plastic pipe.

To understand the role of the dissolution of different minerals in the underground workings in governing the quality of water emerging from the mine, it was necessary to know the chemistry of the input water to the mine. This input water could not be directly sampled at Force Crag due to a lack of access; the mine is unstable and unsafe. Coledale Beck runs adjacent to the mine, and Coledale Beck water immediately upstream of the mine (FC16, Figure 1) was regarded as an approximation of input water to the mine (it represents the least polluted water sampled). Therefore, samples were always collected from FC16 and FC29 on the same day.

After the treatment system was commissioned in March 2014, water-quality and flow monitoring was carried out on average every 2 weeks from April 2014 to September 2019, and up to 130 measurements are available from each sample site. A pre-calibrated Myron L 6P Ultrameter recorded measurements of water temperature, pH, oxidation−reduction potential (ORP) and electrical conductivity during site visits. Total alkalinity was measured with a Hach digital titrator using 0.16 N sulfuric acid and bromcresol-green methyl-red indicator. The collection of water samples was performed in sterile 30 mL polypropylene bottles. Three aliquots were collected for total cations and filtered cations and anions; filtration was carried out using 0.45 μm cellulose nitrate filter paper. Cation samples were acidified with 2% v/v concentrated nitric acid and 1% v/v concentrated hydrochloric acid.

Filtered ion concentrations are often referred to as "dissolved" concentrations, but we use the term "filtered" or "aqueous-phase" rather than "dissolved" because the filtered aliquot may contain colloids <0.45 μm in diameter (colloids are from 1 nm to 1 μm across [29]). Only larger colloids (>0.45 μm) were retained by the filter. "Particulate" concentrations were calculated by subtracting the filtered concentration from the total concentration, albeit this does not include any particulates <0.45 μm in diameter. The great majority of unfiltered samples did not contain particles visible to the naked eye, so the particulates present were very small. Nevertheless, the quantitative distinction between particulate and filterable cation concentrations provides useful insights into the hydrogeochemical behaviour of metals.

All samples were stored at 4 °C prior to laboratory analysis. Cation concentrations were determined using a Varian Vista-MPX Inductively Coupled Plasma-Optical Emission Spectrometer (ICP-OES) or an Agilent Technologies 7700 Series Inductively Coupled Plasma-Mass Spectrometer (ICP-MS). Anion concentrations were measured with a Dionex DX320 Ion Chromatograph (IC). Additional details of sampling and analytical methods are provided elsewhere [13,30], but, in brief, calibration standards were made using certified standards (1000 ppm; accuracy of $\leq\pm1.0\%$; VWR Chemicals) and deionised water (Elga Purelab Ultra 18.2 MΩ at 25.8 °C). Analytical accuracy and precision were checked for every 10 samples using blanks (deionised water) and standards. Triplicate samples were collected periodically as part of a wider QA/QC programme. Mann–Whitney U tests indicated no significant differences ($p > 0.1$) between metals and SO_4 concentrations in replicates. The reliability of sample analysis (for FC16 and FC29) was also tested using charge balance calculations. In the majority of cases (>65% for FC16 (n = 34) and >70% for FC29 (n = 132)), electro-neutrality was within ±5%, and in more than 90% of cases within ±10% for FC16 and within ±10% in all cases for FC29. For FC16, the very low ion content of the water (electrical conductivity of <40 μS/cm) means that the charge balance is very sensitive to even minor analytical errors [30].

Geochemical modelling using the PhreeqC software was used to calculate saturation indices (SI = log (activity product/K_{eq})) for minerals that could be precipitating from the mine water. Minitab v21.2 was used for statistical analysis. For the calculation of correlation coefficients, the non-parametric Spearman's Rho coefficient was used in all cases since data for many variables are non-normally distributed. Because of the covariance of the key variables (flow rate, temperature and water quality), partial correlation coefficients were calculated in Minitab for selected data.

3. Results and Discussion

3.1. Selection of Input Water for Calculation of Water Chemistry Changes within the Mine

Because input water to the Force Crag mine could not be directly sampled, FC16 water (Coledale Beck upstream of the mine; Figure 1) was collected as a reasonable proxy. To assess the suitability of this, FC16 water quality was compared with that of nearby rainfall (Bannisdale, 35 km SE of Force Crag [31]; Table 1, Figure 2), because direct ingress of rainfall to the mine is possible through open voids into the workings. The rainfall data used [31] do not include trace metals, so a one-off sample of rainwater was collected at Force Crag and analysed for trace metals (Table 1).

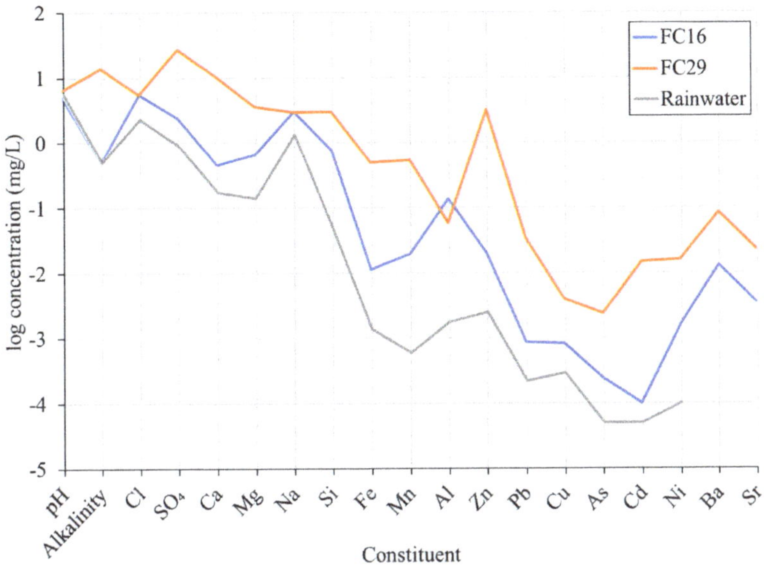

Figure 2. Scholler plot of the median total composition of rainwater, upstream Coledale Beck water (FC16) and output water from the mine (FC29) (for rainwater, Si, As and Cd are reported as half of the detection limit as all were below detection limits of 0.1 mg/L (Si) and 0.1 µg/L (As and Cd); alkalinity in mg/L as $CaCO_3$).

Table 1. Medians of total (0.45 µm filtered) compositions ± standard deviation of Coledale Beck water upstream of the mine (FC16), output water from the mine (FC28/29) and nearby rainfall (Bannisdale; UK-AIR 00114 [31]; n = [a] 124, [b] 137; [c] concentrations in rainfall sample from 2 February 2022); for FC16 and FC28/29 n, number of measurements of total concentration, or filtered if no total measurements = [d] 34, [e] 18, [f] 16, [g] 12, [h] 11, [I] 132, [j] 131, [k] 97, [l] 100, [m] 72, [n] 42. Filtered anion concentrations only. All analyses of K at FC16 and FC28/29 were below detection limit (1 mg/L). See Table S1 for additional details.

	Units	Rainfall	FC16	FC28/29
pH		5.83 ± 0.58 [a]	4.97 ± 0.39 [d]	6.56 ± 0.38 [i]
Alkalinity	mg/L as $CaCO_3$		(<1.0) [d]	(14.3 ± 3.86) [i]
Cl	mg/L	2.32 ± 2.86 [a]	(5.60 ± 1.67) [d]	(5.50 ± 0.46) [i]
SO_4	mg/L	0.89 ± 0.53 [b]	(2.40 ± 0.39) [d]	(27.0 ± 6.09) [i]
Ca	mg/L	0.17 ± 0.96 [b]	0.46 (0.45) ± 0.08 [d]	10.2 (10.1) ± 2.51 [i]
Mg	mg/L	0.14 ± 0.17 [b]	0.67 (0.67) ± 0.13 [d]	3.61 (3.60) ± 0.78 [i]
Na	mg/L	1.33 ± 1.60 [b]	3.00 (3.00) ± 0.59 [d]	2.95 (2.94) ± 0.28 [i]
Si	mg/L	bdl [c]	0.77 (0.77) ± 0.09 [d]	3.00 (2.93) ± 0.53 [i]
Fe	µg/L	1.36 [c]	11.3 (8.76) ± 7.24 [e]	503 (127) ± 127 [j]
Mn	µg/L	0.60 [c]	20.0 (20.0) ± 4.57 [d]	536 (521) ± 131 [i]
Al	µg/L	1.75 [c]	139 (133) ± 46.5 [d]	58.0 (13.3) ± 55.1 [i]
Zn	µg/L	2.49 [c]	20.0 (20.0) ± 4.17 [d]	3190 (3030) ± 723 [i]
Pb	µg/L	0.22 [c]	0.86 (0.64) ± 0.77 [f]	34.0 (1.35) ± 10.7 [k]
Cu	µg/L	0.29 [c]	0.83 (0.77) ± 0.19 [f]	4.00 (2.00) ± 1.83 [k]
Cd	µg/L	bdl [c]	0.05 (0.05) ± 0.04 [f]	14.8 (14.1) ± 3.33 [l]
Ni	µg/L	0.10 [c]	1.61 (1.59) ± 0.22 [f]	16.1 (16.0) ± 3.19 [l]
Ba	µg/L		13.1 (nd) ± 3.79 [g]	84.8 (85.3) ± 26.6 [m]
Sr	µg/L		3.50 (nd) ± 0.34 [h]	23.0 (22.5) ± 3.18 [m]
As	µg/L		0.24 (0.22) ± 0.05 [f]	2.39 (1.45) ± 0.49 [n]

Note: bdl = below detection limits; nd = no data; standard deviation for filtered composition if total composition not available.

The Na and Cl levels in Coledale Beck water at FC16 are 2.3–2.4 times those in rainfall, due to evapotranspiration within the soil as the rainfall infiltrates. However, the concentrations of SO_4, Ca and Mg at FC16 are more elevated than would be expected from evapotranspiration alone, with concentration factors relative to rainfall of 2.7–4.8, and the levels of some metals are also slightly raised (20 μg/L Zn; Table 1). This indicates a minor input to Coledale Beck water from oxidation/dissolution of mineralisation adjacent to the main vein at Force Crag. The low pH of Coledale Beck water (pH 4.97) is probably due to input from organic acids that are common in peaty soils such as those in this area [32]. Thus, upstream Coledale Beck water has been affected by evapotranspiration and the addition of some species, but these processes are equally likely to have affected infiltration into the mine from the hillside above, indicating that FC16 water is a valid proxy for input water to the Force Crag mine. Independent confirmation of this comes from the comparison of Na and Cl concentrations at FC16 with those at FC29 (mine outlet water); they are almost identical (Table 1, Figure 2), indicating that there is no direct rainfall input within the mine, as this would have diluted the Na and Cl levels. The correspondence between these concentrations also shows that there is no significant evaporation within the mine.

3.2. *Mineralogical Controls on Mine Water Chemistry*

To delineate the changes in water chemistry occurring within the mine, median analyses of input water (FC16) and output water (FC29) were compared (Table 1, Figure 2) as both total and filtered (0.45 μm) concentrations. For Zn, Cd, Ni and Ba, there was no significant difference between total and filtered (aqueous-phase) concentrations (Mann–Whitney *U* test; $p > 0.1$ in all cases; Table S1). However, for Pb, Fe, Al and Cu, the filtered concentrations in the mine water (FC29) were significantly lower (<50%) than the total concentrations ($p < 0.001$). This does not affect the following discussion about the mineralogical controls on mine water chemistry but has important implications for the influence of flow and temperature on the composition of mine water, discussed in Sections 3.3 and 3.4.

Using concentrations rather than loadings for comparison assumes that the input and output flows are equal, i.e., there are no gains or losses of water within the system. Na and Cl concentrations at FC16 and FC29 are almost identical (Table 1, Figure 2), indicating that there is no evaporation within the mine and no input of water with a different composition.

3.2.1. Sulphide Oxidation

The concentrations of metals and sulphate in the mine outlet water are much greater than in the input water (Table 1, Figure 2) due to the oxidation of sulphides exposed in the Force Crag vein within the mine. The high concentrations of Zn are due to the oxidation of sphalerite, and hence there is a strong correlation between Zn and SO_4 concentrations in the mine water (Figure 3; $r^2 = 0.95$; $p < 0.001$). Sphalerite oxidation also releases Cd, as shown by the very strong correlation between Zn and Cd (Figure 3; $r^2 = 0.98$; $p < 0.001$). The Zn:Cd weight ratio of 0.0047 indicates ≈0.5 wt% Cd in solid solution within the sphalerite, well within the recorded range of Cd levels in sphalerite [28].

The Ni in the mine water was probably also largely released by sphalerite oxidation because there is a good correlation between Zn and Ni (Figure 3; $r^2 = 0.90$; $p < 0.001$) with a Zn:Ni weight ratio of 0.0042, indicating ≈0.4 wt% Ni in solid solution within the sphalerite. Sphalerite can contain low levels of Ni in solid solution [33,34]. Some Ni may also have been released from chlorite dissolution (discussed below). Ni substitutes for Fe in the octahedral sheet of the chlorite structure, and the clay-rich fraction of Skiddaw Group slates contains on average ≈100 ppm Ni [21].

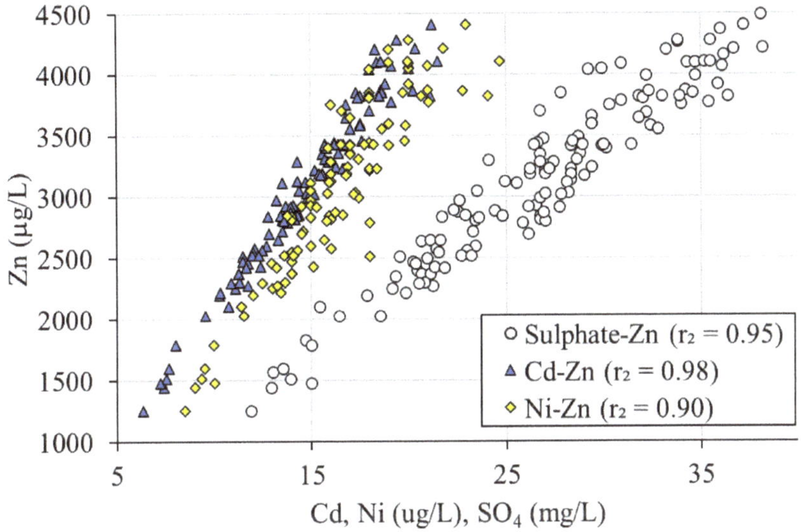

Figure 3. Relationships between sulphate, Zn, Cd and Ni in the Level 1 mine water discharge at Force Crag mine (from analyses of water at location FC29; $p < 0.001$ for all correlation coefficients).

The Pb, Cu, As and Fe in the mine water were released by the oxidation of galena, chalcopyrite and pyrite, which contain these elements and are common in the lower levels of the mine. Pyrite can incorporate large amounts of arsenic (up to 10 wt%) within its structure [35]. The dissolution of supergene Pb/Cu sulphates and oxides may make a minor contribution to the Pb and Cu concentrations. The upper levels of the mine lack sulphides, but pyrite and chalcopyrite may have been previously common because this part of the vein contains reniform masses of goethite [21], which typically forms by oxidation of iron-containing sulphides.

There is approximately 100 times more Zn than Pb in solution in the mine water (as ug/L). This is despite the fact that (1) galena oxidises much more rapidly than sphalerite [36] and (2) there is only twice as much sphalerite as galena in the Force Crag veins, based on production figures and visual estimates of abundance [18,22,28]. This is because two processes attenuate the Pb released by galena oxidation:

(1) Adsorption onto ferrihydrite, which strongly adsorbs Pb at the circumneutral pH typical of the mine water [37–39].

(2) Precipitation of Pb carbonate (cerussite), which is present at Force Crag [24]. This is relatively minor, as the mine waters are consistently undersaturated with respect to cerussite (SI from −2.39 to −0.76, $n = 96$; Figure S1a).

The adsorption of Zn to ferrihydrite will also occur at the pH of the mine water [13,40], but at a lower rate than that for Pb; the order of adsorption affinity of metal ions to ferrihydrite is Pb > Cu > Zn [41]. Zinc is not removed by $ZnCO_3$ (smithsonite) precipitation; this mineral is unknown within the Force Crag mine. Saturation indices for smithsonite in the mine outlet water (FC29) are always negative (range from −1.13 to −3.52; $n = 132$; Figure S1b), reflecting the fact that $ZnCO_3$ is ~1000 times more soluble than $PbCO_3$ (pK_{sp} cerussite = 13.13, pK_{sp} smithsonite = 10.00 [42]). Thus, relatively little Zn is removed from the mine water compared to Pb, and as a result, there is much more Zn than Pb in solution at the mine outlet (FC29; Table 1, Figure 2).

Cd has similar chemical and crystallographic properties to Zn [43], so relatively little Cd is probably removed by adsorption and precipitation. This is verified by the very strong Cd/Zn correlation in the mine water.

The concentrations of Cu and As in the mine water (Table 1) are lower than would be expected from the occurrence of chalcopyrite and pyrite, respectively, in the Force Crag vein. Nevertheless, there is a strong positive correlation between the total Fe concentration and the total ($r^2 = 0.91$; $p < 0.001$) and filtered ($r^2 = 0.72$; $p < 0.001$) As concentrations, suggesting pyrite as the As source. Both Cu and As are removed by adsorption on ferrihydrite [37–39,44], and a small amount of Cu is probably also removed by minor precipitation as supergene Cu minerals such as brochantite and malachite.

3.2.2. Sphalerite/Galena Oxidation Ratio

To determine the ratio in which sphalerite and galena are oxidising, a molar balance calculation was carried out (Tables 2 and S2). Molar balance calculations can reveal the proportions of the various minerals dissolving/oxidising/precipitating within the mine. The format used here was initially developed by Garrels and MacKenzie [45] and was successfully applied to mine drainage [46]. We extend that approach to incorporate bicarbonate concentration to resolve the mass balance for this NMD. This molar mass balance calculation is based on four reactions:

Sphalerite oxidation:

$$ZnS + 2O_2 = Zn^{2+} + SO_4{}^{2-} \tag{1}$$

Galena oxidation:

$$PbS + 2O_2 = Pb^{2+} + SO_4{}^{2-} \tag{2}$$

Pyrite oxidation and ferrihydrite precipitation:

$$FeS_2 + 3.75O_2 + 3.5H_2O = Fe(OH)_3 + 2SO_4{}^{2-} + 4H^+ \tag{3}$$

Calcite dissolution and acid neutralisation:

$$CaCO_3 + H_2CO_3 = Ca^{2+} + 2HCO_3{}^- \tag{4}$$

Table 2. Molar balance calculations for one litre of water, showing the molar proportions of the different mineral dissolution (diss.)/precipitation (precip.) reactions (in bold, relative to sphalerite) within the mine that explain the difference in composition (comp.) between input water (FC16) and outlet water (FC29); see text for explanation. For ferrihydrite, the general formula $Fe(OH)_3$ was used. Cd, Ni and Cu were not taken into consideration as the amounts were too small to make a difference. Mg was not included because the relative proportions contributed by dolomite and chlorite dissolution could not be constrained. Full details of all calculations given in Table S2, which shows calculated FC29 composition to match actual composition for all constituents shown.

	FC16 Comp.	Sphalerite Diss.	Galena Diss.	Pyrite Diss.	Calcite Diss.	Ferrihydrite Precip.	FC29 Comp.
Molar ratio		1.0	2.22	1.0	5.0	0.8	
S	2.50×10^{-5}	4.87×10^{-5}	1.08×10^{-4}	9.94×10^{-5}			2.81×10^{-4}
Ca	1.15×10^{-5}				2.44×10^{-4}		2.55×10^{-4}
Fe	2.03×10^{-7}			4.97×10^{-5}		-4.09×10^{-5}	9.04×10^{-6}
Zn	3.06×10^{-7}	4.87×10^{-5}					4.90×10^{-5}
Pb	4.15×10^{-9}		1.08×10^{-4}			-1.08×10^{-4}	1.64×10^{-7}
$HCO_3{}^-$	8.20×10^{-6}				4.87×10^{-4}		2.86×10^{-4}
H^+	1.07×10^{-5}			1.99×10^{-4}			2.82×10^{-7}

The relative amounts of sphalerite and galena oxidation cannot be determined from the Zn/Pb ratio in the mine water due to the preferential removal of Pb by adsorption on ferrihydrite and secondary mineral precipitation, as previously discussed. Instead, the molar balance calculations used mine input and output pH and bicarbonate concentration data. An initial assumed sphalerite:galena oxidation ratio of 1:2 (based on estimates of sphalerite and galena oxidation rates; see above) was adjusted until the calculated and measured HCO_3^- concentrations in the mine water outlet were equal.

Firstly, the number of moles of Zn (per litre) released by sphalerite oxidation (Equation (1)) was computed from the difference between Zn concentrations in the mine outlet (FC29) and the input (FC16) water. From this, the number of moles of sulphate released by sphalerite oxidation was calculated based on the 1:1 molar Zn:S ratio in sphalerite. The assumed sphalerite-to-galena oxidation ratio (1:2) was then used to calculate the number of moles of Pb and sulphate released by galena oxidation (Equation (2)). The moles of sulphate released by pyrite oxidation (Equation (3)) were calculated as the difference between the sulphate concentration in FC29 and the sum of that in FC16 and the amount released by sphalerite and galena oxidation (chalcopyrite oxidation was ignored because the amount of Cu in solution was too small).

Knowing the moles of sulphate released by pyrite oxidation, it was then possible to calculate the acidity release from this reaction (1 mole of pyrite oxidation generates 4 moles of H^+ if the iron released precipitates as ferric hydroxide; Equation (3)). This would be expected to decrease the pH of the mine outlet water, but instead, there was an increase in pH between FC16 and FC29 (Table 1), indicating that the acidity was consumed within the mine workings. Calcite dissolution was responsible for this, as shown by the increase in Ca concentration at FC29 relative to FC16 (Figure 2). Calcite dissolution also releases bicarbonate: 2 moles for every mole of Ca released at the pH of mine water (pH 5 to 6.5; Equation (4) [47]). This bicarbonate was partly consumed by neutralising the H^+ generated by pyrite oxidation as well as reacting with some of the H^+ already present in the mine input water, which had a pH of 5.

Using the assumed sphalerite-to-galena oxidation ratio of 1:2, the bicarbonate consumed by both these reactions was subtracted from the bicarbonate generated via calcite dissolution to calculate the bicarbonate concentration that should be present in FC29; this was less than the actual bicarbonate concentration in FC29. The sphalerite-to-galena oxidation ratio was then changed iteratively until the calculated FC29 bicarbonate concentration equalled the actual bicarbonate concentration, giving a sphalerite-to-galena oxidation ratio of 1:2.22 (Table 2).

The calculations show that over 99% of the Pb released by galena oxidation is removed within the mine (Table 2). In the molar balance calculations, this loss was assigned to adsorption by ferrihydrite; precipitation as cerussite is probably minor, given the consistent undersaturation of mine waters with respect to this mineral (Figure S1a).

3.2.3. Pyrite/Chalcopyrite Oxidation and Ferrihydrite Precipitation

In the molar balance calculations, the amount of sulphate in excess of that produced by sphalerite and galena oxidation was attributed to the oxidation of pyrite (Table 2). However, sulphate was also released by chalcopyrite oxidation. The relative proportions of pyrite and chalcopyrite oxidation cannot be estimated because of Cu adsorption on ferrihydrite. The adsorption of Zn to ferrihydrite within the mine workings would result in an underestimation of the amount of sphalerite oxidation and an overestimation of the contribution of pyrite/chalcopyrite oxidation, but the effect is minor.

The calculations show that, to explain the composition of the mine outlet water, significant quantities of pyrite and chalcopyrite need to have oxidised (1 mole compared to 3.2 moles of sphalerite and galena). Pyrite and chalcopyrite have been recorded locally in the lower parts of the Force Crag vein; the molar balance calculation suggests that these minerals are more widespread than has been realised, probably because they occur as small, disseminated crystals that are easily overlooked.

The molar balance calculations also show that 82% of the Fe released by pyrite oxidation is removed by ferrihydrite precipitation (Table 2), and this is verified by fine particulate ferrihydrite in the mine outlet water (discussed further in Section 3.3, below), the coating of ferrihydrite ('ochre') up to 1.5 m thick on parts of the mine walls, roof and floor and the orange precipitate on the base of the channel discharging water from Level 1 of the mine (FC30) before it was confined to a pipe [25]. Saturation indices for amorphous iron hydroxide in the mine outlet water are generally negative, indicating that the water should be undersaturated (Figure S1c), but this is clearly incorrect and presumably reflects the difficulty of calculating saturation indices for amorphous minerals such as ferrihydrite, which has a very variable composition [37], so it is difficult to assign an accurate K_{eq} value. Some iron is removed by the precipitation of the minor amounts of iron carbonates (siderite, ankerite) present in the mine, though on all occasions, FC29 water is undersaturated with respect to siderite (SI values of −3.09 to −0.87; n = 131; Figure S1d).

3.2.4. Dissolution of Carbonates

The molar balance calculation showed that the acidity released by pyrite/chalcopyrite oxidation is neutralised by the dissolution of calcite, releasing Ca and alkalinity. Ca and bicarbonate concentrations are strongly correlated in the mine outlet water (r^2 = 0.86; $p < 0.001$). Dissolution of dolomite is probably also occurring (this carbonate has been recorded at Force Crag), but the rate of dolomite dissolution is much lower than that of calcite [48], and the increase in the molar Ca:Mg ratio from ≈1:2.5 at FC16 to 2:1 at FC29 suggests that calcite dissolution is dominant in the subsurface.

The dissolution of calcite also releases Sr, as shown by the correlation between concentrations of Ca and Sr in the mine outlet water (r^2 = 0.77; $p < 0.001$). The Ca/Sr weight ratio in the mine water (0.00115) corresponds to 0.12 wt% Sr in the calcite, well within the range of Sr concentrations in calcite (up to 4% [49]).

For every 1 mole of sphalerite oxidised, 5 moles of calcite/dolomite are dissolved (Table 2). This is substantial and indicates that these carbonates are more abundant in the Force Crag vein than suggested from descriptions of the vein mineralogy, probably because they are typically fine-grained and inconspicuous. There is only one mention of calcite at Force Crag [23].

3.2.5. Dissolution of Silicates

Silicate weathering, probably largely due to the reaction with the acidity released by pyrite oxidation, is responsible for the low levels of Si in the mine water (median of 3.0 mg/L; Table 1). This is not due to the dissolution of quartz because, although it is abundant in the Force Crag vein, quartz has very low solubility at an acid-circumneutral pH [32]. Likewise, dissolution of the K- and Na-rich white micas in the Skiddaw Group slates is not occurring because the levels of K in the mine water are below the detection limit (<1 mg/L) and there is no Na input within the mine workings. However, chlorite, which is common in the Skiddaw slates, is most likely dissolving, releasing Si, Mg and Al (the most common elements in this mineral [50]).

The Al concentration of mine water is lower than in the upstream water (Figure 2; Table 1), despite the release of Al by chlorite dissolution, because the upstream water has a lower pH. Al solubility increases rapidly as pH decreases [38]. The Al in the upstream water is most likely derived from clay dissolution within the peaty soils. The lack of filtered (aqueous-phase) Al within the mine water may also reflect the strong adsorption of Al on ferrihydrite at circumneutral pH [39]; Al can be readily incorporated into the ferrihydrite structure [51].

The weathering of chlorite was not incorporated into the molar balance calculations because the amount is relatively small and Mg is also released by dolomite dissolution. The relative molar amounts of Ca and Si in the mine water (0.255 mmol/L and 0.106 mmol/L, respectively) verify that carbonates are contributing more to the mine water than chlorite, so that although chlorite is more abundant, carbonates are more reactive.

It is notable that because the dissolution of calcite, dolomite and chlorite is largely due to the acid released by the oxidation of pyrite/chalcopyrite, the amounts of Ca, alkalinity, Sr, Mg and Si in the mine water are all strongly correlated (e.g., Ca-Si r^2 = 0.99 ($p < 0.001$)), even though the release of these two species is by different reactions involving different minerals.

3.2.6. Dissolution of Mn Oxides and Barite

The low concentrations of Mn and Ba in the mine water (Table 1) were likely released from the dissolution of Mn oxides and barite in the mine workings, respectively (both are most common in the upper part of the Force Crag vein). These minerals are only sparingly soluble under the temperature, pH and redox conditions of the mine water.

3.3. Flow Control on Mine Water Chemistry

Flow is negatively correlated with the concentrations of many constituents of the mine water, both major species (Ca, alkalinity, Mg, Si and SO_4) and some metals (Zn, Cd, Ni and Mn) (Figures 4a and S2; $p < 0.001$ in all cases). For Zn, Cd, Ni and Mn, total concentration equals filtered concentration (Table S1), so these species are present in dissolved form and perhaps also as very fine (<0.45 µm) colloids. The negative correlation with flow for all these species is due to dilution. Concentrations are ≈1.2 times lower in winter than in summer due to dilution by higher winter flows (though rainfall events throughout the year result in short-term dilution effects). The winter flow at the mine outlet (FC30) is on average 12.1 L/s, 1.2 times higher than the mean summer flow (10.2 L/s), reflecting the higher rainfall in winter (average 100 mm/month compared to 70 mm/month in summer; data from Thornthwaite, 5 km NE of Force Crag mine [52]).

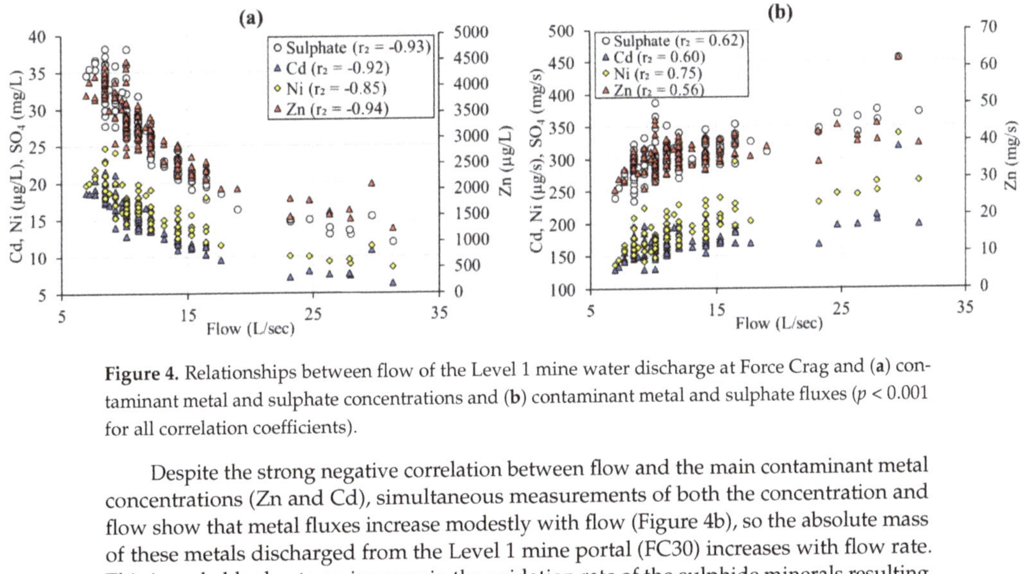

Figure 4. Relationships between flow of the Level 1 mine water discharge at Force Crag and (**a**) contaminant metal and sulphate concentrations and (**b**) contaminant metal and sulphate fluxes ($p < 0.001$ for all correlation coefficients).

Despite the strong negative correlation between flow and the main contaminant metal concentrations (Zn and Cd), simultaneous measurements of both the concentration and flow show that metal fluxes increase modestly with flow (Figure 4b), so the absolute mass of these metals discharged from the Level 1 mine portal (FC30) increases with flow rate. This is probably due to an increase in the oxidation rate of the sulphide minerals resulting from the greater amounts of oxygenated water flowing through the mine. The oxidation rate of sulphides is directly related to the oxygen concentration of the water [53]. This is corroborated by the increase in sulphate flux with flow (Figure 4b).

For Fe, Pb, Al and Cu, the relationship between flow and concentration in the mine water is not straightforward. For all these metals, the filtered (aqueous-phase) concentrations are generally ≤50% of the total concentrations (Table S1), indicating the presence of substantial amounts of fine particulate matter.

Fe in the mine water is present predominantly as finely particulate ferrihydrite (up to ≈80% of total Fe levels). Ferrihydrite has a low density and a large surface area [54], so the freshly precipitated particles are readily entrained by the flowing water within the mine. As flow increases, the concentration of particulate ferrihydrite decreases to approximately one third of the low flow values (Figure 5a); this is too great a reduction to be solely due to dilution. Additional factors probably include the slowing of precipitation rates during cooler temperatures at times of higher winter flows and the reduced time for particles to form when flows are faster. However, during the occasional very high flow events (>20 L/s), the concentrations of particulate ferrihydrite increase (Figure 5a), suggesting the physical flushing of ferrihydrite particles recently deposited on the mine walls.

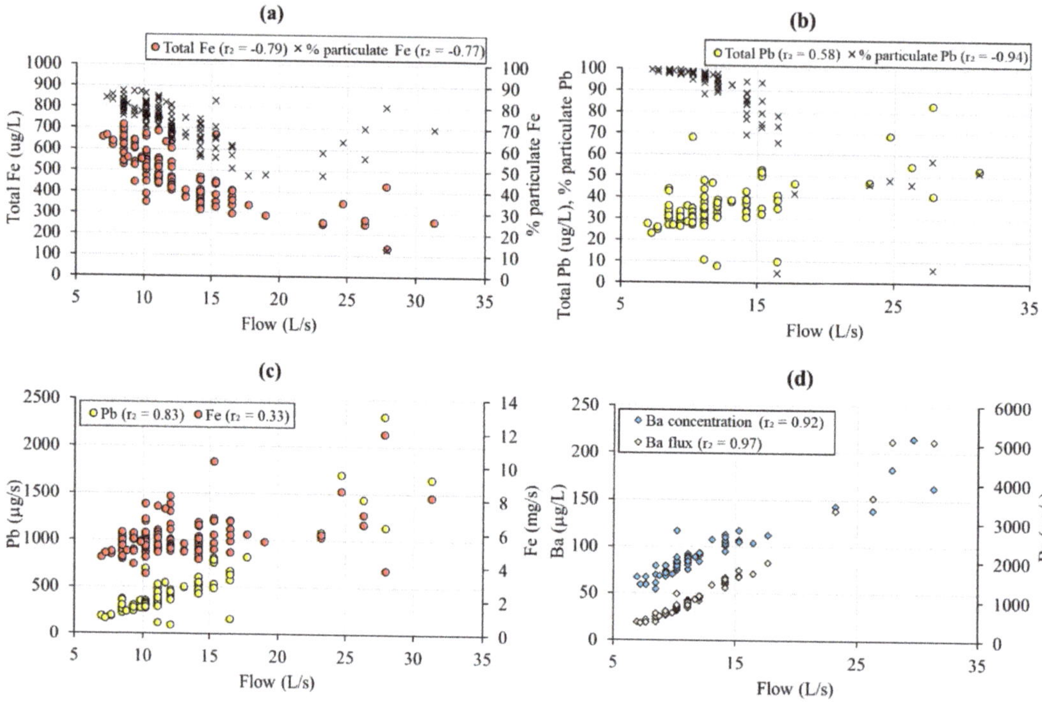

Figure 5. Relationship between Level 1 mine water flow and (**a**) total Fe concentration and percentage particulate Fe, (**b**) total Pb concentration and percentage particulate Pb, (**c**) Fe and Pb flux and (**d**) Ba concentration and flux ($p < 0.001$ for all correlation coefficients).

Pb is predominantly present as fine particles at low flow (96% of the median total Pb concentration; Figure 5b). These particles are unlikely to be a Pb precipitate, as the mine waters are consistently undersaturated with respect to the most likely mineral (cerussite; Figure S1a). Instead, Pb is probably adsorbed onto the ferrihydrite particles described above. As previously discussed, ferrihydrite strongly adsorbs Pb at a circumneutral pH. Particulate Pb concentration is ≥80% of the total concentration on most occasions at flows of <15 L/s (Figure 5b). At low flows, it appears that there are enough ferrihydrite particles to adsorb the majority of the Pb in solution, as shown by the much higher concentrations of particulate Fe (median 376 µg/L) than particulate Pb (median 33 µg/L). However, at flows >15 L/s, the level of particulate ferrihydrite drops sufficiently for a proportion of Pb to remain in solution, so filtered (aqueous-phase) and total Pb concentrations increase at higher flows. As a result, total Pb concentrations are positively correlated with flow, and fluxes of this metal increase with increases in flow (Figure 5c). The lower temperatures at higher flows probably contribute to the higher levels of dissolved Pb because cation adsorption decreases with a drop in temperature in neutral water. This is believed to be at least partly responsible for the widespread diurnal cycles of dissolved concentrations of many trace metals [55]. In addition, the higher flows allow less opportunity for adsorption because of the shorter contact time in the rapidly flowing water. This means that the elevated filtered Pb concentrations at these higher flows are a reflection of the Pb levels released into the mine water by sulphide oxidation before Pb removal onto ferrihydrite.

During the occasional very high flow events (>20 L/s; Figure 5c), Pb fluxes in the mine water show sharp increases, probably due mainly to the physical flushing of Pb-containing ferrihydrite particles recently deposited on the mine walls, as the spikes in Pb flux generally match those in Fe flux. The concentrations of other species in the mine water (e.g., Zn, Cd, Ni), which are relatively unaffected by adsorption, decrease during these very high flow events due to dilution (Figure 4a).

For Ba in the mine water, total concentration equals filtered concentration, but, unlike other species for which this is true (e.g., Zn, Cd, Ni and Mn), Ba concentrations increase rather than decrease with flow (Figure 5d). This takes place because the mine waters are at or close to equilibrium with barite. Saturation indices have a mean of 0.02 and a range of 0.25 to −0.1 (Figure S1e). As flow increases, SO_4 concentrations decrease due to dilution (as previously discussed; Figure 4a), and this causes more barite to dissolve, so Ba concentrations rise. At lower flows, the higher SO_4 concentrations depress barite dissolution. Solution saturation may have some influence on the dissolution/precipitation of other metal phases, but, unlike for Ba, it is not the main cause of fluctuations in concentrations of other metals.

Patterns of change in Cu concentration and flux with flow (Figure 6a) are very similar to those of Pb: increasing concentration with flow (r^2 = 0.63, p < 0.001) and a sharp decrease in the proportion of particulate Cu with flow (r^2 = −0.85, p < 0.001), from ≈85% as particulate Cu on most occasions when flow is <10 L/s to approximately 40% or less as particulate Cu at flows above 15.5 L/s. Cu flux increases with flow (Figure 6c) due to the increase in Cu concentration, likely for the same reasons as for Pb, discussed previously. The somewhat lower proportion of particulate Cu compared to particulate Pb likely reflects the somewhat lower ferrihydrite adsorption affinity of Cu compared to Pb [37,38].

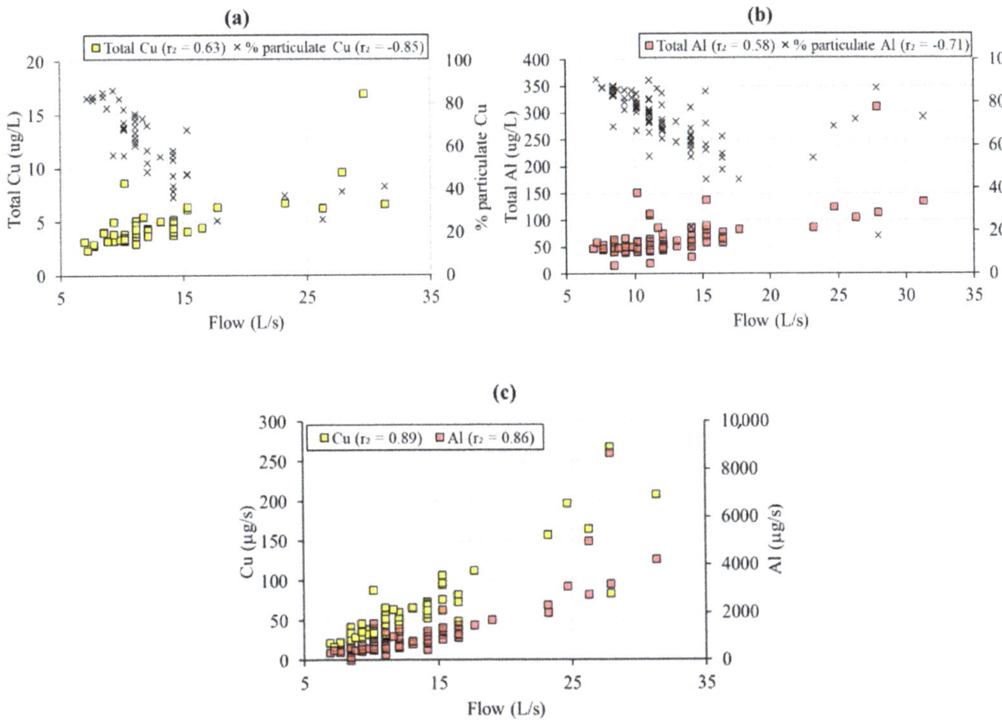

Figure 6. Relationship between Level 1 mine water flow and (**a**) total Al concentration and percentage particulate Al, (**b**) total Cu concentration and percentage particulate Cu and (**c**) Al and Cu flux (p < 0.001 for all correlation coefficients).

Al concentration is also positively correlated with flow (Figure 6b; $r^2 = 0.58$, $p < 0.001$) and Al flux increases with flow (Figure 6c). This is probably for the same reason as for Pb: adsorption to particulate ferrihydrite. Similar to Cu, Al is less strongly adsorbed to ferrihydrite than Pb [37,38], accounting for the smaller percentage of Al present as fine particles (~77% of the median total concentrations; Table S1). However, Figure 6b indicates that the proportion of Al in particulate form decreases up to approximately 20 L/s but then increases again at higher flows. This is likely a consequence of the physical flushing of Al particulates at higher flows and is an almost identical pattern to that seen for Fe (Figure 5a).

3.4. Temperature Control on Mine Water Chemistry

Temperatures in the mine water (FC29) fluctuate seasonally but over a restricted range (8–13.5 °C) compared to air temperatures (−0.2 to 18.8 °C), data from Thornthwaite [52]). The median water temperature (11.2 °C) is ≈3 °C higher than the mean annual surface temperature (8.2 °C). This is because at depths of >10–15 m, temperatures increase with depth in line with the geothermal gradient (≈3.3 °C/100 m in the Lake District [56]). The mine water travels underground for some distance because the mine extends ≈1.2 km horizontally and 350 m vertically, and the water is, overall, warmed by ≈3 °C during its passage through the subsurface due to the warmer temperature at depth. The mine water does not emerge at a constant temperature because it travels through the mine too rapidly to equilibrate with its surroundings. This suggests that residence time within the mine is on the order of hours.

Temperature is negatively correlated with the flow of the mine water ($r^2 = -0.83$; $p < 0.001$; Figure S3, Table S3) because winter flows are higher and cooler than summer flows. This relationship is less clear for the occasional very high flows (>20 L/s), which all have cool temperatures (8–10 °C) irrespective of the season due to the rapid influx and throughflow of relatively cool rainwater.

To investigate the association between temperature and species concentrations in the mine water, the strong inverse relationship between flow rate and temperature must be taken into account. Correlation coefficients between temperature and metal concentrations suggest a negative correlation in the majority of cases, i.e., higher temperatures drive lower metal concentrations (Table S3). However, partial correlation coefficients computed between temperature and the various constituents of the mine water with flow rate held constant show that this is not the case (Table S3). Therefore, the apparent inverse relationship between temperature and the concentrations of these metals is in fact an artefact of the negative correlation between flow rate and temperature.

The partial correlation coefficients calculated with flow rate held constant show a positive correlation between temperature and the concentrations of the main contaminant metals (Zn, Cd, Pb). The strongest partial correlation coefficient is between zinc and temperature ($r_{Zn,T \cdot Q} = 0.55$; $p < 0.001$). This implies that higher water temperatures may be driving more vigorous metal sulphide oxidation within the mine workings, which is consistent with the positive partial correlation coefficient between sulphate and temperature ($r_{SO4.T.Q} = 0.61$; $p < 0.001$). This reflects the fact that reaction kinetics in general are directly impacted by temperature [57]. Although the temperature range is just 8.0–13.5 °C in the FC28/29 water, *Thiobacillus ferrooxidans*, the main organism responsible for microbially mediated sulphide oxidation, becomes inactive at temperatures of <10 °C [58], so perhaps bacterial oxidation is 'switched off' when temperatures drop during times of high flow.

Therefore, the released metal fluxes are a balance between abiotic oxidation reactions (which increase at higher temperatures and, as previously discussed, at higher flows with more highly oxygenated water) and bacterially mediated oxidation (which is depressed at high flow rates when temperatures decrease below 10 °C).

3.5. Mineralogical Control on NMD Composition

The hydrochemical data at Force Crag demonstrate that the major control on the composition of NMD issuing from the mine is mineralogical. The main sulphides present are sphalerite and galena, which do not release acidity when they oxidise, but sufficient amounts of pyrite and chalcopyrite occur in the Force Crag vein that their oxidation generates significant acidity. However, this is neutralised by carbonates present in the mineralised vein and, to a lesser extent, silicates in the host rock, so drainage from the mine is neutral rather than acidic. The iron released by the oxidation of pyrite and chalcopyrite largely precipitates as ferrihydrite, which adsorbs some species, particularly Pb, Cu, As and Al. Therefore, the presence of pyrite and chalcopyrite both causes problems through the release of acidity and provides solutions through the precipitation of ferrihydrite and the adsorption of some contaminant metals.

Force Crag is typical of many mines in the Lake District that have similar ore mineralogy and are emplaced in silicate host rocks [59]. Therefore, the composition of the neutral mine drainage from these mines is determined by the same factors that are applicable at Force Crag, i.e., the mineralogy of the ore deposit and the host rock. The variability in the composition of the mine drainage in the Lake District is likely to be explicable in terms of the relative proportions of key minerals at the sites: iron-containing sulphides (particularly pyrite) and carbonates (particularly calcite and dolomite). These minerals are often overlooked at mine sites in this region because other minerals present are more valuable, more spectacular and/or more sought-after. In contrast, pyrite, calcite and dolomite are typically fine-grained and inconspicuous.

The same conclusions apply to mines discharging water with a near-neutral pH and elevated Zn concentrations throughout the UK. Most of these are Pb/Zn mines hosted in limestone, located in the Pennines from Northumberland to Derbyshire, where iron-containing sulphides (marcasite, pyrite, chalcopyrite and pyrrhotite) widely occur in small proportions associated with lead and zinc sulphides [60,61]. The most important factor is that iron-containing sulphides are not dominant at these sites, so the acidity released when these sulphides oxidise is rapidly neutralised by the limestone, which hosts almost all of these deposits. The dominant control on the composition of the drainage from these mines is, effectively, the same mineralogy as that present at Force Crag.

Zn is the most common contaminant metal in the neutral drainage from mines in the UK because, even if sphalerite is not abundant at a site, Zn is soluble under the hydrochemical conditions present, as it is not removed by precipitation as a supergene mineral and is less susceptible to adsorption to ferrihydrite than other contaminant metals (e.g., Pb). Cd is also soluble under these conditions and is often associated with Zn in ore deposits, usually in solid solution within sphalerite, so it too is a characteristic contaminant of NMD.

The reasons why NMD is so much more common than AMD in the UK are twofold: the dominant sulphides at most mines (sphalerite and galena) do not release acidity when they oxidise, and because most mines are hosted in limestone, there are sufficient carbonates to neutralise the acidity generated by the oxidation of any iron-containing sulphides present. This is true even when the deposits are not hosted by limestone, as the Force Crag vein attests. In contrast, AMD sites in the UK are characterised by abundant iron-containing sulphides and low, or absent, carbonate content. For example, at the Wheal Jane mine in Cornwall, a well-known AMD site, cassiterite mineralisation with abundant pyrite and arsenopyrite is associated with quartz porphyry dykes intruded into lightly metamorphosed mudstones and sandstones [62]. The dominant sulphide minerals present are iron-containing, and carbonates are virtually absent.

3.6. Hydrological Control on NMD Composition

The present study has shown that hydrological factors have a strong influence on the concentrations and fluxes of the constituents of the NMD, due in particular to:

(i) Physical dilution of aqueous-phase metals released by mineral weathering in the subsurface;
(ii) Variation in the degree and duration of rock–water interaction;

(iii) Changes to the rates of reactions, including both sulphide oxidation reactions and aqueous metal attenuation processes.

In particular, it is clear from this study that flow rate plays a critical role in controlling the rates/extent of subsurface (bio)geochemical processes. An understanding of this role is crucial for remedial planning, i.e., fluctuations in flow have a key influence on metal concentration and flux, which in turn have major implications for the design of treatment systems for mine waters. Contaminant concentration and load are crucial criteria for mine-water-treatment system design [47]. The central role of flow rate in determining NMD composition suggests that engineering interventions to limit the ingress of water into mine workings, where feasible, may pay substantial dividends with respect to the size (and therefore cost) of a treatment system by limiting peak metal loadings in mine drainage. In addition, the molar mass balance approach used here revealed that there is substantial ferrihydrite precipitation within the mine workings at the Force Crag mine (which probably removes Pb, Al, As and Cu). Knowing this is important because, in situations where ferrihydrite accumulates within mine workings, there are two potentially important risks: (1) the accumulation of ferrihydrite to the extent that water is re-routed within the workings, potentially emerging to surface at a different location; and (2) the possibility of flushing large volumes of ferrihydrite during extreme flows.

The conclusions of this study also have implications for the impact of climate change on mine water remediation. Warmer temperatures are likely to result in faster sulphide oxidation and, therefore, a deterioration in mine water quality. Furthermore, an increase in the number of extreme flow events will flush more contaminant-laden ferrihydrite particles downstream, increasing pollution in streams below mine sites.

4. Conclusions

The composition of the NMD at Force Crag mine is controlled by a combination of mineralogical and hydrological factors. Mineralogy is the key control on the constituents of the drainage and exerts control on the acidity of the water (the net product of acid-generating and acid-neutralising reactions).

There are strong correlations between the flow rate of the main NMD outflow at Force Crag and the concentrations of contaminant metals and sulphate. There is also a strong inverse correlation between flow and water temperature, due primarily to seasonal fluctuations (lower flows but higher temperatures in the summer months).

The correlation coefficients reveal that Zn, Cd, Fe and SO_4 are strongly inversely correlated with flow, which is indicative of dilution. In contrast, Al, Pb and Cu concentrations are positively correlated with flow due to the flushing of these metals adsorbed on ferrihydrite particles. Although Fe concentrations in the NMD are quite low, a molar mass balance approach demonstrates that (a) pyrite and chalcopyrite oxidations are prevalent in the mine workings, but that (b) there is then substantial precipitation of the iron released as ferrihydrite. The presence of ferrihydrite results in the attenuation, via adsorption, of metals such as Pb, Al, As and Cu within the workings. Thus, concentrations of these metals in the NMD are relatively minor at low flows but then increase at higher flows due to the flushing of the metals in association with ferrihydrite precipitates.

To elucidate the influence of mine water temperature on metal mobilisation within the mine workings, it is necessary to compute partial correlation coefficients. This is because there is a strong inverse relationship between temperature and mine water flow rate, and therefore the role of temperature can only be determined by holding flow rate constant. There are positive correlations between temperature and the main contaminant metals emerging from the Force Crag mine (Zn, Cd, Pb). Since there is also a correlation between temperature and sulphate concentration, this suggests more vigorous sulphide oxidation occurs at higher temperatures, resulting in higher metal concentrations in the mine water discharge.

The important roles of both flow rate and temperature in governing mine water quality have implications for the future management of such mine waters since extreme rainfall events and potentially higher temperatures may exacerbate mine water contamination problems in the UK and elsewhere.

An equivalent approach to that taken here could be used to evaluate mineralogical and hydrological controls on NMD composition at other locations across the UK and around the world. The insights gained from this research have important implications for NMD mine water remediation initiatives wherever they are needed, in particular because they illustrate the critical role of flow rate in controlling the rates and extent of subsurface (bio)geochemical processes and, therefore, in determining the concentrations and loadings of metal contaminants in mine water discharges.

Supplementary Materials: The following supporting information can be downloaded at https://www.mdpi.com/article/10.3390/min13050592/s1, Table S1: Total and filtered (0.45 μm) cation concentrations for FC29 (Level 1 discharge water at inlet to treatment system) and FC16 (Coledale Beck upstream of mine site), showing median (standard deviation), minimum and maximum (n) values; Figure S1: Saturation Indices (calculated in PhreeqC) for minerals in the Force Crag mine Level 1 discharge (at location FC29), showing (a) cerussite, (b) smithsonite, (c) ferrihydrite (amorphous), (d) siderite and (e) barite; Table S2: Molar balance calculations for one litre of water, showing the molar proportions of the different mineral dissolution/precipitation reactions (in bold, relative to sphalerite) within the mine that explain the difference in composition between input water (FC16) and outlet water (FC29); Figure S2: Relationship between flow of the Level 1 mine water discharge at Force Crag with (a) Ca and alkalinity concentrations and (b) Mg, Si and Mn concentrations; Figure S3: Relationship between temperature and flow of water leaving the Force Crag mine (FC28/29); Table S3: Full and partial Spearman's Rho correlation coefficients (r^2) and significance level (p) between mine water temperature and selected variables.

Author Contributions: Conceptualization, A.P.J. and J.A.W.; Data curation, C.J.G.; Formal analysis, A.P.J. and J.A.W.; Funding acquisition, A.P.J.; Investigation, A.P.J., C.J.G. and J.A.W.; Methodology, A.P.J., C.J.G. and J.A.W.; Project administration, A.P.J. and C.J.G.; Resources, A.P.J., C.J.G. and J.A.W.; Software, C.J.G.; Supervision, A.P.J. and J.A.W.; Validation, A.P.J. and J.A.W.; Visualization, A.P.J. and J.A.W.; Writing—original draft, A.P.J. and J.A.W.; Writing—review and editing, A.P.J., C.J.G. and J.A.W. All authors have read and agreed to the published version of the manuscript.

Funding: The research was funded by the UK Coal Authority, primarily via contract No. CA18/2377, under the umbrella of the Water and Abandoned Metal Mines Programme (WAMM). The WAMM programme is a partnership between the Environment Agency, the Coal Authority and the UK Department for Environment, Food and Rural Affairs (Defra).

Data Availability Statement: The data presented in this study are available on request from the corresponding author, subject to the prior permission of the research funder being granted.

Acknowledgments: We express our gratitude to Jane Davis, Patrick Orme (both formerly at Newcastle University) and Clair Roper (Newcastle University) for their invaluable assistance with field work and lab analysis. We are grateful to Hugh Potter (Environment Agency) for many insightful discussions about UK mine water pollution, including Force Crag. We are also indebted to the National Trust, in particular John Malley, for site access and background information. We are grateful for the comments of three anonymous reviewers that helped to improve the manuscript.

Conflicts of Interest: The authors declare no conflict of interest. The funders had no role in the design of the study; in the collection, analyses or interpretation of data; in the writing of the manuscript; or in the decision to publish the results.

References

1. Webb, J.A.; Sasowsky, I.D. The interaction of acid mine drainage with a carbonate terrane: Evidence from the Obey River, north-central Tennessee. *J. Hydrol.* **1994**, *161*, 327–346. [CrossRef]
2. Cravotta, C.A.; Brady, K.B.C.; Rose, A.W.; Douds, J.B. Frequency distribution of the pH of coal-mine drainage in Pennsylvania. In *U.S. Geological Survey Toxic Substances Hydrology Program, Proceedings of the Technical Meeting in Charleston, SC, USA, 8–12 March 1999*; Morganwalp, D.W., Buxton, H., Eds.; U.S. Geological Survey Water Resource Investigations Report 99-4018A; United States Geological Survey: Reston, VA, USA, 1999; pp. 313–324.

3. Younger, P.L. Nature and practical implications of heterogeneities in the geochemistry of zinc-rich, alkaline mine waters in an underground F-Pb mine in the UK. *Appl. Geochem.* **2000**, *15*, 1383–1397. [CrossRef]
4. Iribar, V. Origin of neutral mine water in flooded underground mines: An appraisal using geochemical and hydrogeological methodologies. In Proceedings of the International Mine Water Association Conference 2004, Newcastle upon Tyne, UK, 19–23 September 2004; pp. 169–178.
5. Doulati Ardejani, F.; Rooki, R.; Jodieri Shokri, B.; Eslam Kish, T.; Aryafar, A.; Tourani, P. Prediction of rare earth elements in neutral alkaline mine drainage from Razi coal mine, Golestan province, northeast Iran, using general regression neural network. *J. Environ. Eng.* **2013**, *139*, 896–907. [CrossRef]
6. Pope, J.; Trumm, D. Controls on Zn concentrations in acidic and neutral mine drainage from New Zealand's bituminous coal and epithermal mineral deposits. *Mine Water Environ.* **2015**, *34*, 455–463. [CrossRef]
7. Shahhosseini, M.; Doulati Ardejani, F.; Baafi, E. Geochemistry of rare earth elements in a neutral mine drainage environment, Anjir Tangeh, northern Iran. *Int. J. Coal Geol.* **2017**, *183*, 120–135. [CrossRef]
8. Barago, N.; Pavoni, E.; Floreani, F.; Crosera, M.; Adami, G.; Lenaz, D.; Covelli, S. Hydrogeochemistry of thallium and other potentially toxic elements in neutral mine drainage at the decommissioned Pb-Zn Raibl mine (Eastern Alps, Italy). *J. Geochem. Explor.* **2023**, *245*, 107129. [CrossRef]
9. Heikkinen, P.M.; Räisänen, M.L.; Johnson, R.H. Geochemical Characterisation of Seepage and Drainage Water Quality from Two Sulphide Mine Tailings Impoundments: Acid Mine Drainage versus Neutral Mine Drainage. *Mine Water Environ.* **2009**, *28*, 30–49. [CrossRef]
10. Sracek, O.; Filip, J.; Mihaljevič, M.; Kříbek, B.; Majer, V.; Veselovský, F. Attenuation of dissolved metals in neutral mine drainage in the Zambian Copperbelt. *Environ. Monit. Assess.* **2011**, *172*, 287–299. [CrossRef]
11. EA (Environment Agency). *Prioritisation of Abandoned Non-Coal Mine Impacts on the Environment: The National Picture*; Report of Project SC030136/R14; Environment Agency: Bristol, UK, 2012.
12. Jones, A.; Rogerson, M.; Greenway, G.; Potter, H.A.B.; Mayes, W.M. Mine water geochemistry and metal flux in a major historic Pb- Zn-F orefield, the Yorkshire Pennines, UK. *Environ. Sci. Pollut. Res.* **2013**, *20*, 7570–7581. [CrossRef]
13. Jarvis, A.P.; Davis, J.E.; Orme, P.H.A.; Potter, H.A.B.; Gandy, C.J. Predicting the benefits of mine water treatment under varying hydrological conditions using a synoptic mass balance approach. *Environ. Sci. Technol.* **2019**, *53*, 702–709. [CrossRef]
14. Mayes, W.M.; Potter, H.A.B.; Jarvis, A.P. Riverine flux of metals from historically mined orefields in England and Wales. *Water Air Soil Pollut.* **2013**, *224*, 1425. [CrossRef]
15. de Jonge, M.; Tipping, E.; Lofts, S.; Bervoets, L.; Blust, R. The use of invertebrate body burdens to predict ecological effects of metal mixtures in mining-impacted waters. *Aquat. Toxicol.* **2013**, *142*, 294–302. [CrossRef] [PubMed]
16. Schmidt, T.S.; Clements, W.H.; Zuellig, R.E.; Mitchell, K.A.; Church, S.E.; Wanty, R.B.; San Juan, C.A.; Adams, M.; Lamothe, P. Critical tissue residue approach linking accumulated metals in aquatic insects to population and community-level effects. *Environ. Sci. Technol.* **2011**, *45*, 7004–7010. [CrossRef] [PubMed]
17. Gandy, C.J.; Davis, J.E.; Orme, P.H.A.; Potter, H.A.B.; Jarvis, A.P. Metal removal mechanisms in a short hydraulic residence time subsurface flow compost wetland for mine drainage treatment. *Ecol. Eng.* **2016**, *97*, 179–185. [CrossRef]
18. Young, B.; Cooper, A.H. The geology and mineralisation of Force Crag Mine, Cumbria. *Proc. Cumberl. Geol. Soc.* **1988**, *5*, 5–11.
19. Fortey, N.J. Low grade metamorphism in the Lower Ordovician Skiddaw Group of the Lake District, England. *Proc. Yorks. Geol. Soc.* **1989**, *47*, 325–337. [CrossRef]
20. Bebout, G.E.; Cooper, D.C.; Bradley, A.D.; Sadofsky, S.J. Nitrogen-isotope record of fluid-rock interactions in the Skiddaw Aureole and granite, English Lake District. *Am. Mineral.* **1999**, *84*, 1495–1505. [CrossRef]
21. Merriman, R.J.; Breward, N.; Stone, P.; Green, K.; Kemp, S. *Element Mobility and Low-Grade Metamorphism of Mudrocks in British Caledonian Basins*; British Geological Survey Geology and Landscape Programme Internal Report OR/09/017; British Geological Survey: Keyworth, UK, 2009; p. 27.
22. Eastwood, T. *The Lead and Zinc Ores of the Lake District*; Memoirs of the Geological Survey, Special report on the mineral resources of Great Britain 22; Her Majesty's Stationery Office (HMSO): London, UK, 1921; p. 42.
23. Mindat, nd. Force Crag Mine. Available online: https://www.mindat.org/loc-1468.html (accessed on 3 December 2019).
24. Green, D.I.; McCallum, D.; Wood, M. Supergene Cu, Pb, Zn and Ag minerals from Force Crag Mine, Coledale, Cumbria. *UK J. Mines Miner.* **1997**, *18*, 10–14.
25. Bell, F.G.; Donnelly, L.J. *Mining and Its Impact on the Environment*; Taylor and Francis: London, UK, 2006; p. 543.
26. Bigham, J.; Schwertmann, U.; Traina, S.; Winland, R.; Wolf, M. Schwertmannite and the chemical modeling of iron in acid sulfate waters. *Geochim. Cosmochim. Acta* **1996**, *60*, 2111–2121. [CrossRef]
27. Oswald, A.; Pearson, T. *Force Crag Mine, Cumbria*; English Heritage Archaeological Investigation Report Series AI/1/1999; Historic England: Portsmouth, UK, 1999; p. 109.
28. Dewey, M. Mines in Cumbria. Available online: http://www.cbdc.org.uk/CumbriaLGS/Leaflets/3_046.pdf (accessed on 3 December 2019).
29. Stumm, W.; Morgan, J.J. *Aquatic Chemistry: Chemical Equilibria and Rates in Natural Waters*, 3rd ed.; John Wiley & Sons Inc.: New York, NY, USA, 1996; p. 1022.

30. Mayes, W.M.; Perks, M.T.; Large, A.R.G.; Davis, J.E.; Gandy, C.J.; Orme, P.A.H.; Jarvis, A.P. Effect of an extreme flood event on solute transport and resilience of a mine water treatment system in a mineralised catchment. *Sci. Total Environ.* **2021**, *750*, 141693. [CrossRef]
31. UKEAP (UK Eutrophying & Acidifying Network). Bannisdale; UK-AIR 00114. © Crown 2019 copyright Defra via uk-air.defra.gov.uk, licenced under the Open Government Licence (OGL). 2019. Available online: https://uk-air.defra.gov.uk/networks/site-info?uka_id=UKA00114 (accessed on 3 December 2019).
32. Drever, J.I. *The Geochemistry of Natural Waters: Surface and Groundwater Environments*, 3rd ed.; Prentice Hall: Upper Saddle River, NJ, USA, 1997.
33. Cook, N.J.; Ciobanu, C.L.; Pring, A.; Skinner, W.; Shimizu, M.; Danyushevsky, L.; Saini-Eidukat, B.; Melcher, F. Trace and minor elements in sphalerite: A LA-ICPMS study. *Geochim. Cosmochim. Acta* **2009**, *73*, 4761–4791. [CrossRef]
34. Wu, P.; Kershaw, R.; Dwight, K.; Wold, A. Growth and characterization of nickel-doped ZnS single crystals. *Mater. Res. Bull.* **1989**, *24*, 49–53. [CrossRef]
35. Blanchard, M.; Alfredsson, M.; Brodholt, J.; Wright, K.; Catlow, C.R.A. Arsenic incorporation into FeS_2 pyrite and its influence on dissolution: A DFT study. *Geochim. Cosmochim. Acta* **2007**, *71*, 624–630. [CrossRef]
36. Rimstidt, J.D.; Chermak, J.A.; Gag, P.M. Rates of reaction of galena, sphalerite, chalcopyrite, and arsenopyrite with Fe(III) in acidic solutions. In *Environmental Geochemistry of Sulfide Oxidation*; Alpers, C.N., Blowes, D.W., Eds.; ACS Symposium Series; American Chemical Society: Washington, DC, USA, 1993; pp. 1–13.
37. Dzombak, D.A.; Morel, F.M.M. *Surface Complexation Modelling: Hydrous Ferric Oxide*; Wiley-Interscience: New York, NY, USA, 1990; p. 430.
38. Schultz, M.F.; Benjamin, M.M.; Ferguson, J.F. Adsorption and desorption of metals on ferrihydrite: Reversibility of the reaction and sorption properties of the regenerated solid. *Environ. Sci. Technol.* **1987**, *21*, 863–869. [CrossRef]
39. McDonald, D.M.; Webb, J.A.; Taylor, J. Chemical stability of acid rock drainage treatment sludge and implications for sludge management. *Environ. Sci. Technol.* **2006**, *40*, 1984–1990. [CrossRef]
40. Trivedi, P.; Axe, L. Modeling Cd and Zn sorption to hydrous metal oxides. *Environ. Sci. Technol.* **2000**, *34*, 2215–2223. [CrossRef]
41. Martinez, C.E.; McBride, M.B. Dissolved and labile concentrations of Cd, Cu, Pb, and Zn in aged ferrihydrite-organic matter systems. *Environ. Sci. Technol.* **1999**, *33*, 745–750. [CrossRef]
42. Aqion. Solubility Product Constants Ksp at 25 °C. Available online: https://www.aqion.de/site/16 (accessed on 15 April 2020).
43. Schwartz, M.O. Cadmium in zinc deposits: Economic geology of a polluting element. *Int. Geol. Rev.* **2000**, *42*, 445–469. [CrossRef]
44. Hering, J.; Chen, P.; Wilkie, J.; Elimelech, M.; Liang, S. Arsenic removal by ferric chloride. *J. Am. Water Work. Assoc.* **1996**, *88*, 155–167. [CrossRef]
45. Garrels, R.M.; MacKenzie, F.T. Origin of the chemical compositions of some springs and lakes. In *Equilibrium Concepts in Natural Water Systems*; Stumm, W., Ed.; ACS Publications, Advances in Chemistry: Washington, DC, USA, 1967; Volume 67, pp. 222–241.
46. Nordstrom, D.K. Hydrogeochemical processes governing the origin, transport and fate of major and trace elements from mine wastes and mineralized rock to surface waters. *Appl. Geochem.* **2011**, *26*, 1777–1791. [CrossRef]
47. Younger, P.L.; Banwart, S.A.; Hedin, R.S. *Mine Water: Hydrology, Pollution, Remediation*; Environmental Pollution Series (Vol. 5); Kluwer Academic Publishers: Dordrecht, The Netherlands, 2002; p. 442.
48. Pokrovsky, O.S.; Schott, J. Surface chemistry and dissolution kinetics of divalent metal carbonates. *Environ. Sci. Technol.* **2002**, *36*, 426–432. [CrossRef]
49. Tucker, M.E.; Wright, V.P. *Carbonate Sedimentology*; Blackwell Science Ltd.: Hoboken, NJ, USA, 1990. [CrossRef]
50. Deer, W.A.; Howie, R.A.; Zussman, J. *An Introduction to the Rock-Forming Minerals. Mineralogical Society of Great Britain and Ireland*; GeoScienceWorld: Tysons Corner, VA, USA, 2013. [CrossRef]
51. Hansel, C.M.; Learman, D.R.; Lentini, C.J.; Ekstrom, E.B. Effect of adsorbed and substituted Al on Fe(II)-induced mineralization pathways of ferrihydrite. *Geochim. Cosmochim. Acta* **2011**, *75*, 4653–4666. [CrossRef]
52. Climate-data.org. Thornthwaite Climate. Available online: https://en.climate-data.org/europe/united-kingdom/england/thornthwaite-70265/ (accessed on 3 December 2019).
53. Bilenker, L.D.; Romano, G.Y.; McKibben, M.A. Kinetics of sulfide mineral oxidation in seawater: Implications for acid generation during in situ mining of seafloor hydrothermal vent deposits. *Appl. Geochem.* **2016**, *75*, 20–31. [CrossRef]
54. Mendez, J.C.; Hiemstra, T. Surface area of ferrihydrite consistently related to primary surface charge, ion pair formation and specific ion adsorption. *Chem. Geol.* **2020**, *532*, 119304. [CrossRef]
55. Nimick, D.A.; Gammons, C.H.; Parker, S.R. Diel biogeochemical processes and their effect on the aqueous chemistry of streams: A review. *Chem. Geol.* **2011**, *283*, 3–17. [CrossRef]
56. Busby, J. Geothermal prospects in the United Kingdom. In Proceedings of the World Geothermal Congress, Bali, Indonesia, 25–30 April 2010; pp. 1–7.
57. Schott, J.; Pokrovsky, O.S.; Oelkers, E.H. The Link Between Mineral Dissolution/Precipitation Kinetics and Solution Chemistry. *Rev. Mineral. Geochem.* **2009**, *70*, 207–258. [CrossRef]
58. Dixon, D.G. Analysis of heat conservation during copper sulphide heap leaching. *Hydrometallurgy* **2000**, *58*, 27–41. [CrossRef]
59. Stanley, C.J.; Vaughan, D.J. Copper, lead, zinc and cobalt mineralization in the English Lake District: Classification, conditions of formation and genesis. *J. Geol. Soc.* **1982**, *139*, 569–579. [CrossRef]
60. Dunham, K.C. *Geology of the Northern Pennine Orefield*; Memoirs of the Geological Survey (Economic) HMSO: London, UK, 1990.

61. Ford, T.D.; Worley, N.E. Mineralization of the South Pennine Orefield, UK—A review. *Proc. Yorks. Geol. Soc.* **2016**, *61*, 55–86. [CrossRef]
62. Jackson, N.J.; Willis-Richards, J.; Manning, D.A.C.; Sams, M.S. Evolution of the Cornubian ore field, Southwest England; Part II, Mineral deposits and ore-forming processes. *Econ. Geol.* **1989**, *84*, 1101–1133. [CrossRef]

minerals

MDPI

Article

Categorization of Mining Materials for Restoration Projects by Means of Pollution Indices and Bioassays

Inmaculada Ferri-Moreno [1], Jose Ignacio Barquero-Peralbo [2], Oscar Andreu-Sánchez [3,*], Pablo Higueras [2], Luis Roca-Pérez [3], Mari Luz García-Lorenzo [1,*] and Jose María Esbrí [1]

1 Departamento de Mineralogía y Petrología, Facultad de Ciencias Geológicas, Universidad Complutense de Madrid, 28040 Madrid, Spain
2 Instituto de Geología Aplicada, Universidad de Castilla-La Mancha, Almadén, 13400 Ciudad Real, Spain
3 Departamento de Biología Vegetal, Facultad de Farmacia, Área de Edafología y Química Agrícola, Universitat de València, 46100 Valencia, Spain
* Correspondence: oscar.andreu@uv.es (O.A.-S.); luzgarcia@ucm.es (M.L.G.-L.); Tel.: +34-913-94-5014 (M.L.G.-L.)

Abstract: Sulfide mining wastes may lead to severe environmental and human health risks. This study aims to use geochemical and ecotoxicological indicators for the assessment of the ecological risks of potentially toxic elements (PTEs) in the San Quintín mining group to categorize wastes prior to mining restoration. Ecotoxicity was evaluated using crustacean (*Dahpnia magna, Thamnocephalus platyurus*) and algae (*Raphidocelis subcapitata*) bioassays. The geochemical and mineralogical results suggested that the mining residues underwent intense weathering processes, with active processes of acidity generation and metal mobility. Total PTEs concentrations indicated that the mining materials were extremely polluted, with Pb, Zn and Cd geoaccumulation index (I_{geo}) values higher than 5 in more than 90% of the samples. The pollution load index (*PLI*) showed average values of 18.1, which classifies them as very highly polluted. The toxicity tests showed a higher toxicity for plants than crustaceans, being the highest values of toxicity related to toxic elements (Pb, Cd and Zn), electrical conductivity and to pH. This paper presents for the first time the combination of indices in the categorization of mining waste prior to its restoration. The combination of them has made it possible to categorize the waste and adapt the restoration and remediation procedures.

Keywords: tailings; dumps; potentially toxic elements; *Daphnia magna; Raphidocelis subcapitata; Thamnocephalus platyurus*; ecotoxicity; mining restoration

1. Introduction

Potentially toxic elements (PTEs) are naturally occurring, ubiquitous substances in the human environment which typically originate from the weathering of parent materials. Nevertheless, due to a variety of human activities, PTEs have substantially accumulated in the global environment in recent years, particularly in soil and sediment environments. Potentially toxic elements include metals and non-metals such as arsenic (As) and selenium (Se) [1].

The presence of PTEs supposes a risk in areas affected by ancient mining activities [2,3]. These zones, such as the San Quintín Mining Group (SQMG), have large amounts of wastes with nonrecoverable sulfides that are dumped at disposal sites and exposed to weathering. Acid mine drainage (AMD) occurs when sulfides are exposed to atmospheric, hydrological or biological weathering (oxygen, water and bacteria), becoming oxidized products and producing low pH and high concentrations of PTEs and soluble sulfates in surficial water and groundwater. This environmental concern is especially relevant in areas with historical metallic mining activities, where the application of less efficient mineral extraction and processing techniques and the lack of environmental awareness and controls have left huge environmental liabilities [4–6].

There are many derelict mining areas with these conditions that require the application of adapted restoration projects. A mining restoration project must seek to design restoration actions focused on minimizing erosive processes and preventing the generation of acidity and immobilizing dangerous PTEs contained in mining residue and soils affected by mining. Likewise, edaphic support must be provided to the roof for the revegetation of the restored materials. Restoration project design can be challenging in mining areas with a long history of exploitation that have dispersed mining materials (dumps and tailings) with very different metal loads and toxicities.

Traditionally, soil toxicity has been evaluated through quantifying total and/or extractable PTEs contents [7]. Once the PTEs contents are determined, one of the most used ways to assess the environmental risk in soils is the calculation of pollution indices based on the comparison of an element concentration in soil samples and the background level of this element in the study area. Among all indices, the geoaccumulation index (I_{geo}) and the pollution index (*PI*) are widely used as single indices and the pollution load index (*PLI*) is used among the integrated ones [8].

In SQMD, some studies have been carried out on the geochemical and mineralogical characterization of soils and waters [3–6]. Even if these procedures are sensitive, they fail to provide information about the bioavailability of contaminants and their synergistic or antagonistic interactions between pollutants. For that reason, and to estimate the environmental risk of contaminants, chemical analytical techniques need to be complemented with the use of bioassays [9].

Taking into account that the aquatic environment usually represents the destination of contaminants, organisms such as algae (i.e., *Raphidocelis subcapitata*) or crustaceans (i.e., *Daphnia magna* or *Thamnocephalus platyurus*) are very important to be included in the ecotoxicological evaluation procedure. Primary producers are the first link of aquatic systems, and adverse impacts on them may have important consequences for the health of the whole aquatic ecosystem. On the other hand, primary consumers are filter-feeding organisms and can be useful indicators of the bioavailability of particle-bound contaminants. *D. magna* and *R. subcapitata* have been commonly used for the assessment of the ecotoxicity of environmental samples such as water, soil leachates and sediments, but recently *T. platyurus* has gained interest as a potential reference organism for environmental monitoring [10]. The results of these tests (e.g., the effect concentration EC50 or the lethal concentration LC50) are often recalculated to toxic units (TU) in order to classify the potential toxicity. Persoone et al. [11] introduced the classification of the water and wastewater toxicity, becoming the most commonly used system in the last two decades [12].

The aim of the present research was to assess the polluting potential and geochemical behavior of PTEs in mine wastes accumulated by the SQMG (Ciudad Real, Spain), evaluating the mineralogical and geochemical composition of 18 mining waste samples. Additionally, we analyzed the ecotoxicity of PTEs using three environmentally relevant test species of different trophic levels and biological complexity: algae (*Raphidocelis subcapitata*) and crustaceans (*Thamnocephalus platyurus* and *Daphnia magna*). The suitability of those assays for the assessment of environmental risk in mine sites strongly polluted by PTEs is discussed and evaluated to determine if they could be applied for residue categorization prior to mining restoration.

2. Study Area

The SQMG is part of the so-called Alcudia Valley Mining District [13]. It is located within the so-called Meseta Sur (the Spanish southern mesa), which has a continental Mediterranean climate with contrasted seasonal variations in mean temperatures: 6–8 °C (winter) and 26–28 °C (summer). The rain concentrates in late autumn and early spring, with an annual total of 500–700 mm. The mining group is located within a region morphologically characterized by WNW trending valleys and sierras, within a landscape ranging in altitude between 600–700 m above sea level [4].

This mining district is formed by a large number of mines distributed along the southwest area of Ciudad Real province. In particular, SQMG represents the largest one and comprises two mining zones (Figure 1) delimited and separated by 800 m: San Quintín East (affecting an area of 12 Ha) and San Quintín West (affecting an area of 40 Ha) [13].

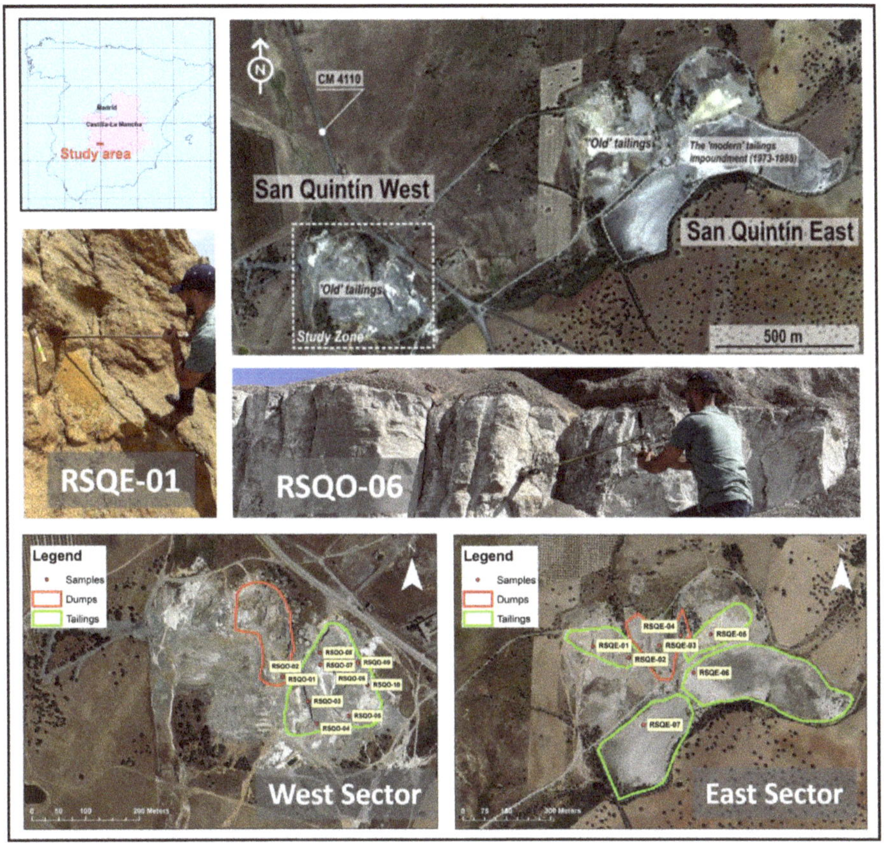

Figure 1. Situation map and sample location.

The main soils are entisols, with localized development of anthrosols. Other local soil-like materials include spolic technosols related to abandoned tailings deposits.

The SQMG (East and West) (Figure 1) was active between 1888 and 1923, when about 515,000 tons of galena concentrates were produced (via gravity concentration with the aid of primitive jigs). Subsequently, between 1973 and 1988, the *Sociedad Minero-Metalúrgica de Peñarroya España* carried out a second mineral recovery operation of the old tailings (this time using froth flotation), which was estimated at around 3 million tons and showed high Zn content—a metal with hardly any interest at the time of the older exploitation. However, this operation did not affect the tailings left at San Quintín West. From the older period, there are many waste dumps of barren rock whose environmental risk is low; thus, the main hazard relates to the tailings, which in this case are constituted by residues of mineral concentration via gravity, a process that was inefficient at the time and left behind residues very rich in galena (PbS) and sphalerite (ZnS). These old and chaotic tailing deposits are a separate case and present a significant environmental hazard in relation to the heavy metals (Pb, Zn-Cd) and metalloids (As). These tailings not only contain high concentrations of Pb and Zn (up to 48,600 mg kg^{-1} Pb and 34,000 mg kg^{-1} Zn), but also are a source of AMD.

Additionally, the lack of vegetation cover on these unstable wastes reveals other hazards associated with this type of infrastructure: slope instability, gully production, collapses and great sediment mobilization [13].

3. Materials and Methods

3.1. Sampling Design

The main criterion for the selection of the sampling points has been the access to the interior of the sludge pools for the sampling of humid material, neither aerated nor in an active oxidation process. To achieve this, the vertical profiles of the sludge ponds have been drilled with an Ejkelkamp sampler until access to the fine materials was processed in the corresponding sink. In the large ponds in the East, samples have been taken from the upper deck to avoid taking samples from the containment dikes.

The selection of sampling points has been made to try to ensure the representativity of mining wastes in the studied area (Figure 1). A total of 18 samples were collected: 7 in San Quintín East and 11 in San Quintín West.

3.2. Geochemical Analysis of Samples

Samples were air-dried and sieved through a 2-mm mesh. The pH and electrical conductivity (EC) were measured in a 1:5 (w/v) suspension of an aliquot of the sample in deionized water.

Organic matter (OM) was determined following the Standard Test Methods for the Moisture, Ash, and OM of Peat and Organic Soils [14]. In order to apply the step specifications of the standard method, the samples were powdered by an agate mortar and were sieved through a 53-μm mesh. Then, they were dried in a laboratory oven for 24 h and cooled down in a desiccator. Afterwards, the samples were heated up in a furnace following temperature intervals determined by the ASTM [14].

Elemental concentration data in samples were achieved by means of Energy Dispersive X-ray Fluorescence Spectrometry (EDXRF) using a Bruker, S2 Ranger spectrometer with a Pd detector at the Unidad de Técnicas Geológicas, Universidad Complutense de Madrid. Samples were quartered and an aliquot of 10g was ground to fine powder using an agate ball mill and passed through a 100-μm sieve prior to analysis. Sieved samples were placed into cylindrical sample holders with a 3.6-μm Mylar filter and then introduced in the EDXRF spectrometer. An extended analysis time (23 min) was used to achieve the minimum standard deviation of each data. From the total list of major, minor and trace elements analyzed, As, Cd, Fe, Pb and Zn were specially chosen for this study because of their abundance in these types of mine wastes and because most of them are included in the priority contaminant list of environmental protection agencies [15]. To ensure the quality of the analyses, duplicate samples were analyzed to check precision and a certified reference material (SRM 2711) was analyzed for accuracy. The precision results of the studied elements were Zn (87.3%), As (82.9%), Pb (93.1%) and Fe_2O_3 (88.4%). The recovery percentages for the certified elements in SRM 2711 were Fe (97.7–98.0), Zn (98.8–99.9), As (79.1–83.9) and Pb (111.0–112.7).

3.3. Mineralogical Analysis of Samples

Main mineral phases were determined by means of X-ray diffraction (XRD) using a Bruker D8 Advance® diffractometer equipped with a Cu anticathode at CAI-Unidad de Técnicas Geológicas, Universidad Complutense de Madrid. An aliquot of each sample was grinded in an automatic agate mortar and manually until all the grain size was less than 53 μm. Work conditions of the diffractogram obtention were: 2θ angles (2°–65°), 0.02 stepping intervals and 1 s per step. The semiquantitative analysis was carried out according to the Chung method [16–18] using EVA from Bruker software.

3.4. Samples Leaching Procedure and Soluble PTE Content

The leaching process was carried out for each sample to obtain the aqueous extract used on bioassays with aquatic organisms. The method followed was in accordance with the Spanish legislation of contaminated soils, the Royal Decree (RD) 9/2005, which establishes a list of potentially polluting soil activities and the criteria and standards for the declaration of contaminated soils. The method proposed by the RD is the DIN 38414-S4 [19]. It specifies to use 100 g of dry soil sample (with a grain size <10 mm), mix it with 1000 mL of deionized water (leaching mass relation: 10 Liquid/Solid) and submit it to an upside-down agitation, in a Reax20 rotary agitator (Heidolph®), during 24 h at room temperature. The solid and liquid phases are separated via sedimentation at 4 °C overnight. Liquid one was filtered with a 0.45-μm pore size fiber membrane (Pall®). To avoid the growth of microorganisms during the storage, the samples were stored at 4 °C in darkness until assayed.

The Pb, Cu, Fe and Cd contents in leachates were determined via electrothermal atomic absorption spectrometry (ETAAS) using an ICE 3300 spectrometer (Thermo Fisher, USA). The reliability of the results was verified by means of certified reference material (EnviroMAT SS-2 Soil Standard), obtaining recoveries of Pb (101.5%), Fe (108.4%) and Cd (87.9%).

3.5. Bioassays

3.5.1. Immobilization Test with *Daphnia magna*

The assay with *Daphnia magna* (DM) was carried under the OCDE Technical Guideline 202 [20]. To determine the 48 h EC50 (sample dilution required to immobilize 50% of *Daphnia magna* population after 48 h exposure), *D. magna* organisms aged less than 24 h were obtained from dormant eggs (ephippia) supplied in a commercial kit, Daphtoxkit F™ (Microbiotests Inc., Ghent, Belgium). Ephippia were hatched after 72 h of incubation in culturing media at 21 ± 1 °C and 4000 lux of continuous illumination in a climatic chamber. Afterwards, twenty neonates (five neonates per four replicates) were exposed to different tested concentrations and a blank control [21]. The mortality rate in Daphnia magna was 0% in the controls. Concentrations were obtained via dilution of the leachates with the standard culture medium supplied with the Daphtoxkit™. The concentrations, expressed as percentages of the sample in the final dilution, were 50%, 25%, 12.5%, 6.25% and 3.12% (v/v). The assays were conducted in multiwell plates (30 wells/15 mL) and incubated in a climatic chamber at 21 ± 1 °C in darkness. After 48 h of incubation, immobility was checked in order to determine the mortality of the exposed population. Daphnids were considered immobilized if they were not swimming within the observation period of 15 s [22]. The toxicity was expressed as the EC50 (%; v/v), causing lethality on 50% of the exposed organisms.

3.5.2. Immobilization Test with *Thamnocephalus platyurus*

The *Thamnocephalus platyurus* (TP) test was performed using the Thamnotoxkit F™ (Microbiotests Inc., Ghent, Belgium) following the standard operational procedure provided in the kit [18], with some modifications [23]. The test follows the ISO 14380:2011 standard [24]. Dormant eggs of *T. platyurus* were incubated 24 h prior to the start of the toxicity test under continuous illumination (4000 lux) at 24 ± 1 °C in a climatic chamber (model ARTI-150L, Microbiotests Inc, Gent, Belgium). The assays were performed in 24-well plates (10 organisms per well in triplicate). Five dilutions (100%, 50, 25, 12.5, 6.25 and 3.12%; v/v) were prepared from each leachate by using the standard fresh water (U.S. EPA moderately hard reconstituted water) supplied in the Toxkit™. Ten milliliters of each dilution were added to each well; the fresh medium was used as a control. The number of dead organisms was used as an endpoint, and the EC50 (%; v/v) causing lethality on 50% of the exposed organisms was calculated. Exposed organisms were considered immobilized (dead) if they were not swimming within the observation period of 10 s [22]. The mortality rate in Thamnocephalus platyurus was 0% in the controls.

3.5.3. Freshwater Algae—Growth Rate Inhibition Test with *Raphidocelis subcapitata*

Raphidocelis subcapitata (RS) (formerly *Selenastrum capricornutum*) cells from the commercial Algaltoxkit F™ (Microbiotests Inc., Ghent, Belgium) were used. This kit test conformed with ISO 8692:2012 and OECD TG 201 [25]. The alginate beads containing algae were disaggregated according to OECD guidelines (Annex 3, TG 201, 2011). The cells were exposed to undiluted samples and five serial dilutions (50%, 25%, 13.5%, 6.25% and 3.12%, *v/v*) and one negative control (100% of growth). Sample dilutions were conducted with an ISO algal culturing medium. Each sample was run in triplicate. The algae growth assays were carried out in 100-mm long spectrophotometric cuvettes. Initially, the initial biomass concentration was 10^4 cells mL^{-1}. The cuvettes were incubated at 25 ± 1 °C for 72 h under continuous illumination (4000 lux) in a climatic chamber with model ARTI-150L (Microbiotests Inc., Ghent, Belgium), equipped with four white light led lamps. Growth inhibition rates relative to negative controls were determined using measurements of optical density (OD_{670}) in a spectrophotometer (model Auris 2021, CECIL Instruments™, Cambridge, UK) equipped with a holder for 10-cm path-length cells. The inhibition growth (% I) in the tested dilutions versus the control growth is based on the determination of the average growth rates (μ) after transformation of the OD_{670} values into cell numbers according to the manufacturer protocol [26]. The toxicity was expressed as a 72 h ErC50, the concentration (%; *v/v*) causing a 50% on the growth rate inhibition.

3.6. *Assessment of Potential Environmental Risk*

The potential environmental risk in SQMD was assessed by using the geoaccumulation index (I_{geo}), the pollution index (*PI*) and the pollution load index (*PLI*).

I_{geo} index [27] is calculated by conforming to Equation (1):

$$I_{geo} = log_2\left(\frac{C_i}{1.5B_i}\right) \tag{1}$$

where C_i the total concentration of the metal *i* in the sample and B_i the background concentration of this metal in the San Quintín mine. Those regional background values have been previously reported by Gallego et al. [28]. Based on the values obtained for I_{geo}, samples can be classified as [29]: <0, practically unpolluted; 0–1, unpolluted-to-moderately polluted; 1–2, moderately polluted; 2–3, moderately–strongly polluted; 3–4, strongly polluted; 4–5, strongly–extremely polluted and >5, extremely polluted.

PI is an index single that evaluates the pollution of soils and sediments. Its equation of calculation is defined using Equation (2):

$$PI = \frac{C_i}{B_i} \tag{2}$$

where *PLI* is the pollution load index and n is the number of metals evaluated. *PI* is the pollution index of each metal.

Finally, the *PLI* is calculated using the nth root of the multiplication of the contamination factors of the investigated elements, as in the following expression (3):

$$PLI = (PI_1 \times PI_2 \times PI_n \times \ldots)^{1/n} \tag{3}$$

where PI_1, PI_2 and PI_n are the pollution indexes of the elements 1, 2 and *n*, respectively. This index classifies the soil or water into four categories, which are: $PLI < 2$ (unpolluted-to-moderately polluted); $2 \leq PLI \leq 4$ (moderately polluted); $4 \leq PLI < 6$ (highly polluted) and $PLI > 6$ (very highly polluted) [30].

3.7. *Statistical Analysis*

A multivariate analysis was applied for the geochemical and mineralogical variables, including a correlation matrix and principal component and factor analysis, using Statgraphics 19 software package (Statgraphics Technologies, The Plains, VA, USA) [31]. The multivariate analysis was used to determine potential relationships between the PTE total contents and the mineralogical composition of the original samples and between the PTE contents in the leachates and ecotoxicological results. Factorial analysis permits a statistical approximation for analyzing interrelations between a large number of variables. The factorial analysis was carried out via the principal component extraction method, using a varimax normalized rotation of the factors. Toxicity was expressed as the percentage of effect as the median effective concentration (EC50) and the 10% effect concentration (EC10), as an alternative to the No Observed Effect Concentration (NOEC) [32]. Whenever possible, the 95% confidence limit values were set. ECx was calculated by means of Probit regression [33] in EPA Probit software (v1.5).

4. Results

4.1. Geochemistry and Mineralogy of Samples

The pH values ranged from 2.0 to 6.5, with a mean value of 4.7 in SQ east and 5.1 in the west area. Samples RSQO-01, RSQE-01 and RSQE-05 are the most acidic, with pH values of 2.0, 2.3 and 2.8, respectively (Table S1). As regards the EC values, SQ east showed similar values (average of 2200 μS cm^{-1}) to the west area, with average values of 2380 μS cm^{-1}. The highest EC values were found in RSQE-01 and RSQO-01, which showed 5470 and 4100 μS cm^{-1}, respectively. The obtained results suggested that samples with lower pH values correspond to samples with the highest of EC.

All samples showed a low OM content, with an average value of 1.65% in the west sector and 1.62% in the east sector. The highest OM value was 7.8% and was found in the west zone.

Total and soluble PTE contents are summarized in Tables S2 and S3. Average values for target elements were 130 mg kg^{-1} for Cd, 42,900 mg kg^{-1} for Pb, 15,100 mg kg^{-1} for Zn, 63,600 mg kg^{-1} for Fe and 47 mg kg^{-1} for As. Except for Fe, determined total PTE contents are higher in the west area (Table S2).

Compared with background values in the study region (Table 1), all samples exceed the average of the local background levels established by Gallego et al. [28] for the mining area. The mean concentrations are 85 times higher than the background for Cd and 268 times for Pb. The average values are also higher or similar than other studies, such as [4,13,28,34]. The results expose the need for treatment of the tailings and the restoration of the mining area. The determination of the potential risk of the pollutants is required to determine their influence on human life and associated ecosystems.

Table 1. Main PTE content in samples from the San Quintín mine, local background levels established in the San Quintín area [24] and other values of PTEs in other studies of the area [4,13,28,34]. All values are expressed as mg kg^{-1}.

	Cd	Pb	Zn	Fe	As	References
Mean SQMD	130	42,900	15,100	63,600	47	This work
Mean value east zone	58	32,400	6,700	71,300	7	
Mean value west zone	170	49,600	20,400	58,700	72	
Minimum	25	1800	600	36,300	0.05	
Maximum	440	144,900	47,700	202,600	549	
Background levels San Quintín mining area						
Local Background	1.5	160	87	28,963	8	[28]
Other studies in San Quintín mining area						
Mean	38	18,036	8825	42,597	88	[4]
Mean	-	21,892	11,242	54,573	-	[13]
Mean	-	11,260	8549	42,170	-	[34]

PTE content in leachates (Table S3) showed an average value of 9.7 mg kg^{-1} for Cd, 20 mg kg^{-1} for Pb, 580 mg kg^{-1} for Zn and 230 mg kg^{-1} for Fe. Considering both subzones, in the SQ west zone the mean values are much higher than those determined in the SQ east zone.

The semi-quantitative mineralogical composition is summarized in Table S4. In most cases, samples showed similar mineral composition, mainly composed of quartz (23%–51%), phyllosilicates (in some cases with chamosite of note contain) (15%–44%), feldspars (2%–8%), plagioclase (3%–10%), gypsum (3%–19%) and, to a lesser extent, iron oxides such as goethite, magnetite or hematite; sulfates and/or hydroxysulfates such as jarosite, baryte or anglesite and also alunite, sphalerite, siderite, pyrite, cerussite and dolomite/ankerite, all as minor phases.

The mineralogy observed is coherent with the geological and mining context. The higher Pb concentrations are explained with the greatest amount of minerals phases such as jarosite or anglesite. In some cases, the content of these minerals reached 22% (RSQO-01). The presence of galena, sphalerite, and pyrite in some of the waste samples should be highlighted, even though they appeared in a low content percentage in most of the cases. Nevertheless, there are samples where the quantity of these ore, such as with FeS_2 and ZnS, rise to 6% (RSQO-6A) and 9% (RSQO-6B), respectively. The presence of pyrite is especially important in the context of mining restoration since it implies a very high capacity to generate acidity from mining waste. In this sense, the results suggest that the tailings in the western area, which were concentrated in old mineralurgical facilities with inefficient technologies, are more reactive and will need more aggressive amendments during restoration work. This oxidation process is evidently active in some areas of the western zone, where the presence of jarosite reaches concentrations of 22%, although it appears to be active in a low proportion in both areas and in almost all tailings, regardless of the efficiency of the oxidation concentration technology used.

4.2. Bioassays

The results of the ecotoxicity bioassays (48 h EC50 and 48 h EC10) are shown in Table S5; the toxicity values are shown by the percentage of the water extract accompanied by the confidence interval and the toxic units (T.U.). There is also a classification of samples based in R.D. 9/2005 and in [11].

For *Daphnia magna* organisms, the toxicity levels (EC50) are between 0.2% and 14.9% of dilution. In three of the samples, the toxicity values have not been determined because the levels are over 50% of dilution, so they are not considered assoils that are toxic toward *D. magna* (RSQE-03, RSQE-07 and RSQO-03). The higher mortality is attributed to RSQO-6A, which showed 458.7 toxic units. On the other hand, EC10 presents values between 0.1% and 6.5% of dilution (Table S6).

According to the hazard classification system for wastes discharged into the aquatic environment carried out by Persoone et al. [11], 27.7% of the samples showed very high acute toxicity. Based on R.D. 9/2005, 28% (EC50) of the samples are contaminated, corresponding to RSQE-01, RSQE-05, RSQO-02, RSQO-05 and RSQO-6A in both cases.

The toxic effect toward *Thamnocephalus platyurus* showed the greater mortality in RSQE-01, RSQE-04, RSQE-06, RSQO-05, RSQO-06A and RSQO-08. RSQE-01 showed the most mortal levels, reaching 2500 toxic units.

For *T. platyurus* organisms, 33.3% of the samples presented high acute toxicity, slightly greater than the *D. magna* tests [11]. In addition, 33% of the samples were contaminated, 5% higher than in the *D. magna* test. In the case of EC10, 44% of the samples exceed levels (R.D. 2005) (Table S6).

Tests based on *Raphidocelis subcapitata* showed the largest mortality data against PTE content, where 50% of the samples were under 1% of dilution for ErC50 values and 78% for EC10 (Table S6). For that reason, 50% is considered as a very high acute toxic sample [11] and also as a contaminated sample, which manifests the sensibility of these organisms to PTEs with respect to *D. magna* and *T. platyurus* results. The most toxic sample was RSQO-05. The bioassays reveal that a major part of the samples presented at least acute toxicity [11] for *D. magna*, *T. platyurus* and *R. subcapitata*, in particular the samples whose higher element concentrations were Pb and Zn. This relationship between mortality and PTE contents is shown in Table 2. There is also a relation between the EC, pH and the bioassay results; EC especially seems to be an important factor for life development. However, the higher mortality in all the cases was not attributed to the same samples, therefore the effect of the characteristics of the leachates was different for each organism.

Table 2. Pearson's correlation coefficients between leachates, PTE content, general geochemical parameters and bioassay's results (n = 18). * values are $p < 0.05$.

	Cd	Pb	Zn	Fe	pH	EC	EC50 *D.m.*	EC50 *T.p.*
Pb	0.3							
Zn	0.3	0.1						
Fe	−0.2	0.1	0.0					
pH	−0.4	−0.3	−0.7 *	−0.6 *				
EC	0.5	0.3	0.5 *	0.5	−0.8 *			
EC50 *D.m.*	−0.5 *	−0.7 *	−0.4	−0.2	0.5 *	−0.7 *		
EC50 *T.p.*	−0.5 *	−0.6 *	−0.4	−0.1	0.4	−0.5 *	0.8 *	
EC50 *R.s.*	-0.4	−0.8 *	−0.4	−0.2	0.5	−0.6 *	0.9 *	0.8 *

The sensitivity of *D. magna* against these parameters is confirmed in other works [9,35,36] that also show a correlation between Pb and Zn content.

A cluster analysis has differentiated very clearly the group of Fe and OM from the rest of the variables, suggesting an absolute absence of influence of these factors on the toxicity of mining materials (Figure 2A). However, the relationship between the other variables appears to be very high, although two subgroups can be distinguished: the subgroup that includes pH with toxicity in crustaceans and algae, and the subgroup that includes PTEs (Cd, Pb and Zn) together with EC. The factorial analysis (Figure 2B) helped to discriminate between the samples that did not present acute toxicity—the B subgroup (RSQE-03; RSQE-07 and RSQO-03)—from the rest of the samples with acute toxicities in at least one of the tests carried out. In this group, the A1 subgroup (RSQE-02, RSQE-05, RSQO-05 and RSQO-06A) is related to high EC and Cd contents, the A2 subgroup (RSQO-04, RSQO-6B, RSQO-07, RSQE-04 and RSQE-06) shows low Cd concentrations and the last, the A3 subgroup (RSQO-02, RSQO-08, RSQO-09 and RSQO-10), presents slightly acute-to-acute toxicities and medium concentrations of Cd, Pb and Zn.

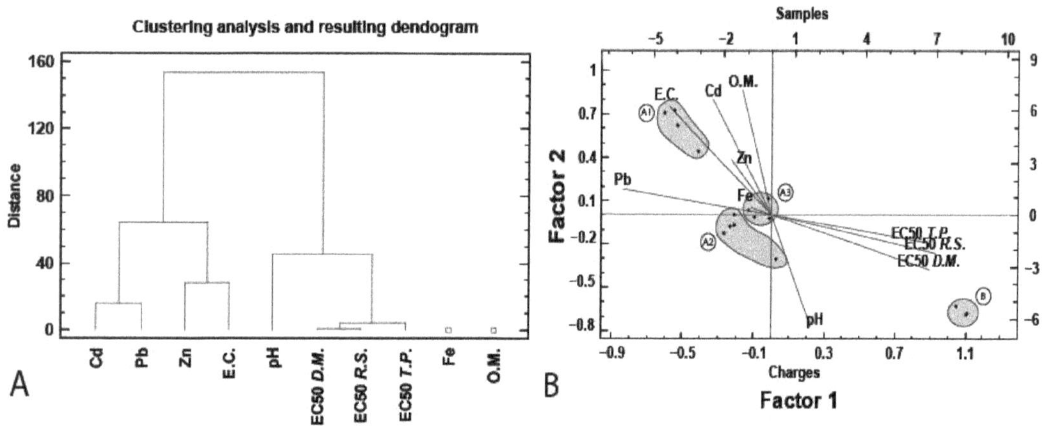

Figure 2. (**A**): Cluster and factor analysis for the studied samples. The lines correspond to the punctuations for the first and second factors of the considered values. (**B**): Principal component analysis (PCA) for the PTE, bioassays values, pH, EC and OM. Groups in the factor analysis graph were obtained in the cluster analysis.

4.3. Assessment of Potential Environmental Risk

Cd and Pb I_{geo} results are high in most of the samples, showing a high degree of contamination that is interpreted as being from strongly to extremely polluted. The highest one for Cd corresponds to RSQO-6B, whereas lead showed the greatest value in the RSQO-05 sample. As Cd I_{geo} and Pb I_{geo} values, Zn I_{geo} results are elevated; nevertheless, two samples (RSQE-03 and RSQO-01) showed slightly lower values than the rest; thus, they are considered as strongly polluted. For Fe and As, the results are quite different. Most of the samples were moderately polluted according to Fe I_{geo}. Arsenic values were the lowest ones, indicating unpolluted mining materials; however, there are highlighted samples that exceed the low values determined for the rest: RSQE-02, RSQO-05 and RSQO-6B. The I_{geo} values reveal the potential risk that the area poses to the surroundings as well to the human and environmental life.

According to the *PLI* calculated for the selected elements, only RSQE-03 and RSQE-07 can be categorized as moderately polluted–unpolluted and moderately polluted, respectively, whereas the rest are highly polluted-to-very highly polluted. The maximum *PLI* value is 100.16 and corresponds to RSQO-05, located in one of the ponds of San Quintín West. Seventy-eight percent of the samples are classified as very highly polluted ($PLI > 6$).

In relation to the I_{geo} index and *PI* and *PLI* results, we could make a comparison with other works in the San Quintín mine [4,37]. In case of the I_{geo} [37], the results concerning Pb and Zn are quite similar, manifesting extremely contaminated wastes. For Cd values, results show differences, moderately contaminated for [37] and extremely contaminated in our case. The same phenomena appear in the comparison of [4] the *PLI* results. The data from this work estimate larger values, except for As and Fe, that are similar to those determined by García-Lorenzo et al. [4]. Thus, the *PLI* is higher in our case. In any case, all the works interpret the result as being a very highly polluted residue. Differences may be due to the background values selected.

5. Discussion

In the characterization of the risk produced by mining waste from the San Quintín mine, data on basic pedological parameters (pH, EC, OM) as well as mineralogical and multielemental data have been obtained. Based on the latter, both the I_{geo} of the most important PTEs in the area have been obtained, as well as the *PLI* of a combination of these, to evaluate the synergies that can be established between them. Finally, the toxicity of the residues has been evaluated by means of tests with crustaceans (*Daphnia magna* and *Thamnocephalus platyurus*) and algae (*Raphidocelis subcapitata*), obtaining a categorization of the wastes, both from the multi-elemental point of view (I_{geo} and *PLI*) and from a toxicological point of view (based on the EC50 of the three biotests) (Table 3).

Table 3. Categorization of bioassays, I_{geo} values and *PLI* results, according to [11,29,30], respectively. The color scale is shown for each classification. DM: *Daphnia magna*; TP: *Thamnocephalus platyurus*; RS: *Raphidocelis subcapitata*. Toxicity results are shown as TU. Toxicity values less than 2 are indicated without color in the cell because they cannot be assigned to a single class.

Sample	*DM* TU	*TP* TU	*RS* TU	I_{geo} Cd	I_{geo} Pb	*PI* Zn	I_{geo} Fe	I_{geo} As	*PLI*
RSQE-01	100	2500	430	5.9	8.9	6.9	3.1	−6.1	9.0
RSQE-02	76	68	490	6.1	9.7	6.8	1.9	3.3	32
RSQE-03	<2	20	<2	4.6	4.1	3.4	1.1	−6.0	1.8
RSQE-04	7	145	<2	6.5	7.7	8.3	1.3	−6.0	7.9
RSQE-05	320	<2	4760	6.3	8.4	6.4	1.9	−6.0	7.0
RSQE-06	15	1560	250	5.9	7.7	7.0	1.5	−6.0	6.2
RSQE-07	<2	<2	<2	5.0	5.0	5.1	1.2	−6.0	2.8
RSQO-01	38	10	59	5.1	7.9	3.5	3.4	−6.0	4.6
RSQO-02	180	<2	2080	6.8	8.5	8.6	1.6	−6.0	10
RSQO-03	<2	<2	<2	6.0	6.5	7.8	0.9	−6.0	5.6
RSQO-04	66	19	90	7.3	9.0	8.3	0.9	−6.0	10
RSQO-05	270	660	190	8.3	10	9.2	1.5	6.8	100
RSQO-06A	460	240	130	7.4	10	7.9	1.4	−6.0	12
RSQO-6B	38	44	110	8.8	10	9.7	1.2	5.5	84
RSQO-07	90	66	150	6.9	7.7	8.1	0.9	−6.0	7.7
RSQO-08	48	110	28	7.4	7.7	8.2	1.2	−6.0	8.8
RSQO-09	54	50	8	6.8	7.3	8.7	1.0	−6.0	7.9
RSQO-10	26	55	49	7.5	7.5	8.4	1.0	−6.0	8.6

The values obtained from I_{geo} that categorize a greater number of samples as extremely polluted correspond to Pb, Zn and Cd, the elements present in the mineral phases of paragenesis (galena and sphalerite). Other PTEs related with pyrite or arsenopyrite (Fe, As) appear to be less pollutant in terms of I_{geo}, surely due to the presence of Fe oxides in the local lithologies. Three samples with low I_{geo} values have been found for almost all PTEs, a sample from an old mineralurgical facility (RSQE-03), another from a modern tailing of coarse (sandy) material (RSQE-07) and a sample from a waste material from the western zone with abundant signs of acidity generation and high proportions of OM (RSQO-01). When we consider the PTEs as a whole through the *PLI*, all the samples are categorized as very highly polluted except for the three aforementioned samples, which coincide with the Cd, Pb and Zn I_{geo}. Iwasaki et al. [32] have applied a similar approach to categorize the evolution of the risk of contaminated soils in the San Quintin area through the use of I_{geo} and *PLI*. Their findings are in accordance with our I_{geo} data for Pb and Zn, with differences in relation to Cd and Cu.

From a toxicological point of view, mining waste has shown a great variety of categories according to [11]. Some of the samples have shown toxicities typical of classes I, II or III (<2 TU) for the three tested organisms and low toxicity values determined in samples of processed materials (tailings) from the two areas of the mining group. A remarkable fact is the values of the RSQE-05 sample, which has shown low toxicity values (<2 TU) for *Thamnocephalus platyurus*, but high values for the other two organisms tested.

Differences can be observed between crustaceans and algae, with a higher number of samples with acute toxicity in the case of algae, which suggests that the phytotoxicity of these mining residues could be of concern. These findings are in accordance with

García-Lorenzo et al. (2019) [4], who described an inhibition of the germination of *Sorghum saccharatum* (Sorgho), *Lepidium sativum* (Garden cress) and *Sinapis alba* (mustard) for samples with high contents of Pb, As and Cd, in addition to root inhibition in the samples with high Pb contents in the San Quintin area. However, the non-toxicity for the TS of two samples that have acute toxicity for DM and RS does not seem to be explained by the data obtained in the present study. It could be a synergistic effect between inorganic contaminants (PTEs) and major elements, such as those described for Cd-Zn-Cu in some plant species [38].

The combination of "inorganic" indices is often used in the characterization of risks in mining areas, such as gold mines in Ghana [39] or Brazil [40], or coal mines in Australia [41] and India [42], but the combination with toxicological risk categorization using bioassays does not have many published references. For instance, Varga et al. [43] describes the differences in sensitivity of different plant species in the categorization of risk for sediment and pore water in a polluted channel in Serbia.

This paper presents for the first time the combination of these indices in the categorization of mining waste prior to its restoration, taking into account not only the presence of PTEs in high concentrations compared to local geochemical backgrounds, but also their toxicity through tests with organisms that live in an aquatic environment (crustaceans and algae). In the case of the San Quintín area, the combination of these indices shows that practically all wastes can be considered as polluted (moderately–extremely) based on their Pb, Zn and Cd content, but that the risk derived from the presence of these PTEs does not imply acute toxicity in a group of five samples from tailings and dumps nor a risk of slight acute toxicity in a large group of samples from tailings in the western area of the mining group. It has also been possible to establish that phytotoxicity is higher than the toxicity to small organisms (crustaceans), in accordance with data described by García-Lorenzo et al. [4], which could have implications for the management of these wastes during restoration. The fact that the toxicity of the mining materials is less than their contaminated levels offers the possibility of not completely confining these residues or allowing the access of the roots of trees and shrubs to the residues through the technosol designed to cover the mining materials. Some recent studies suggest that Pb and Zn are less toxic for local plants than Cd [44,45]; therefore, it would be possible to categorize wastes with non-acute toxicity risk and a low Cd content (samples RSQE-03 and RSQE-07) as susceptible to lighter restoration treatments and thinner technosol covers, which would produce an improvement in the optimization of restoration treatments. This optimization will be especially interesting in the case of the RSQE-07 sample, which represents tailings with a large volume of materials and a thickness of 8–12 m. It would be excessively expensive to apply a limestone amendment to prevent the generation of acidity, so the possibility of applying the amendment to the first meter of waste and covering it with a 50-cm technosol makes it possible to restore this large volume of waste.

6. Conclusions

This work reveals that the San Quintín mine area is highly polluted by PTEs, especially the western sector, where the mineral concentration treatment was not effective enough. The geochemical and mineralogical characterization shows high contents of Pb, Cd, Zn, Fe and also As. The most elevated ones are Pb, Zn and Fe, reaching values of 202,600 mg kg^{-1}, 144,900 mg kg^{-1} and 47,700 mg kg^{-1}, respectively, all of them located in the western area. The mineralogical phases are constituted principally by quartz, phyllosilicates, oxides, sulfates, hydroxysulfate and others in less percentage. The presence of ores such as pyrite and sphalerite in the west area of San Quintin has to be highlighted; indeed, this has to be considered in future restorations and also in acid drainage amendments.

Bioassays reveal that algae organisms (*R. subcapitata*) are more sensitive to PTE contents; thus, their toxicity values are higher than the results for the crustaceans (*D. magna* and *T. platyurus*). The lowest EC50 data from *R. subcapitata*, *D. magna* and *T. platyurus* were 0.20, 0.22 and 0.04, respectively. Statistical information disclose that the highest values of toxicity are related to PTEs (Pb, Cd and Zn), EC and especially to pH.

The joint use of biological, geochemical and mineralogical techniques can provide useful information for environmental management and also to improve the restoration procedure in terms of cost–benefit.

Supplementary Materials: The following supporting information can be downloaded at: https://www.mdpi.com/article/10.3390/min13040492/s1, Table S1: pH, EC (μS cm^{-1}) and OM values (%). Table S2: Total PTEs values determined for target elements (mg kg^{-1}); Table S3: PTEs content in leachates (mg kg^{-1}); Table S4: Mineralogical composition (%); Table S5. EC50 values calculated (Probit regression) and their respective confidence intervals (95%) for *Daphnia magna, Thamnocephalus platyurus* and *Raphidocelis subcapitata*. The values are expressed as a percentage of sample in the final dilution. * TU: toxic units, 100/EC50. ** R.D. 9/2005 (Annex III); <1% C: contaminated; >1% NC: not contaminated. *** [11] Hazard classification system (HC); <0.4: Class I, no acute toxicity; 0.4 < TU < 1: Class II, slight acute toxicity; 1 < TU < 10: Class III, acute toxicity; 10 < TU < 100: Class IV, high acute toxicity; TU > 100: Class V; very high acute toxicity. ND: undetermined, HC between I–III. Table S6: EC50 and EC10 values calculated and their respective confidence interval (95%) for *Daphnia magna, Thamnocephalus platyurus* and *Raphidocelis subcapitata*. The values are expressed in percentage of dilution used in the bioassay. * TU: toxic units, 100/EC50. ** R.D. 9/2005 (Annex III); <1% C: contaminated; >1% NC: not contaminated. *** Persoone et al., 2003 Hazard classification system (HC); <0.4: Class I, no acute toxicity; 0.4 < TU < 1: Class II, slight acute toxicity; 1 < TU < 10: Class III, acute toxicity; 10 < TU < 100: Class IV, high acute toxicity; TU > 100: Class V; very high acute toxicity.

Author Contributions: Conceptualization M.L.G.-L., J.M.E., O.A.-S., I.F.-M.; Data curation M.L.G.-L., J.M.E., O.A.-S., I.F.-M.; Formal analysis M.L.G.-L., J.M.E., O.A.-S., I.F.-M.; Funding acquisition M.L.G.-L., J.M.E.; Investigation M.L.G.-L., J.M.E., O.A.-S., I.F.-M., J.I.B.-P., P.H.; Methodology M.L.G.-L., J.M.E., O.A.-S., I.F.-M., L.R.-P., J.I.B.-P., P.H., L.R.-P.; Project administration M.L.G.-L., J.M.E.; Resources M.L.G.-L., J.M.E.; Software I.F.-M., O.A.-S.; Supervision M.L.G.-L., J.M.E., O.A.-S.; Validation M.L.G.-L., J.M.E., O.A.-S., I.F.-M.; Visualization I.F.-M.; Roles/Writing—original draft M.L.G.-L., O.A.-S., I.F.-M., J.M.E.; Writing—review & editing M.L.G.-L., J.M.E., O.A.-S., I.F.-M. All authors have read and agreed to the published version of the manuscript.

Funding: This has been funded by the Spanish "Ministerio de Ciencia e Innovación" project, TED2021-130498B-I00.

Data Availability Statement: Data will be made available by request.

Conflicts of Interest: The authors declare no conflict of interest.

References

1. Pan, L.; Fang, G.; Wang, Y.; Wang, L.; Su, B.; Li, D.; Xiang, B. Potentially Toxic Element Pollution Levels and Risk Assessment of Soils and Sediments in the Upstream River, Miyun Reservoir, China. *Int. J. Environ. Res. Public Health* **2018**, *15*, 2364. [CrossRef] [PubMed]
2. Bird, G. The influence of the scale of mining activity and mine site remediation on the contamination legacy of historical metal mining activity. *Environ. Sci. Pollut. Res.* **2016**, *23*, 23456–23466. [CrossRef] [PubMed]
3. Martín Crespo, T.; Gómez Ortiz, D.; Esbrí, J.M.; Monescillo, C.I.; García-Noguero, E.V. Caracterización geoquímica de la balsa de lodos de la mina de San Quintín, Ciudad Real. *Geogaceta* **2009**, *46*, 143–146.
4. García-Lorenzo, M.L.; Crespo-Feo, E.; Esbrí, J.M.; Higueras, P.; Grau, P.; Crespo, I.; Sánchez-Donoso, R. Assessment of Potentially Toxic Elements in Technosols by Tailings Derived from Pb–Zn–Ag Mining Activities at San Quintín (Ciudad Real, Spain): Some Insights into the Importance of Integral Studies to Evaluate Metal Contamination Pollution Hazards. *Minerals* **2019**, *9*, 346. [CrossRef]
5. Martín-Crespo, T.; Gómez-Ortiz, D.; Martín-Velázquez, S.; Ebrí, J.M.; de Ignacio-San José, C.; Sánchez-García, M.J.; Montoya-Montes, I.; Martín-González, F. Abandoned mine tailings in cultural itineraries: Don Quixote Route (Spain). *Eng. Geol.* **2015**, *197*, 82–93. [CrossRef]
6. Rodríguez, L.; Ruiz, E.; Alonso-Azcárate, J.; Rincón, J. Heavy metal distribution and chemical speciation in tailings and soils around Pb-Zn mine in Spain. *J. Environ. Manag.* **2009**, *90*, 1106–1116. [CrossRef]
7. Hussain, F.; Ashun, E.; Jung, S.P.; Kim, T.; Lee, S.-H.; Kim, D.-J.; Oh, S.-E. A direct contact bioassay using immobilized microalgal balls to evaluate the toxicity of contaminated field soils. *J. Environ. Manag.* **2022**, *321*, 115930. [CrossRef]
8. Ferreira, S.L.C.; da Silva, J.B.; Ferreira dos Santos, I.; de Oliveira, O.M.C.; Cerda, V.; Queiroz, A.F.S. Use of pollution indices and ecological risk in the assessment of contamination from chemical elements in soils and sediments—Practical aspects. *Trends Environ. Anal. Chem.* **2022**, *35*, e00169. [CrossRef]

9. Andreu-Sánchez, O.; García-Lorenzo, M.L.; Esbrí, J.M.; Sánchez-Donoso, R.; Iglesias-Martínez, M.; Arroyo, X.; Crespo Feo, E.; Ruiz Costa, N.; Roca Pérez, L.; Castiñeiras, P. Soil and freshwater bioassays to assess ecotoxicological impact in soils affected by mining activities in the Iberian Pyrite Belt. *Toxics* **2022**, *10*, 353. [CrossRef]
10. Palma, P.; Penha, A.M.; Novais, M.H.; Fialho, S.; Lima, A.; Catarino, A.; Mourinha, C.; Alvarenga, P.; Iakunin, M.; Rodrigues, G.; et al. Integrative toolbox to assess the quality of freshwater sediments contaminated with potentially toxic metals. *Environ. Res.* **2023**, *217*, 114798. [CrossRef]
11. Persoone, G.; Marsalek, B.; Blinova, I.; Törökne, A.; Zarina, D.; Manusadzianas, L.; Nalecz-Jawecki, G.; Tofan, L.; Stepanova, N.; Tothova, L.; et al. A Practical and User-Friendly Toxicity Classification System with Microbiotests for Natural Waters and Wastewaters. *Environ. Toxicol.* **2003**, *18*, 395–402. [CrossRef] [PubMed]
12. Liwarska-Bizukojc, E. Evaluation of Ecotoxicity of Wastewater from the Full-Scale Treatment Plants. *Water* **2022**, *14*, 3345. [CrossRef]
13. Sánchez Donoso, R.; Martín-Duque, J.F.; Crespo, E.; Higueras, P.L. Tailing's geomorphology of the San Quintin mining site (Spain): Landform catalogue, aeolian erosion and environmental implications. *Environ. Earth Sci.* **2019**, *78*, 166. [CrossRef]
14. ASTM. Section four Construction Volume 04.08 and 04.09, Soil and Rock(I). D 2974, D 4972. In *Annual Book of ASTM STANDARDS*; ASTM International: Conshohocken, PA, USA, 2004.
15. US EPA. *Clean Water Act. Section 503. U.S.*; Environmental Protection Agency: Washington, DC, USA, 1993; Volume 58.
16. Chung, F.H. Quantitative interpretation of X-ray diffraction patterns. I. Matrix-flushing method of quantitative multicomponent analysis. *J. Appl. Cryst.* **1974**, *7*, 519–525. [CrossRef]
17. Chung, F.H. Quantitative interpretation of X-ray diffraction patterns. II. Adiabatic principle of X-ray diffraction analysis of mixtures. *J. Appl. Cryst.* **1974**, *7*, 526–531. [CrossRef]
18. Chung, F.H. Quantitative interpretation of X-ray diffraction patterns. III. Simultaneous determination of a set of reference intensities. *J. Appl. Cryst.* **1975**, *8*, 17–19. [CrossRef]
19. Deutsches Institut für Normung (DIN). *38414-S4: Determination of Leachability by Water (S4)*; DIN: Belin, Germany, 1984.
20. Organisation for Economic Cooperation and Development (OECD). *Test No. 202: Daphnia sp. Acute Immobilisation Test*; OECD: Paris, France, 2004. [CrossRef]
21. DaphtoxkitF™ Standard Operating Procedure. 2000. Available online: https://www.microbiotests.com/wp-content/uploads/2019/07/daphnia-toxicity-test_daphtoxkit-f_standard-operating-procedure.pdf (accessed on 26 December 2022).
22. ThamnotoxkitF™. Standard Operating Procedure. Crustacean Toxicity Screening Test for Freshwater, 1995. Available online: https://www.microbiotests.com/wp-content/uploads/2019/05/folder-Thamnotoxkit-F-Thamnocephalus-toxicity-test.pdf (accessed on 26 December 2022).
23. Davoren, M.; Fogarty, A.M. A test battery for the ecotoxicological evaluation of the agri-chemical Environ. *Ecotoxicol. Environ. Saf.* **2004**, *59*, 116–122. [CrossRef]
24. *ISO 14380:2011*; Water Quality-Determination of the Acute Toxicity to Thamnocephalus Platyurus (Crustacea, Anostraca). International Organization for Standardization: Geneva, Switzerland, 2011.
25. Organisation for Economic Cooperation and Development (OECD). *Test No. 201: Freshwater Alga and Cyanobacteria, Growth Inhibition Test*; OECD: Paris, France, 2011. [CrossRef]
26. Algaltoxkit F™ Standard Operating Procedure. 2015. Available online: https://www.microbiotests.com/wp-content/uploads/2019/07/freshwater-algae-toxicity-test_algaltoxkit-f_standard-operating-procedure.pdf (accessed on 26 December 2022).
27. Muller, G. Index of Geoaccumulation in Sediments of the Rhine River. *Geojournal* **1969**, *2*, 108–118.
28. Gallego, S.; Esbrí, J.M.; Campos, J.A.; Peco, J.D.; Martin-Laurent, F.; Higueras, P. Microbial diversity and activity assessment in a 100-year-old lead mine. *J. Hazard. Mater.* **2021**, *15*, 410. [CrossRef]
29. Qing, X.; Yutong, Z.; Shenggao, L. Assessment of heavy metal pollution and human health risk in urban soils of steel industrial city (Anshan), Liaoning, Northeast China. *Ecotoxicol. Environ. Saf.* **2015**, *120*, 377–385. [CrossRef]
30. Jorfi, S.; Maleki, R.; Jaafarzadeh, N.; Ahmadi, M. Pollution load index for heavy metals in Mian-Ab plain soil Khuzestan, Iran. *Data Brief* **2017**, *15*, 584–590. [CrossRef]
31. Statgraphics 19 Software. Available online: https://www.statgraphics.com/ (accessed on 26 December 2022).
32. Iwasaki, Y.; Kotani, K.; Kashiwada, S.; Masunaga, S. Does the Choice of NOEC or EC10 Affect the Hazardous Concentration for 5% of the Species? *Environ. Sci. Technol.* **2015**, *49*, 9326–9330. [CrossRef] [PubMed]
33. Finney, D.J. *Probit Analysis*, 2nd ed.; Cambridge University Press: Cambridge, UK, 1971.
34. Gómez Ortiz, D.; Martín Crespo, T.; Esbrí, J.M. Geoenvironmental characterization of the San Quintín mine tailings, Ciudad Real (Spain). *Dyna* **2010**, *77*, 131–140.
35. Alveranga, P.; Palma, P.; de Varennes, A.; Cunha-Queda, A.C. A contribution towards the risk assessment of soils from São Domingos Mine (Portugal): Chemical, microbial and ecotoxicity indicators. *Environ. Pollut.* **2012**, *161*, 50–56. [CrossRef] [PubMed]
36. Loureiro, S.; Ferreira, A.; Soares, A.; Nogueira, A. Evaluation of the toxicity of two soils from Jales Mine (Portugal) using aquatic bioassays. *Chemosphere* **2005**, *61*, 168–177. [CrossRef] [PubMed]
37. Rodriguez, L.; González-Corrochano, B.; Medina-Díaz, H.; López-Bellido, F.; Fernández-Morales, F.J.; Alonzo-Azcárate, J. Does environmental risk really change in abandoned mining areas in the medium term when no control measures are taken? *Chemosphere* **2022**, *391*, 133129. [CrossRef]

38. Sharma, S.S.; Schat, H.; Vooijs, R.; van Heerwaadens, L.M. Combination toxicology of copper, zinc, and cadmium I binary mixtures: Concentration dependent antagonistic, nonadditive, and synergistic effects on root growth in *Silene vulgris*. *Environ. Toxicol. Chem.* **1999**, *18*, 348–353. [CrossRef]
39. Mensah, A.K.; Marschner, B.; Antoniadis, V.; Stemn, E.; Shaheen, S.M.; Rinklebe, J. Human health risk via soil ingestion of potentially toxic elements and remediation potential of native plants near an abandoned mine spoil in Ghana. *Sci. Total Environ.* **2021**, *798*, 149272. [CrossRef]
40. Santos, M.V.S.; da Silva Júnior, J.B.; de Carvalho, C.E.V.; dos Santos Vergílio, C.; Hadlich, G.M.; de Santana, C.O.; de Jesus, T.B. Geochemical evaluation of potentially toxic elements determined in surface sediment collected in an area under the influence of gold mining. *Mar. Pollut. Bull.* **2020**, *158*, 111384. [CrossRef]
41. Ali, A.; Strezov, V.; Davies, P.J.; Wright, I. River sediment quality assessment using sediment quality indices for the Sydney basin, Australia affected by coal and coal seam gas mining. *Sci. Total Environ.* **2018**, *616*, 695–702. [CrossRef]
42. Masto, R.E.; George, J.; Rout, T.K.; Ram, L.C. Multi element exposure risk from soil and dust in a coal industrial area. *J. Geochem. Explor.* **2017**, *176*, 100–107. [CrossRef]
43. Varga, N.; Krcmar, D.; Dalmacija, B.; Gvozdenac, S.; Tricković, J.; Roncevic, S.; Prica, M. Assessment of sediment pollution using chemical and biological trait approach. *Carpathian J. Earth Environ. Sci.* **2018**, *13*, 359–368. [CrossRef]
44. Peco, J.D.; Higueras, P.; Campos, J.A.; Olmedilla, A.; Romero-Puertas, M.C.; Sandalio, L.M. Deciphering lead tolerance mechanisms in a population of the plant species Biscutella auriculata L. from a mining area: Accumulation strategies and antioxidant defenses. *Chemosphere* **2020**, *261*, 127721. [CrossRef]
45. Peco, J.D.; Campos, J.A.; Romero-Puertas, M.C.; Olmedilla, A.; Higueras, P.; Sandalio, L.M. Characterization of mechanisms involved in tolerance and accumulation of Cd in *Biscutella auriculata* L. *Ecotoxicol. Environ. Saf.* **2020**, *201*, 110784. [CrossRef] [PubMed]

minerals

MDPI

Review

Polymer/Clay Nanocomposites as Advanced Adsorbents for Textile Wastewater Treatment

Adel Mokhtar [1,2,*], Boubekeur Asli [2], Soumia Abdelkrim [1,3], Mohammed Hachemaoui [1,4], Bouhadjar Boukoussa [1,5], Mohammed Sassi [1], Gianluca Viscusi [6] and Mohamed Abboud [7]

1 Laboratoire de Chimie des Matériaux L.C.M, Université Oran1 Ahmed Ben Bella, BP 1524, El Mnaouer, Oran 31000, Algeria; abdelkrim_soumia@yahoo.fr (S.A.); mohamedb.f@live.fr (M.H.); bbouhdjer@yahoo.fr (B.B.); sassim2006@yahoo.fr (M.S.)

2 Département Génie des Procédés, Faculté des Sciences et Technologies, Université de Relizane, Relizane 48000, Algeria; bbk.asli@gmail.com

3 Institut des Sciences Techniques et Appliquées (ISTA), Université Oran1 Ahmed Ben Bella, BP 1524, El-Mnaouer, Oran 31000, Algeria

4 Département de Chimie, Faculté des Sciences et Technologies, Université de Relizane, Relizane 48000, Algeria

5 Département de Génie des Matériaux, Faculté de Chimie, Université des Sciences et de la Technologie Mohamed Boudiaf, BP 1505, El-Mnaouer, Oran 31000, Algeria

6 Department of Industrial Engineering, University of Salerno, Via Giovanni Paolo II, 132, 84084 Fisciano, SA, Italy; gviscusi@unisa.it

7 Catalysis Research Group (CRG), Department of Chemistry, College of Science, King Khalid University, P.O. Box 9004, Abha 61413, Saudi Arabia; abboud_med@yahoo.fr

* Correspondence: mokhtar.adel80@gmail.com or adel.mokhtar@univ-relizane.dz

Abstract: This review explores the removal of textile dyes from wastewater using advanced polymer/clay composites. It provides an in-depth analysis of the chemical and physical properties of these composites, emphasizing how the combination of polymers and clays creates a synergistic effect that significantly improves the efficiency of dye removal. The structural versatility of the composites, derived from the interaction between the layered clay sheets and the flexible polymer matrices, is detailed, showcasing their enhanced adsorption capacity and catalytic properties for wastewater treatment. The review outlines the key functional groups present in both polymers and clays, which are crucial for binding and degrading a wide range of dyes, including acidic, basic, and reactive dyes. The role of specific interactions, such as hydrogen bonding, ion exchange, and electrostatic attractions between the dye molecules and the composite surface, is highlighted. Moreover, the selection criteria for different types of clays such as montmorillonite, kaolinite, and bentonite and their modifications are examined to demonstrate how structural and surface modifications can further improve their performance in composite materials. Various synthesis methods for creating polymer/clay composites, including in situ polymerization, solution intercalation, and melt blending, are discussed. These fabrication techniques are evaluated for their ability to control particle dispersion, optimize interfacial bonding, and enhance the mechanical and chemical stability of the composites. Furthermore, the review introduces advanced characterization techniques, such as X-ray diffraction (XRD), scanning electron microscopy (SEM), Fourier-transform infrared spectroscopy (FTIR), and thermogravimetric analysis (TGA), to help researchers assess the morphological, structural, and thermal properties of the composites, aligning these features with their potential application in dye removal. Additionally, the review delves into the primary mechanisms involved in the dye removal process, such as adsorption, photocatalytic degradation, and catalytic reduction. It also provides an overview of the kinetic and thermodynamic models commonly used to describe the adsorption processes in polymer/clay composites. The environmental and operational factors influencing the efficiency of dye removal, such as pH, temperature, and composite dosage, are analyzed in detail, offering practical insights for optimizing performance under various wastewater conditions. In conclusion, this review not only highlights the promising potential of polymer/clay composites for textile dye removal but also identifies current challenges and future research directions. It underscores the importance of developing eco-friendly, cost-effective, and scalable solutions to address the growing concerns related to water pollution and sustainability in wastewater management.

Keywords: clay; polymer; textile; composite; pollutant; adsorption; catalyst; wastewater

1. Introduction

Artificial dyes have long captivated attention due to their affordability, aesthetic appeal, and the wide spectrum of colors they offer [1]. Overextension of their use in several areas including textiles, medicine, food and drink, paper, and leather industries to provide color to substrates has a detrimental effect on the environment. Currently, there are more than 10,000 dyes with various chemical structures [2,3], based on functional groups such as the chromophoric group (—NR_2, —NHR, —NH_2, —COOH, and —OH) and auxochromes (—N_2, —NO, and —NO_2) [3,4]. This variety in the functional groups is one of the challenges that researchers face in their elimination. Therefore, several scientific researchers have adapted composite adsorbents and catalysts for the adsorption or the reduction of these toxic substances from wastewater [5–8]. Figure 1 displays a schematic diagram of the preparation steps, chemical properties, and various environmental application fields of polymer/clay composite.

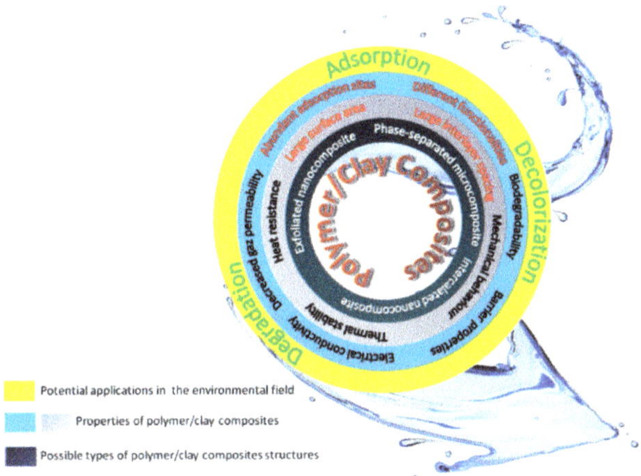

Figure 1. Schematic sketch showing the preparation steps, properties, characteristics, and environmental application fields of polymer/clay composite.

Several materials for water treatment have been developed based on grapheme, diatomite, zeolites, hydrated layered polysilicates, mesoporous materials, or double-layered hydroxides. Consequently, the resultant composites generate a synergistic effect, having the possibility for elimination of a variety of organic pollutants. Unfortunately, these materials have some drawbacks either in their availability, preparation, stability, cost, biodegradability, or toxicity; for example, graphene and its modified forms have shown effectiveness in capturing several dye types from water, but they are very expensive to prepare. Exploitation of clays in organic dye removal could bring great economic and environmental profit to wastewater industries.

Clay minerals are relatively abundant in nature, have low toxicity, are inexpensive, eco-friendly, and have good physical and chemical stability [9,10]. This class of solids has been widely used for medicinal and environmental applications throughout human history due to their adsorption and fixing effect on pollutants, micro-pollutants, and other harmful substances [11–17]. They have been used for the elimination of bacteria as well as organic and inorganic substances due to their high cation exchange capacity, high surface area, high swelling capacity, high water dispersibility, stability, and non-toxic properties [18–20].

Their negative charge allows them to exchange with positively charged substances on their surfaces, which can provide a variety of forms of hybrid materials. Nevertheless, using natural clays alone is limited by their low specific surface area, their difficult regeneration and recovery protocol in aqueous systems, their limited potential for micro-pollutant removal, and the formation of aggregates in aqueous media. On the other hand, their combination with polymers can overcome these drawbacks.

To overcome the drawbacks mentioned above, composites, nanocomposites, or even resins have been developed based on polymers and clays. These resulting materials have attracted more and more attention lately for wastewater treatment [21]. The elaboration of polymers and inorganic clay minerals has resulted in a class of new materials of a micro-composite and nanocomposite nature (depending on the modification procedure used), the clay mineral being well dispersed in the polymer matrix, adsorbed on the surface, immobilized, or interposed. The resulting composites show dramatic improvements in several important characteristic properties including physical–chemical properties, mechanical strength, heat resistance, and reduced gas permeability. Consequently, clay–polymer nanocomposites are exceedingly promising, not only because they are inexpensive, but also because they are very effective in removing different types of organic pollutants from aqueous solutions. In addition, some of these clay–polymer hybrids are durable materials as they can be made from abundant resources such as biopolymers [8,22].

2. The Structure of Clays

There are many types of clay minerals, each of them having unique and attractive chemical and physical properties that arise from the structure (see Figure 2). Almost all clays contain only two basic components, which occur in different arrangements; these two elements are the silica tetrahedron and the aluminum octahedron. They are classified into two types, 1:1 and 2:1. Class 1:1 clay consists essentially of a tetrahedral silica sheet stacked on an octahedral gibbsite or brucite sheet. For example, kaolinite consists essentially of tetrahedral silica and an octahedral gibbsite sheet. For class 2:1, the best-known example is montmorillonite, which consists of two sheets of tetrahedral silica and one sheet of octahedral gibbsite. Taking the case of montmorillonite, this structure has been widely used in several fields of application in adsorption, ion exchange, and in catalysis to a much greater extent. Montmorillonite has been classified as an important clay mineral due to its industrial applications; its chemical composition is $(Na,Ca)_{0.3}(Al,Mg)_2Si_4O_{10}(OH)_2$ nH_2O. The octahedral aluminum layer centered between two tetrahedral sheets of silicon forms a crystalline structure. Montmorillonite is characterized by three layers interconnected via oxygen bridges. It is possible to have isomorphic substitution of octahedral aluminum (III) ions by metal ions such as iron (II) and magnesium (II). Thus, it is possible to have an isomorphic substitution between silicon (IV) and aluminum ions (III). One of the characteristics of this structure is that the substitution generates the formation of a negatively charged surface, which can be balanced by different compensating cations (like Na^+ and Ca^{2+}). As a result of this characteristic, montmorillonite exhibits a significant cation exchange capacity and excellent swelling properties. It is also possible to immobilize polar molecules or cationic surfactants inside the galleries, which generates new surface properties. This variety of clay structure has made it possible to produce a variety of composites by combining them with polymers with the objective of generating novel functional hybrid materials with various properties for the treatment of textile effluents [23]. The elaboration of aluminosilicate clay 2:1 stratified structure with acrylamide and acrylic acid through polymerization, hydrogen bonds, amination, and electrostatic interactions to form the three-dimensional reticular-structured hydrogel for methylene blue removal [24] allowed an efficient removal of 97% at a high concentration of 200 mg/L using a small dose of hydrogel of about 0.5 g/L within a short time of 20 min.

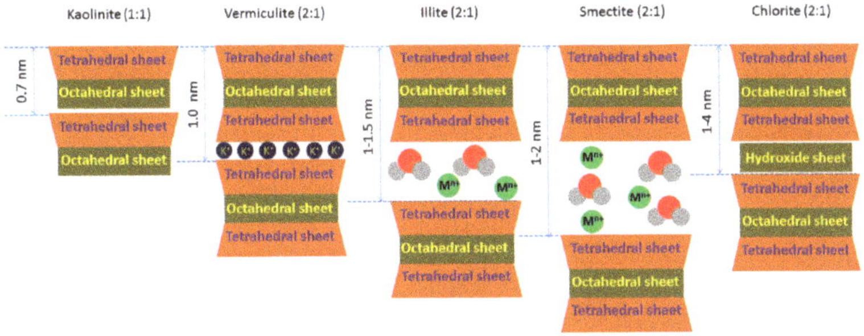

Figure 2. Schematic representation of the structure of the clays.

3. The Most Common Clays in Wastewater Treatment

Natural and synthetic clays have long been employed in wastewater treatment due to their distinct physicochemical properties, which make them versatile for a variety of environmental applications [25]. Their widespread use, particularly in adsorption and catalytic reduction processes, has been driven by their high surface area, layer charge, and their ability to modify their structure to enhance functionality. In recent years, advancements in nanotechnology have led to the modification of clay minerals into low-dimensional materials such as nanosheets and nanotubes, increasing their utility in developing multifunctional composites for more efficient wastewater treatment. Structurally, clays exhibit a range of configurations, including 2:1 layered (such as montmorillonite), 1:1 layered (such as kaolinite), elongated discontinuous octahedral sheets (such as in halloysite), and multilayered nanotubular formations. These structural characteristics significantly influence their surface properties, reactivity, and interaction with pollutants. For instance, the 2:1 layered clays, where two tetrahedral sheets sandwich an octahedral sheet, offer a high degree of flexibility in terms of swelling and ion exchange capacity. This flexibility allows them to effectively trap pollutants between their layers, making them highly suitable for adsorption processes. Figure 3 illustrates the six classes of clay minerals most commonly used in environmental applications, each with distinct structural and surface chemistry properties. Montmorillonite, a smectite mineral, is particularly favored in wastewater treatment due to its high cation exchange capacity (CEC) and large interlayer space, which facilitate the adsorption of various contaminants, including heavy metals, dyes, and organic molecules. The presence of hydroxyl groups on the surface and edges of montmorillonite enhances its reactivity, allowing for both physical and chemical interactions with pollutants. These clay minerals are favored not only for their abundance but also for their ease of functionalization. By incorporating polymers, metal oxides, or other nanomaterials, clay minerals can be transformed into advanced composites that exhibit superior adsorption capacity, enhanced mechanical stability, and selective pollutant targeting. For example, the formation of polymer–clay nanocomposites allows the introduction of functional groups capable of interacting with specific contaminants, such as acidic or basic dyes, through electrostatic interactions, hydrogen bonding, or complexation mechanisms. Furthermore, the high surface area of these materials provides numerous active sites for adsorption, while the layered structure enables intercalation of pollutants into the clay galleries. This makes clays particularly effective for removing inorganic contaminants like heavy metals as well as organic pollutants such as dyes and pharmaceutical residues from wastewater. Their tunable properties and ability to undergo various modifications make them versatile tools for tackling a wide range of environmental pollutants.

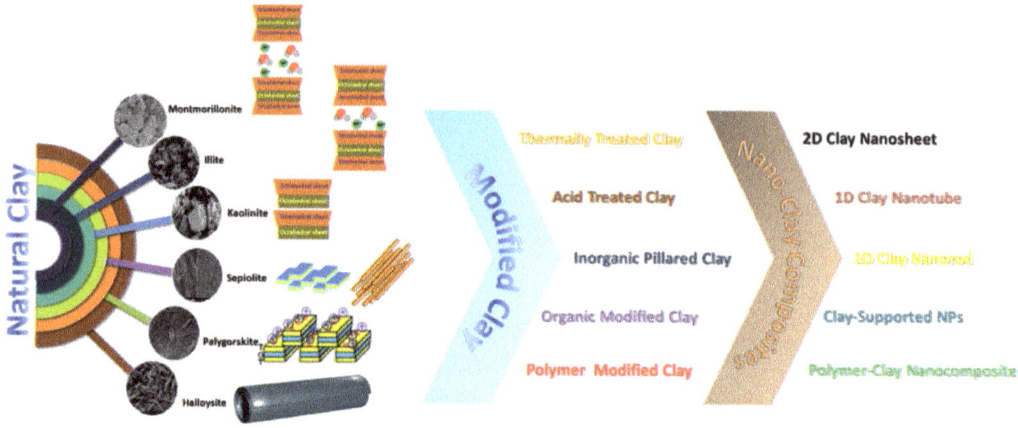

Figure 3. Schematic diagram of the modification of the most common clays used in wastewater treatment.

4. The Most Common Polymers in Wastewater Treatment

A variety of porous and layered materials, such as activated carbon, natural clays, zeolites, and metal–organic frameworks (MOFs), are commonly employed for wastewater treatment due to their low cost, availability, ease of use, and environmentally friendly nature. These materials exhibit high efficiency in removing pollutants like dyes, heavy metals, and organic contaminants from wastewater. However, despite their widespread use, they also possess several limitations, including difficult recovery from aqueous solutions, poor selectivity toward certain dyes, low reusability, and a tendency to aggregate in water. These drawbacks hinder their broader application in practical wastewater treatment processes. To overcome these challenges, the integration of these materials into polymer or biopolymer matrices has emerged as a promising strategy. Polymers provide structural support, enhance dispersibility in aqueous media, and increase the reusability of adsorbent materials. This association not only prevents aggregation but also improves the selectivity and stability of the treatment materials in diverse wastewater environments. Figure 4 presents the most frequently studied natural and synthetic polymers for wastewater treatment [1]. Among the most commonly researched polymers are chitosan [26], polyacrylamide [27], and polyvinyl alcohol (PVA) [28,29], which account for 23%, 11%, and 10% of studies, respectively. These polymers are chosen based on their solubility, biocompatibility, and the variety of functional groups they possess, which play a crucial role in the adsorption of pollutants. Chitosan, for instance, is widely used due to its solubility in acidic solutions (low pH), its biocompatibility as a natural biopolymer, and the presence of reactive amino and hydroxyl groups. These functional groups allow chitosan to form hydrogen bonds or electrostatic interactions with a wide range of contaminants, enhancing its adsorption capacity.

The diversity of functional groups found in polymers such as hydroxyl, carboxylic, amine, aromatic, aliphatic, and amide groups further contributes to their effectiveness in wastewater depollution. For instance, carboxylic and amine groups can bind to heavy metal ions through ion exchange mechanisms, while hydroxyl and amide groups can form hydrogen bonds with organic pollutants like dyes. Polymers such as polyacrylamide and PVA, with their versatile chemical structures, are ideal for forming composite materials with enhanced adsorption properties [30]. These composites not only increase the adsorption surface area but also improve the interaction between the adsorbent and the pollutants, resulting in higher removal efficiencies. In addition to their adsorption capacity, polymers offer the advantage of tunable properties. By functionalizing polymer matrices with specific chemical groups, it is possible to tailor their selectivity toward targeted pollutants. This makes polymer-based materials particularly effective in addressing a broad spectrum of contaminants, ranging from organic dyes to inorganic heavy metals. Moreover, the reusability of polymer-based composites is significantly improved due to their structural stability and resistance to chemical degradation, allowing for multiple cycles of wastewater treatment with minimal loss of efficiency.

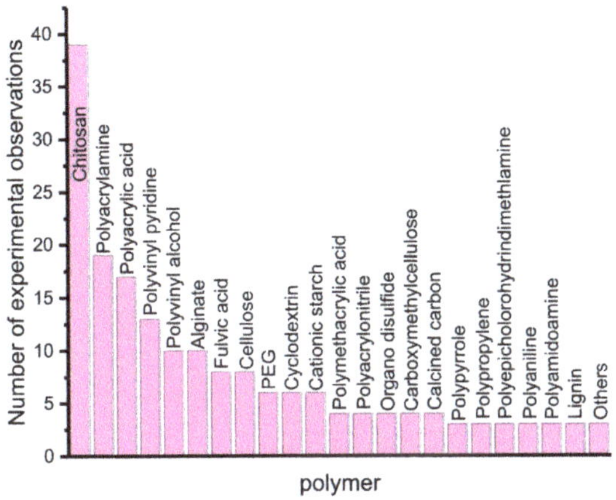

Figure 4. The most frequently used polymers in water treatment.

Overall, the incorporation of polymers into wastewater treatment materials addresses the limitations of traditional adsorbents, offering a more robust, selective, and sustainable solution for water purification.

5. Polymer/Clay Composites

5.1. Advanced Modification of Clay Particles

Due of their hydrophilic character and the covalent connections that form between their interlayers, clays are not suited for reacting or dispersing with most polymer matrices [26,27]. Moreover, the clay layer stacks are tightly held together by electrostatic forces. To address this difficulty, modified clay particles are first distributed in polymer matrices. In essence, surfactant intercalation increases the spacing between the clay layered silicate sheets throughout the modification process. By making such modifications, it is possible to give clay minerals a hydrophobic quality that will enable the modified clay to disperse extremely finely in polymer matrices. There are two main methods for altering the structure of clay: (i) physical modification and (ii) chemical modification. The physical modification procedure involves only the adsorption of modifying chemicals on the surface of the clay. Although the clay particles are not altered by this method, the resultant polymer composites do benefit slightly from it. Because clay and the modifying chemicals have weak van der Waals forces, there may be a modest change in properties. On the other hand, the stacked silicates and the modifier initiate chemical reactions with the help of chemical modification. Moreover, the molecules are altered during the ion exchange process by the addition of cationic or anionic functional groups. Consequently, the clay's capacity to disperse in the polymer matrix is enhanced. The increased surface area and higher aspect ratio of clay particles when dispersed in polymeric composites, along with the nano-thick layers present in the clay material, are responsible for the enhanced mechanical, thermal, optical, and barrier properties of polymer/nanoclay composite systems [28–33].

5.2. Fabrication of Polymer/Clay Composites

Generally, the direct use of clays as a material for removing organic dyes is limited due to many inconvenient factors such as smaller surface area, low adsorption tendency to organic species, difficulty in recovering clay particles from aqueous solutions, and their reduction after regeneration. In order to overcome these limitations, researchers have developed clay–polymer nanocomposites [34–37]. Clays have a large field of application due to their attractive properties. A small amount of these substances can change the overall physical, chemical, and mechanical properties of polymers. Materials of type conventional composites or nanocomposites can be fabricated relying on the nature of the constituents, the treatment condition, and the resistance of the interface interactions between the polymer and clay layer surfaces (Figure 5). For the fabrication of conventional composites, the clay particles with their original aggregated state block the diffusion of the polymer between the interlayer spaces of the clay. These conventional composites generate properties similar to those of micro-composites with an advantageous improvement in modulus, consequently resulting in a lack of other properties such as resistance or toughness. If the clay is intercalated and exfoliated, the layers of the latter retain the multilayered structure of polymer and clay alternating in intercalated nanocomposites, whereas in exfoliated nanocomposites, the clay layers are well separated and randomly distributed in the continuous polymer matrix [37–40].

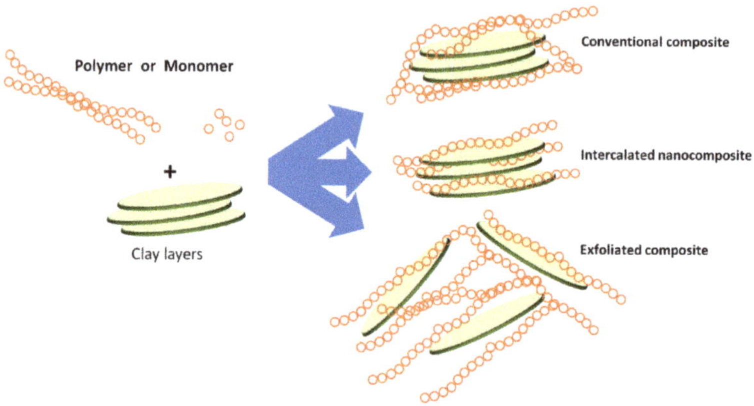

Figure 5. Synthesis of different types of polymer/clay nanocomposites.

In order to improve the chemical, physical, and mechanical properties, various approaches are available for the polymer/clay nanocomposite synthesis, namely solution-blending, melt-blending, and in situ polymerization [41–44]. As a result, the nanocomposite involves uniform dispersion of clay, abundant functional groups, good compatibility, and thermal stability. Figure 6 shows a simple diagram which gives the path of the three main methods of incorporating clay into the polymer matrix [43,45]. Concerning the solution-blending procedure, the polymer is dissolved in a suitable solvent with the dispersing of the clay in the same solvent if possible or a different one in the aim to obtain a miscible mixture. The polymer solution is then intercalated or incorporated between the clay dispersion. In situ polymerization targets swollen clay layer sheets using a monomer solution where polymerization starts and propagates [45–47]. The polymerization initiating agent can be heat or a suitable chemical. This increase in the molecular weight between layers increases the basal spacing d001, sometimes resulting in an almost fully exfoliated morphology. In order obtain a thorough dispersion of clay, the melt-blending process is used. Clay is simultaneously incorporated into the annealed polymer matrix in a melt mixing device, and the resulting composite is kneaded to obtain an even distribution [45,48,49]. Organic modification by using polymer can create active functional groups and reform the surface charge on clays, raising the ability to eliminate the organics pollutants [25,31,50,51].

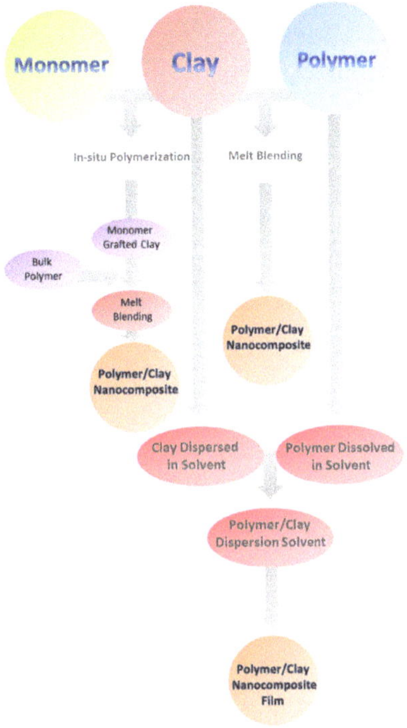

Figure 6. Main techniques for polymer/clay nanocomposite preparation.

Polymer composites based on clay, which integrate the special qualities of clay materials into the polymer matrix, improve overall performance. These composites exhibit enhanced mechanical strength, thermal stability, flame retardancy, and barrier qualities by using the clay's vast surface area, high aspect ratio, and ion exchange capacity. Many polymer composites based on clay have been reported to date; Table 1 provides a complete list of the remaining noteworthy findings, and a review of these important studies is given below.

In order to improve dye adsorption, a study by Akbari and colleagues concentrated on surface-functionalizing halloysite nanotubes (HNTs) with carboxyl and amino-dendritic groups [67]. The excellent synthesis and characterization of functionalized HNTs demonstrated the efficient elimination of the dyes methylene blue and acid red 1. Electrostatic interactions and host–guest encapsulation were highlighted during the exploration of the adsorption mechanisms. The investigation also looked at the modified nanoclay's stability and thermodynamic and kinetic behaviors, providing information about effective dye pollutant adsorption utilizing functionalized clay minerals.

Nambiar et al., in 2022, used solution casting to create composite films of poly(vinyl alcohol) (PVA) and halloysite nanotubes (HNTs) [68], with glutaraldehyde acting as a biodegradable crosslinker to improve the adsorption of methylene blue dye. After the films were characterized, the ideal adsorption conditions were found, and 555 mg/g of monolayer adsorption was attained. Thermodynamic investigation revealed that the adsorption process was endothermic and favorable, and it followed the pseudo-second-order kinetic model and the Langmuir isotherm. The composite proved to be successful in treating wastewater since it was highly recyclable and maintained its adsorption efficacy for at least six cycles.

Table 1. Preparation methods of polymer/clay composites and targeted dyes.

Adsorbents	Preparation Method	Dye	References
Poly(AA-co-AM)/MMT	Radical polymerization	Methylene blue	[52]
Polymethylmethacrylate/organo-MMT nanocomposites (PMMA/OMT nanocomposite)	Exfoliated structure by in situ polymerization	Azo dyes	[42]
CMC-graft-poly(methacrylic acid-co-acrylamide)/kaolin	Free-radical polymerization	Nile blue	[53]
Polymer–clay hybrid composite (KG-g-PMETAC/MMT)	Microwave-assisted; free-radical polymerization	Methylene blue, toluidine blue, crystal violet, and azure B	[54]
Kaolin/chitosan beads	Sol–gel method	Fuchsine and methylene blue	[55]
Carboxymethyl cellulose-g-Poly(AM)/MMT	Free-radical polymerization method; ultra-sonication	Malachite green	[56]
PVA/CMC/halloysite nanoclay membrane	Crosslinking	Methylene blue	[57]
Poly(AA-co-AMPSA)/MMT	Simple free-radical copolymerization	Methylene blue	[58]
poly(MAA-co-AAm)/Bentonite nanocomposite hydrogels	Free-radical polymerization	Methyl violet	[59]
Crosslinked chitosan-epichlorohydrin/kaolin (Chi-ECH/KA) composite	Incorporating kaolin into chitosan and crosslinking with epichlorohydrin	Reactive blue 19	[60]
Sodium alginate/halloysite/hemp hurd	Ionotropic gelation	Methylene blue	[61]
Na-MMT/PNIPAm-Na-AMPS	Crosslinking polymerization	Methylene blue	[62]
Lignin-g-p(AM-co-NIPAM)/MMT	Graft polymerization; ultra-sonication	Methylene blue	[63]
CMC-g-poly (AM-co-ITA)/Clay nanocomposite	Graft free-radical co-polymerization	Brilliant Green	[64]
MMT@HTAB@PEG	Intercalated with HTAB surfactant, then incorporated into PEG through sonication	Trypan blue and methylene blue	[31]
Alginate–montmorillonite–polyaniline nanocomposite	Chemical oxidative polymerization	Reactive orange 13	[65]
Polyaniline/poly(vinyl alcohol)/montmorillonite-hybrid aerogels	In situ polymerization method	Reactive blue 5, methyl orange, and safranin	[66]

Nguyen et al. (2021) [69] offer a simple casting procedure for agar/maltodextrin/poly(vinyl alcohol) walled montmorillonite (AMP–MMT) composites, addressing the urgent issue of textile pollutants such as methylene blue in aquatic systems. By using physicochemical characterization, the composite's structure was confirmed, and several parameters affecting dye adsorption were examined. Possible processes, including cation exchange and interactions, were discovered. The composite demonstrated strong MB absorption to 71.51 mg/g over a minimum of five cycles, as well as simple desorption using a green eluent made of ethanol and water. This study encourages the use of recyclable, eco-friendly, and biodegradable hydrogel membranes for water filtration.

Overall, these investigations collectively underline the tremendous potential of polymer/clay composites in improving dye adsorption for wastewater treatment. These materials have been successfully synthesized, characterized, and tested, demonstrating their efficacy, recyclable nature, and environmental sustainability. These composites are more in line with the expanding need for environmentally friendly and sustainable water filtration solutions because of the incorporation of biodegradable materials and cutting-edge functionalization procedures. Subsequent investigations might concentrate on extending the reach of these methods and investigating their use for a wider variety of contaminants.

5.3. Characterization of Polymer/Clay Composites

Polymer/clay nanocomposites can be characterized by different methods depending on the desired applications. For example, analysis through zeta potential measurements is frequently used when employing polymer/clay nanocomposites in adsorption or catalysis fields. The surface properties change depending on the nature of the polymer used (a polymer with basic or acidic sites), as it allows the determination of the zero-charge point. Additionally, pH significantly influences zeta potential; at low pH, the surface may become positively charged, favoring the adsorption of anionic dyes, while at high pH, the surface becomes negatively charged, enhancing cationic dye adsorption [31]. Then, all the modifications of these nanocomposites by functional organic group can be studied by FTIR analysis [31].

But the analyses most used to confirm the nature of the nanocomposites (intercalated, exfoliated, or immiscible) are XRD and TEM, often combining TGA thermal analysis to confirm their stability. The structural properties of nanocomposites have been widely studied using XRD analysis, from which the different forms of nanocomposites can be determined. Regarding the immiscible form, in general, no shift of the characteristic peak of clay will occur. This is sometimes due to the size of the polymer and also related to its structure, in which the immobilization of a linear polymer in the galleries of the clay is better compared to the branched form [69]. Then, it is sometimes even difficult to immobilize the polymer in the galleries of the clay, which implies the use of other procedures. Among the most used procedures is the immobilization of surfactants with different carbon chains, which leads to the swelling of the interlayer distance. This is followed by an ion exchange process between the metal cations existing in the structure of the clay and the surfactants. Then, another method involves the use of ultrasonic radiation with a solution containing clay or organophilic clay and the polymer. According to the work of Cherifi et al. [70], it was found that this method generates the formation of exfoliated or intercalated nanocomposites with a good dispersion of polymer in the clay galleries (see Figure 7). From this figure, it is clear that the nature of the surfactant, its concentration, and the method used during the preparation significantly influence the structural properties of the obtained nanocomposites. Intercalation results an expansion of 2–3.5 nm of clay sheets, and they remain parallel to each other. Extensive penetration of polymer chains into the interlayer spaces of clay results in exfoliation of the clay sheets [71]. They are separated/dispersed during exfoliation by more than 10 nm. Among the best techniques that can be combined with XRD analysis is TEM analysis. From Figure 8, the TEM images of Nano-So4 and Nano-So7 nanocomposites are totally different due to several modifications conducted for each sample. Nano-So4 nanocomposite is a material based on poly(GMA) and CTA-Magh(2CEC), while Nano-So7 contains poly(GMA) and TBA-Magh(1CEC); the TEM images of Nano-So4 show very fine lines with similar shapes corresponding to the intercalated form, while Nano-So7 nanocomposite shows non-uniform scattered lines corresponding to the exfoliated form, which is in agreement with the XRD analysis [70]. So, for TGA analysis, this technique allows us to know the different stages of degradation as well as the hydrophilic/hydrophobic character and the stability of the nanocomposite. As it is known, the loss of mass of physisorbed water is generally observed at temperatures below 150 °C; at this stage, it is easier to determine the character of the resulting material. So, to establish whether the stability of the nanocomposite has been improved, it is very important to test the parent polymer without any modification. For example, as shown in Figure 9 with nanocomposite prepared from an organophilic clay and cellulose acetate, there was a significant change in the degradation of the nanocomposite, from which it was confirmed that the nanocomposite has a greater thermal stability compared to the parent polymer [72]. To achieve this stability, there is an optimal content of clay that should not be exceeded. From this figure, it is clear that the CA/Mag 5% nanocomposite was the most stable material in this whole series of materials [72].

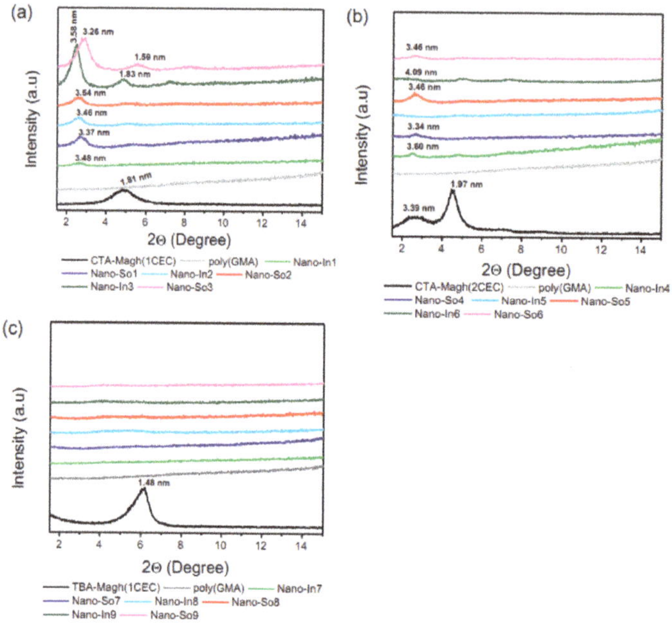

Figure 7. XRD patterns of nanocomposites obtained by different organoclays. (**a**) XRD patterns of nanocomposite prepared by an in situ method (symbolized by In) or by direct dispersion of poly(glycidylmethacrylate) (Poly(GMA)) in organophilic clay galleries using ultrasound radiation (symbolized by So). CTA-Magh(2CEC) is an Algerian MMT clay modified by CTA+ surfactant using 1CEC. (**b**) XRD patterns of poly(GMA)/organophilic clay nanocomposite (containing 2CEC of CTA+). (**c**) XRD patterns of poly(GMA)/organophilic clay nanocomposite (containing 1CEC of TBA+). Reproduced with permission from Ref. [71].

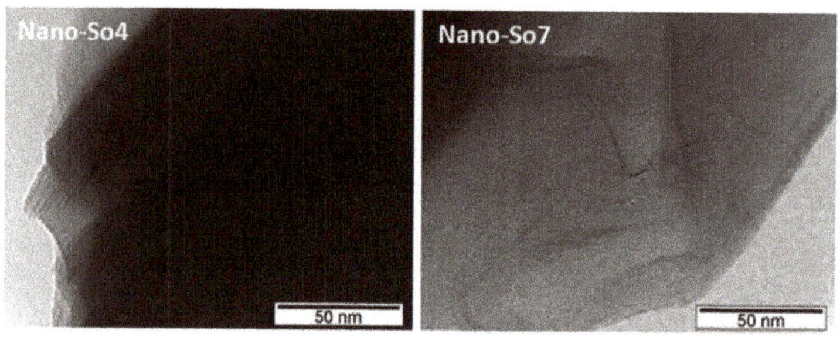

Figure 8. TEM images of nanocomposites Nano-So4 and Nano-So7 (the abbreviation of these samples is well detailed in the XRD part). Reproduced with permission from Ref. [71].

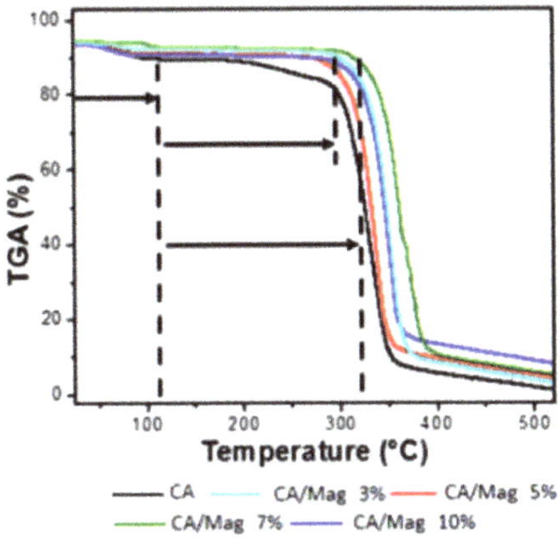

Figure 9. Cellulose acetate/organophilic clay. CA: cellulose acetate; Mag: Maghnite (MMT) modified by CTA+; CA/Mag (3%–10%): nanocomposite obtained at different weights of Maghnite. Reproduced with permission from Ref. [72].

6. Dye Elimination

6.1. The Most Common Studied Dyes in the Literature

These colored substances are groups of atoms called chromophores whose light absorption causes color. A soluble dye is referred to as a dye; otherwise, it is a pigment [73]. In several application techniques, such as pharmaceuticals or food, a colorant can contain a combination of soluble molecules and insoluble particles. In biology, we speak of pigments for the molecules which naturally give color to the organisms that we study, and of dyes for the products that we add to the preparations to be studied. There are three types of colorings: natural ones, synthetic colors produced by the chemical industry but which exist identically in nature, and artificial colors, i.e., produced by the art of man and which have no equivalent in nature [74]. Synthetic dyes, commonly used in textiles, cosmetics, and the food industry, often find their way into wastewater, causing serious environmental pollution. These compounds are non-biodegradable and persist in aquatic systems, leading to toxic effects on aquatic organisms and potential bioaccumulation. For example, dyes like crystal violet and methylene blue are harmful to fish and other aquatic species, while some, such as rhodamine and Congo red, exhibit carcinogenic and mutagenic effects in humans [75–78]. Addressing these harmful impacts is a critical focus of dye removal studies. Figure 10 present the chemical structures of synthetic dyes most frequently studied in the literature.

The main methods of classifying dyes are based either on their chemical constitution (see Figure 11) or on their methods of application to the various substrates such as textile fibers, paper, leather, plastics [79,80], etc. Classification according to chemical structure is based mainly on the nature of the chromophore, which constitutes the backbone necessary for the coloring of the molecule [81,82]. All the chromophores have in common the presence within the molecular structure of a series of conjugated double bonds, which most often include aromatic or pseudo-aromatic rings. Auxochromes, for their part, are ionizable groups of NH_2, OH, COOH, and SO_3H types. They make it possible to ensure solubility in water, but above all to create a solid bond between the dye and the substrate. This is how auxochromes define the tinctorial class: the presence of amino groups characterizes a basic dye; that of sulfonic groups characterizes acid dyes. These two modes of classification are far from being identical; the proof is the case of reactive dyes, which are distinguished by the possibility of establishing a true covalent bond with the substrate, but whose chromophore can be derived from several chemical categories of dyes, in particular the azo, anthraquinone, and phthalocyanine families. This variety of functional groups in these pollutants has prompted researchers to develop low-cost and eco-friendly composites in the aim to reduce or eliminate these substances. In 2023, Kankana et al. [83] created an organogel PAni/MMT composite that effectively eliminated harmful contaminants such as Congo red and chlorophenol, showcasing exceptional adsorption abilities. With a 99.44% removal efficiency, the gel was able to absorb 349.72 mg/g of Congo red and 20.84 times its own weight in chlorophenol. The adsorption procedure followed the pseudo-second-order and Langmuir isotherm models. The gel showed promise for environmentally friendly applications because it retained 83% of the dye even after five cycles. Olad et al. [84] conducted investigations that indicated that the removal mechanism of biopolymer/clay/polymer nanocomposite was based on both the adsorption and electrostatic attraction between nanocomposite and dye molecules. The removal results of these composites demonstrate the effectiveness of the hybrid system (clay and polymer) as an efficient adsorbent for removal of several dye types from textile effluents.

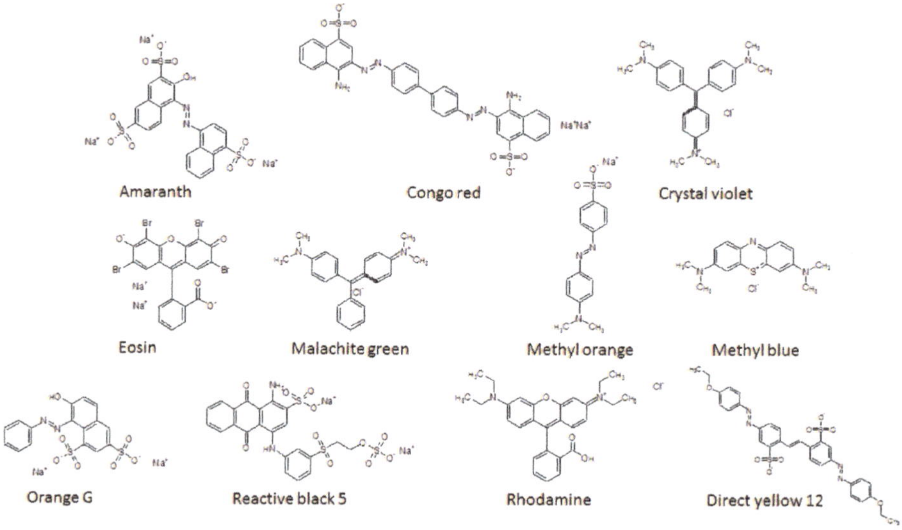

Figure 10. The chemical structures of synthetic dyes most frequently studied.

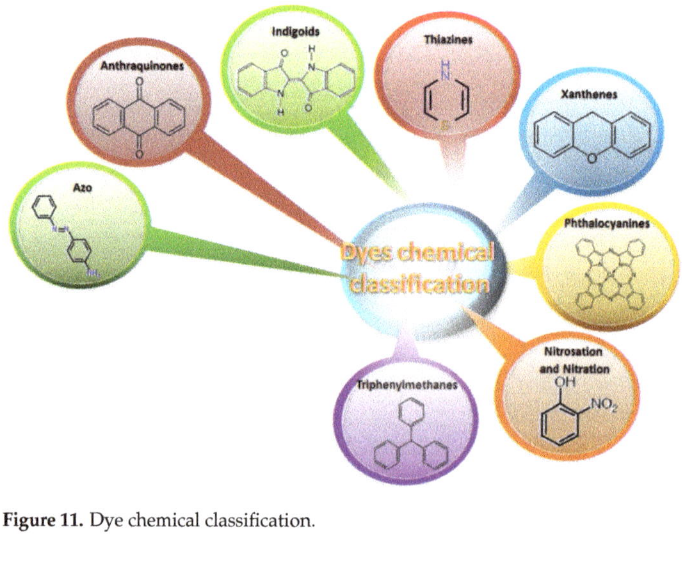

Figure 11. Dye chemical classification.

6.2. Polymer/Clay-Composite-Based Adsorbents

Polymer-based materials are organic compounds possessing a large specific surface area, microporosity, and a three-dimensional network [85]. This gives them an advantage of use as adsorbents such as ion exchange resins, membranes, molecularly imprinted polymers, microporous organic polymers, etc., while their elaboration with other polymers or metals, clays, and zeolites for the formation of novel polymeric composite materials is important in the environmental field [5,8,85,86]. The tendency to develop low-cost adsorbents with high removal capacity has always been the focus of researchers for dye removal. Among the most powerful additives are clays, activated alumina, metal oxides, and zeolites. The advantages of clays are numerous compared to other additives; clays are cheap, abundant, have a porous surface, a high exchange capacity, and are respectful of nature. Their modification by certain inorganic or organic agents can overcome some disadvantages. Their layered structure allows penetration of polymer chains or in situ polymerization, and this results in equal or better performance in dye adsorption. These modifications made it possible to improve the surface of the composite with structural characteristics, also offering the opportunity for abundant functional groups [5,6]. The resulting composite exhibiting a variety of complex elimination mechanisms towards different types of dye O, H, S, N, and F groups. Elimination is achieved either via electrostatic attraction, surface hydrophobicity, π-π interaction, or hydrogen bonding. Figure 12 shows a simple diagram which groups together all the possible mechanisms for the elimination of dyes by polymer/clay composite. Over the years, this type of composite (polymer/clay) based on the inorganic (clay) and organic (polymer) substances has always been necessary, and it is very interesting to apply it as an adsorbent for wastewater treatment due to its compatibility, low cost, and lower toxicity. Figure 13 shows the number of publications per year on these polymer/clay nanocomposites as adsorbents.

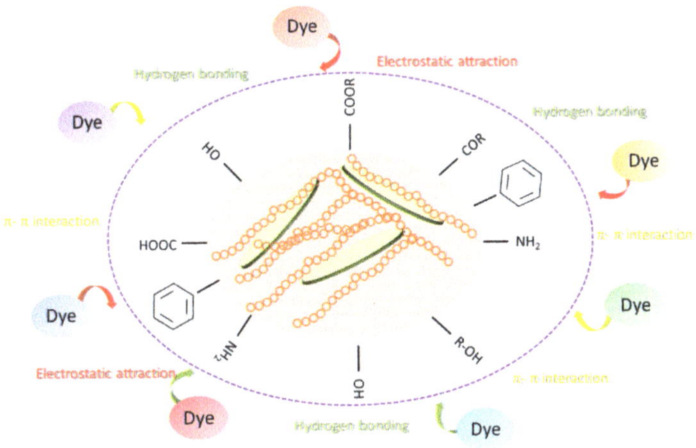

Figure 12. Schematic representations of mechanisms for the elimination of various dyes using polymer/clay composite via electrostatic attraction, surface hydrophobicity, π-π interaction, and hydrogen bonding.

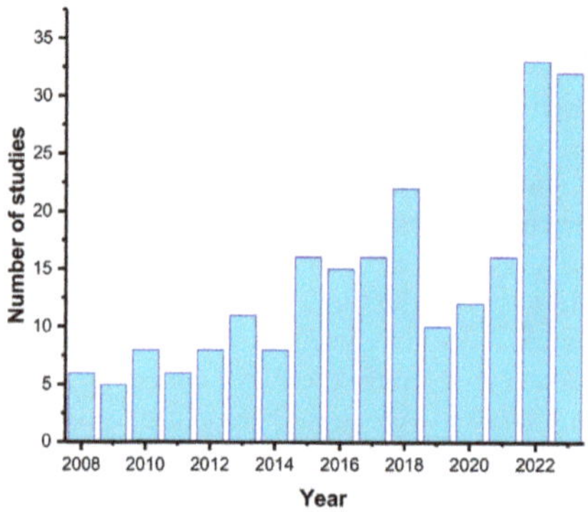

Figure 13. Number of publications per year on polymer/clay nanocomposites as adsorbents.

6.3. *Magnetic-Polymer/Clay-Nanocomposite-Based Adsorbents*

Research on effective and sustainable treatment technologies has accelerated due to growing environmental concerns about the discharge of untreated textile wastewater, which is high in synthetic dyes and other dangerous chemicals. Polymer/clay composites have proven to be one of the most successful materials for the adsorption and removal of dyes from aqueous solutions among the numerous technologies investigated [31,36,87]. Particularly, the potential of magnetic polymer/clay nanocomposites as superior adsorbents for dye removal has been highlighted by recent advancements in the field. These hybrid materials combine the distinctive qualities of polymers, like flexibility and customizable surface functionality, with the high adsorption capacity of clay minerals and the magnetic responsiveness provided by embedded magnetic nanoparticles. This combination makes it simple to apply an external magnetic field to separate and collect the adsorbent from treated water, which improves process efficiency and lowers costs [88]. Magnetic clay–polymer nanocomposites are usually prepared by a combination of techniques such as surface modification, sol–gel procedures, and in situ polymerization [36,89,90]. The structure and characteristics of the composite material can be precisely controlled using these techniques. To ensure a uniform distribution of all components, in situ polymerization includes polymerizing monomers in the presence of clay and magnetic nanoparticles. In contrast, the sol–gel method entails the hydrolysis and condensation of precursors to create a network resembling a gel that contains magnetic and clay particles. Clay particles can also have their surfaces functionalized by surface modification procedures, which will improve their interaction with magnetic particles and the polymer matrix [89]. These preparation techniques allow magnetic nanoparticles, like Fe_3O_4, to be incorporated into the clay–polymer matrix more easily and help to produce a stable and strong composite structure. This improves the adsorption efficiency of the composite and makes it possible for the adsorbent to separate from the liquid phase quickly, which is essential for large-scale water treatment applications. Anionic and cationic dyes, as well as a variety of other dyes, have been effectively removed from textile wastewater by magnetic clay–polymer nanocomposites. The clay, polymer, and magnetic components work in concert to give these materials their high adsorption capacity and selectivity. The adsorption performance is mostly determined by the specific surface area, pore structure, type of polymer, and clay utilized. For the purpose of removing methylene blue, Agnès Bée et al. [88] created a magnetic clay/chitosan composite, which demonstrated a strong adsorption capacity of 82 mg/g and efficient performance over a wide pH range. The adsorption of the substance is propelled by electrostatic interactions between the clay and dye, and its magnetic characteristics facilitate effortless recovery. To effectively remove organic dyes from water, Edwin et al. [91] prepared and characterized a CMC-cl-pAA hydrogel and its magnetic nanocomposite. The study demonstrated high adsorption capacity, improved thermal stability, and excellent regeneration stability. Bisphenol A was adsorbed and separated from aqueous solutions using magnetic molecularly imprinted polymers based on kaolinite/Fe_3O_4 composites [92], which showed excellent selectivity and efficiency. The adsorption process was carried out using a pseudo-second-order kinetic model and the Langmuir isotherm model. Additionally, these nanocomposites' magnetic qualities make it easier for them to be recovered from the aqueous solution, enabling efficient dye removal without leaving behind leftover adsorbents in the treated water. Following adsorption, the nanocomposites can be recovered by straightforward procedures such as washing in appropriate solvents, allowing for repeated cycles without noticeably degrading their function. This reusability minimizes the wastewater treatment process's environmental impact in addition to lowering costs.

6.4. Polymer/Clay-Composite-Based Catalysts

The utilization of metallic species as nanoparticles or oxides in the degradation or the reduction of various colorant-contaminated wastewaters is a very promising, simple, rapid, and efficient treatment method. These species as catalysts have unique electrical, optical, and thermal properties that are ideal, as well as having a high surface area and easy fabrication of their selectivity and specificity. Using the noble metals (Ag, Au, Pt, and Pd) for catalytic reduction is very common and widely investigated for wastewater depollution. The application of these species alone is unsuitable due to their agglomeration/aggregation along with the reduction in their performance and the difficulty of their reuse in aqueous solutions [93,94]. In order to minimize the inconveniences, several researchers used low-cost, abundant, and eco-friendly materials in the aim to immobilize/disperse these metals on their surface. Clays possess a high specific surface area and sheet-like structure with surface negative charges, which make them a suitable challenger as support for the immobilization/dispersion of metallic species using a simple method such as ion exchange. By using a chemical, biological, or thermal treatment, the nanoparticle species can be formed on the surface of the clay. To obtain an effective nanocomposite catalyst, polymerization or in situ polymerization can also increase other factors such as specific surface area and surface adsorption efficiency, minimize leaching of metals, and facilitate their use and reuse in aqueous solution. Figure 14 presents a simple diagram for catalytic reduction/degradation of dye using polymer/clay nanocomposite. To limit environmental pollution, effective and environmentally friendly treatment strategies have been developed. Among these strategies is the application of advanced oxidation processes. These are oxidative treatment processes which require the presence of metallic catalytic support. After the diffusion of the pollutant in the nanocomposite surface, the metallic nanoparticles supported on the nanocomposite (polymer/clay) facilitate the dye reduction/degradation reaction. The catalytic reductions require the presence of a reducing agent such as $NaBH_4$, NaOH, or $LiBH_4$, whereas the phot ocatalytic reaction requires the presence of UV irradiation. In addition to using clay-based nanocomposites, these elimination pollutant techniques are preferable to other techniques due to their shorter elimination time, less toxic product, and their interesting reduction products as raw molecules, as well as their low cost, ease of design, and ease of process.

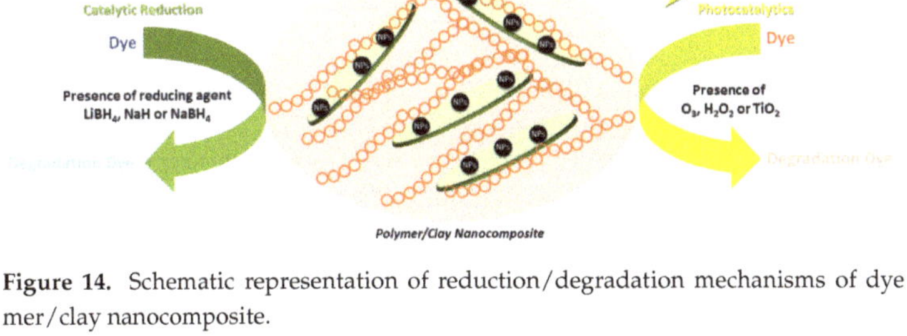

Figure 14. Schematic representation of reduction/degradation mechanisms of dye using polymer/clay nanocomposite.

7. Response Surface Methodology

Response surface methodology (RSM) is a statistical method for enhancing procedures when many input variables affect an output variable (response). RSM considers every aspect simultaneously, in contrast to methods that consider one factor at a time. The most popular types of response surface design were Central Composite Design (CCD), Doehlert design, and Box–Behnken design (BBD). With its five-level experimental design, which efficiently identifies optimal circumstances with a small number of tests, the CCD is a crucial part of RSM. By using this approach, less testing is necessary to obtain the maximum response values [60,95–98]. To remove anionic (Remazol brilliant blue R, RBBR) and cationic (malachite green) dyes from aqueous solutions [98], a pH-sensitive chitosan/sepiolite clay/algae biocomposite was optimized for adsorptive performance using a Box–Behnken design. Response surface plots were employed in the investigation to illustrate the correlations between the variables that affected dye adsorption, including biocomposite dosage, solution pH, and process time. The results demonstrated the importance of pH in improving dye removal effectiveness; a pH of 8 was found to be ideal for MG removal, whereas a pH of about 4 was best for RBBR removal. The biocomposite's strong affinity for both dyes was confirmed by the adsorption, which proceeded according to pseudo-second-order kinetics and Freundlich isotherm models via processes such electrostatic interactions, H-bonding, and n-π interactions. In 2024, Abdallah and colleagues created an environmentally friendly adsorbent by functionalizing kaolin clay with peanut shell cellulose [99]. They then assessed the adsorbent's efficacy in eliminating both anionic and cationic dyes from water. The BBD was used to optimize the adsorption process and pinpoint important variables like the pH of the solution, cellulose loading, and adsorbent dosage. According to the data, the adsorption proceeded according to the Freundlich model and pseudo-second-order kinetics, while the Langmuir model estimated large monolayer capacities for both dyes. This demonstrates that the adsorbent has the potential to be a workable treatment option for wastewater contaminated with dyes.

8. Assessing Economic and Environmental Aspects

Polymer/clay composites have shown great promise as materials in a range of industrial applications owing to their special blend of improved barrier qualities, mechanical strength, and thermal stability. These composites are made by combining clay minerals—such as kaolin, bentonite, montmorillonite, or halloysite—with a polymer matrix. The product is a material that performs better than conventional polymers. By lowering the demand for non-renewable resources, the combination of the polymer and clay components not only enhances the physical qualities but also has a major positive impact on the environment. In addition, there is growing interest in the economic viability of large-scale production of these composites, which makes them a viable option for upcoming technical developments. Thankfully, chitosan is the most widely used polymer for creating polymer/clay composites. Due to its expanding commercial market, it is an excellent substitute for synthetic polymers and a naturally occurring polymer that is widely accessible at a reasonable cost [100]. Compared to most synthetic polymers, the cost of making 1 kg of chitosan is currently between USD 5 and USD 14 in nations like Ecuador, Colombia, and Spain. This makes the material economically more viable [101]. Owing to its market pricing, which ranges from USD 0.04 to USD 0.12 per kilogram, montmorillonite and its modified forms are the most utilized clay for the synthesis of chitosan–clay composites [28,102]. As of 2021, the market prices for kaolin and bentonite are USD 0.160 and USD 0.094, respectively [28]. The costs of specific clays that are used to create clay–polymer composites differ greatly. The cost of a ton of bentonite is about USD 43.59 [83], but the cost of a ton of nano-Ca-bentonite can range from EUR 4400 to EUR 10,000. The cost of montmorillonite ranges from USD 40 to USD 120 per ton, while the cost of kaolin is approximately USD 106.91 per ton. The cost of palygorskite is USD 200 per ton. An additional crucial aspect that needs to be considered before using polymer/clay composites on a broad scale is the assessment of the environmental impact of doing so. Because they decay slowly, have a short lifespan, or may never biodegrade, synthetic polymers like epoxy and polyurethane could be harmful to the environment [103]. For this reason, it is necessary to promote and seek out low-cost and environmentally friendly polymers for current and future research. As a result, natural and biodegradable polymers, such as chitosan, Kappa-Carrageenan, and Moringa olifera, as discussed in this research, are garnering more attention and ought to be promoted further to create innovative adsorbents made of polymer and clay composites. Both materials have a lot to offer when evaluating the price, adsorption effectiveness, and reusability of clay/chitosan and clay/cellulose composites. Because chitosan has relatively low production costs, clay/chitosan composites are more cost-effective and have great adsorption effectiveness, especially for heavy metals and dyes. However, because of cellulose's inherent resilience and structural integrity, clay/cellulose composites, though marginally more expensive, offer excellent reusability, which makes them perfect for applications needing several adsorption cycles. In the end, which of these composites is best depends on the environmental and economic needs of the intended use.

9. Conclusions and Prospects

The ongoing development of polymer/clay composites for the removal or degradation of textile effluents continues to show great promise. The synergistic interactions between the organic polymer matrices and inorganic clay components generate composites with a broad range of functionalities, offering significant selectivity towards various organic pollutants. However, considering the extensive diversity of dyes in textile wastewater, current polymer/clay composites can remove only a limited subset of organic dyes. This highlights the urgent need for the development of new, multifunctional composites that can effectively target a wider range of pollutants, including complex mixtures of dyes and other organic contaminants. One of the key challenges in advancing this field lies in the design of composites that can address a broader spectrum of pollutants while maintaining economic and practical viability for large-scale applications. Future research should focus on creating materials that not only have enhanced adsorption and catalytic properties but are also cost-effective, easily recoverable, and capable of maintaining high efficiency after multiple cycles of use. Despite the significant progress made in the application of polymer/clay composites for wastewater treatment, several limitations persist. These include difficulties in recovering the composites from aqueous solutions after use, the decline in adsorption efficiency after repeated cycles, low selectivity for certain types of dyes, and challenges in desorbing the adsorbed dyes from the composite matrix. Furthermore, the formation of by-products during dye degradation or reduction processes may introduce new concerns regarding secondary pollution. To overcome these challenges, it is crucial to adopt a multi-disciplinary approach that integrates experimental and theoretical methods. Computational modeling and simulation techniques can be employed to predict the interaction between pollutants and composite materials, thereby guiding the design of more efficient polymer/clay composites. Additionally, the development of novel surface modification techniques and the incorporation of functional nanoparticles could significantly enhance the selectivity and reusability of these composites. In summary, while polymer/clay composites have demonstrated remarkable potential in wastewater treatment, especially in the removal of textile dyes, further advancements are required to address their current limitations. The future direction of this research should focus on the development of composites that are not only more versatile in pollutant removal but also environmentally sustainable, economically viable, and scalable for industrial applications. Continued efforts in both material innovation and theoretical understanding will be essential in achieving more effective and sustainable solutions for wastewater management.

Author Contributions: A.M.: Conceptualization, methodology, formal analysis, investigation, and writing—original draft. B.A.: Formal analysis, investigation, and writing—original draft. S.A.: Resources and methodology. M.H.: Resources and methodology. B.B.: Review and editing, methodology, and validation. M.S.: Polishing the language, conceptualization, supervision, and review and editing. G.V.: Review and editing, methodology, and validation. M.A.: Conceptualization, validation, formal analysis, review and editing, and funding acquisition. All authors have read and agreed to the published version of the manuscript.

Funding: The authors extend their appreciation to the Deanship of Scientific Research at King Khalid University for funding this work through a large group research project under grant number RGP2/280/45.

Data Availability Statement: All the data and materials of this paper are available for reader.

Acknowledgments: The authors gratefully acknowledge the support provided by the Directorate General for Scientific Research and Technological Development (DGRSDT), Algeria, for facilitating this research.

Conflicts of Interest: The authors declare no conflicts of interest.

References

1. Shabtai, I.A.; Lynch, L.M.; Mishael, Y.G. Designing Clay-Polymer Nanocomposite Sorbents for Water Treatment: A Review and Meta-Analysis of the Past Decade. *Water Res.* **2021**, *188*, 116571. [CrossRef] [PubMed]
2. Adeyemo, A.A.; Adeoye, I.O.; Bello, O.S. Adsorption of Dyes Using Different Types of Clay: A Review. *Appl. Water Sci.* **2017**, *7*, 543–568. [CrossRef]
3. Kausar, A.; Iqbal, M.; Javed, A.; Aftab, K.; Nazli, Z.; Nawaz, H.; Bhatti, H.N.; Nouren, S. Dyes Adsorption Using Clay and Modified Clay: A Review. *J. Mol. Liq.* **2018**, *256*, 395–407. [CrossRef]
4. Kishor, R.; Purchase, D.; Saratale, G.D.; Saratale, R.G.; Ferreira, L.F.R.; Bilal, M.; Chandra, R.; Bharagava, R.N. Ecotoxicological and Health Concerns of Persistent Coloring Pollutants of Textile Industry Wastewater and Treatment Approaches for Environmental Safety. *J. Environ. Chem. Eng.* **2021**, *9*, 105012. [CrossRef]
5. Mokhtar, A.; Abdelkrim, S.; Djelad, A.; Sardi, A.; Boukoussa, B.; Sassi, M.; Bengueddach, A. Adsorption Behavior of Cationic and Anionic Dyes on Magadiite-Chitosan Composite Beads. *Carbohydr. Polym.* **2020**, *229*, 115399. [CrossRef] [PubMed]
6. Mokhtar, A.; Abdelkrim, S.; Hachemaoui, M.; Adjdir, M.; Zahraoui, M.; Boukoussa, B. Layered Silicate Magadiite and Its Composites for Pollutants Removal and Antimicrobial Properties: A Review. *Appl. Clay Sci.* **2020**, *198*, 105823. [CrossRef]
7. Hachemaoui, M.; Mokhtar, A.; Mekki, A.; Zaoui, F.; Abdelkrim, S.; Hacini, S.; Boukoussa, B. Composites Beads Based on Fe_3O_4@MCM-41 and Calcium Alginate for Enhanced Catalytic Reduction of Organic Dyes. *Int. J. Biol. Macromol.* **2020**, *164*, 468–479. [CrossRef]
8. Mokhtar, A.; Abdelkrim, S.; Zaoui, F.; Sassi, M.; Boukoussa, B. Improved Stability of Starch@Layered-Materials Composite Films for Methylene Blue Dye Adsorption in Aqueous Solution. *J. Inorg. Organomet. Polym. Mater.* **2020**, *30*, 3826–3831. [CrossRef]
9. Jayrajsinh, S.; Shankar, G.; Agrawal, Y.K.; Bakre, L. Montmorillonite Nanoclay as a Multifaceted Drug-Delivery Carrier: A Review. *J. Drug Deliv. Sci. Technol.* **2017**, *39*, 200–209. [CrossRef]
10. Awasthi, A.; Jadhao, P.; Kumari, K. Clay Nano-Adsorbent: Structures, Applications and Mechanism for Water Treatment. *SN Appl. Sci.* **2019**, *1*, 1076. [CrossRef]
11. Unuabonah, E.I.; Ugwuja, C.G.; Omorogie, M.O.; Adewuyi, A.; Oladoja, N.A. Clays for Efficient Disinfection of Bacteria in Water. *Appl. Clay Sci.* **2018**, *151*, 211–223. [CrossRef]
12. Williams, L.B.; Metge, D.W.; Eberl, D.D.; Harvey, R.W.; Turner, A.G.; Prapaipong, P.; Poret-Peterson, A.T. What Makes a Natural Clay Antibacterial? *Environ. Sci. Technol.* **2011**, *45*, 3768–3773. [CrossRef] [PubMed]
13. Gomes, C.F.; Gomes, J.H.; da Silva, E.F. Bacteriostatic and Bactericidal Clays: An Overview. *Environ. Geochem. Health* **2020**, *42*, 3507–3527. [CrossRef] [PubMed]
14. Zoveidavianpoor, M. *Current Topics in the Utilization of Clay in Industrial and Medical Applications*; IntechOpen: London, UK, 2018.
15. Hundáková, M.; Dêdková, K.; Martynková, G.S. Decoration of Inorganic Substrates with Metallic Nanoparticles and Their Application as Antimicrobial Agents. In *Metal Nanoparticles in Pharma*; Springer: Berlin/Heidelberg, Germany, 2017.
16. Biswas, B.; Warr, L.N.; Hilder, E.F.; Goswami, N.; Rahman, M.M.; Churchman, J.G.; Vasilev, K.; Pan, G.; Naidu, R. Biocompatible Functionalisation of Nanoclays for Improved Environmental Remediation. *Chem. Soc. Rev.* **2019**, *48*, 3740–3770. [CrossRef] [PubMed]
17. Dorioz, J.M.; Robert, M.; Chenu, C. The Role of Roots, Fungi and Bacteria on Clay Particle Organization. An Experimental Approach. *Geoderma* **1993**, *56*, 179–194. [CrossRef]
18. Özdemir, G.; Yapar, S. Preparation and Characterization of Copper and Zinc Adsorbed Cetylpyridinium and N-Lauroylsarcosinate Intercalated Montmorillonites and Their Antibacterial Activity. *Colloids Surf. B Biointerfaces* **2020**, *188*, 110791. [CrossRef] [PubMed]
19. Ghadiri, M.; Chrzanowski, W.; Rohanizadeh, R. Biomedical Applications of Cationic Clay Minerals. *RSC Adv.* **2015**, *5*, 29467–29481. [CrossRef]
20. Li, G.L.; Zhou, C.H.; Fiore, S.; Yu, W.H. Interactions between Microorganisms and Clay Minerals: New Insights and Broader Applications. *Appl. Clay Sci.* **2019**, *177*, 91–113. [CrossRef]
21. Unuabonah, E.I.; Taubert, A. Clay-Polymer Nanocomposites (CPNs): Adsorbents of the Future for Water Treatment. *Appl. Clay Sci.* **2014**, *99*, 83–92. [CrossRef]
22. Mokhtar, A.; Bennabi, F.; Abdelkrim, S.; Sardi, A.; Boukoussa, B.; Souna, A.; Bengueddach, A.; Sassi, M. Evaluation of Intercalated Layered Materials as an Antimicrobial and Drug Delivery System: A Comparative Study. *J. Incl. Phenom. Macrocycl. Chem.* **2020**, *96*, 353–364. [CrossRef]

23. El-Zahhar, A.A.; Awwad, N.S.; El-Katori, E.E. Removal of Bromophenol Blue Dye from Industrial Waste Water by Synthesizing Polymer-Clay Composite. *J. Mol. Liq.* **2014**, *199*, 454–461. [CrossRef]
24. Wang, W.; Wang, J.; Zhao, Y.; Bai, H.; Huang, M.; Zhang, T.; Song, S. High-Performance Two-Dimensional Montmorillonite Supported-Poly(Acrylamide-Co-Acrylic Acid) Hydrogel for Dye Removal. *Environ. Pollut.* **2020**, *257*, 113574. [CrossRef] [PubMed]
25. Zhang, T.; Wang, W.; Zhao, Y.; Bai, H.; Wen, T.; Kang, S.; Song, G.; Song, S.; Komarneni, S. Removal of Heavy Metals and Dyes by Clay-Based Adsorbents: From Natural Clays to 1D and 2D Nano-Composites. *Chem. Eng. J.* **2020**, *420*, 127574. [CrossRef]
26. Biswas, S.; Fatema, J.; Debnath, T.; Rashid, T.U. Chitosan–Clay Composites for Wastewater Treatment: A State-of-the-Art Review. *ACS EST Water* **2021**, *1*, 1055–1085. [CrossRef]
27. Zhang, J.; Wang, A. Study on Superabsorbent Composites. IX: Synthesis, Characterization and Swelling Behaviors of Polyacrylamide/Clay Composites Based on Various Clays. *React. Funct. Polym.* **2007**, *67*, 737–745. [CrossRef]
28. Hnamte, M.; Pulikkal, A.K. Clay-Polymer Nanocomposites for Water and Wastewater Treatment: A Comprehensive Review. *Chemosphere* **2022**, *307*, 135869. [CrossRef]
29. Mok, C.F.; Ching, Y.C.; Muhamad, F.; Abu Osman, N.A.; Hai, N.D.; Che Hassan, C.R. Adsorption of Dyes Using Poly(Vinyl Alcohol) (PVA) and PVA-Based Polymer Composite Adsorbents: A Review. *J. Polym. Environ.* **2020**, *28*, 775–793. [CrossRef]
30. Sanchez, L.M.; Shuttleworth, P.S.; Waiman, C.; Zanini, G.; Alvarez, V.A.; Ollier, R.P. Physically-Crosslinked Polyvinyl Alcohol Composite Hydrogels Containing Clays, Carbonaceous Materials and Magnetic Nanoparticles as Fillers. *J. Environ. Chem. Eng.* **2020**, *8*, 103795. [CrossRef]
31. Sardi, A.; Bounaceur, B.; Mokhtar, A.; Boukoussa, B.; Abbes, M.T.; Chaibi, W.; Nacer, A.; Khadidja, K.B.; Issam, I.; Iqbal, J.; et al. Kinetics and Thermodynamic Studies for Removal of Trypan Blue and Methylene Blue from Water Using Nano Clay Filled Composite of HTAB and PEG and Its Antibacterial Activity. *J. Polym. Environ.* **2023**, *31*, 5065–5088. [CrossRef]
32. van Oss, C.J.; Giese, R.F. The Hydrophilicity and Hydrophobicity of Clay Minerals. *Clays Clay Miner.* **1995**, *43*, 474–477. [CrossRef]
33. Kausar, A. A Review of Fundamental Principles and Applications of Polymer Nanocomposites Filled with Both Nanoclay and Nano-Sized Carbon Allotropes—Graphene and Carbon Nanotubes. *J. Plast. Film Sheeting* **2020**, *36*, 209–228. [CrossRef]
34. A Review Featuring the Fundamentals and Advancements of Polymer/CNT Nanocomposite Application in Aerospace Industry | Polymer Bulletin. Available online: https://link.springer.com/article/10.1007/S00289-019-03096-0 (accessed on 26 August 2024).
35. Hassan, T.; Salam, A.; Khan, A.; Khan, S.U.; Khanzada, H.; Wasim, M.; Khan, M.Q.; Kim, I.S. Functional Nanocomposites and Their Potential Applications: A Review. *J. Polym. Res.* **2021**, *28*, 36. [CrossRef]
36. Abulyazied, D.E.; Ene, A. An Investigative Study on the Progress of Nanoclay-Reinforced Polymers: Preparation, Properties, and Applications: A Review. *Polymers* **2021**, *13*, 4401. [CrossRef] [PubMed]
37. Rajeshkumar, G.; Seshadri, S.A.; Ramakrishnan, S.; Sanjay, M.R.; Siengchin, S.; Nagaraja, K.C. A Comprehensive Review on Natural Fiber/Nano-Clay Reinforced Hybrid Polymeric Composites: Materials and Technologies. *Polym. Compos.* **2021**, *42*, 3687–3701. [CrossRef]
38. Singha, S.; Hedenqvist, M.S. A Review on Barrier Properties of Poly(Lactic Acid)/Clay Nanocomposites. *Polymers* **2020**, *12*, 1095. [CrossRef] [PubMed]
39. Amari, A.; Alzahrani, F.M.; Katubi, K.M.; Alsaiari, N.S.; Tahoon, M.A.; Rebah, F. Ben Clay-Polymer Nanocomposites: Preparations and Utilization for Pollutants Removal. *Materials* **2021**, *14*, 1365. [CrossRef]
40. Usmani, M.A.; Khan, I.; Ahmad, N.; Bhat, A.H.; Sharma, D.K.; Rather, J.A.; Hassan, S.I. Modification of Nanoclay Systems: An Approach to Explore Various Applications. In *Nanoclay Reinforced Polymer Composites: Nanocomposites and Bionanocomposites*; Springer: Berlin/Heidelberg, Germany, 2016; pp. 57–83.
41. Babu Valapa, R.; Loganathan, S.; Pugazhenthi, G.; Thomas, S.; Varghese, T.O. Chapter 2—An Overview of Polymer–Clay Nanocomposites. In *Clay-Polymer Nanocomposites*; Jlassi, K., Chehimi, M.M., Thomas, S., Eds.; Elsevier: Amsterdam, The Netherlands, 2017; pp. 29–81. ISBN 978-0-323-46153-5.
42. Terchi, S.; Hamrit, S.; Ladjal, N.; Bachari, K.; Ben Rhaiem, H. Synthesize of Exfoliated Poly-Methylmethacrylate/Organomontmorillonite Nanocomposites by in Situ Polymerization: Structural Study, Thermal Properties and Application for Removal of Azo Dye Pollutant. *J. Therm. Anal Calorim.* **2024**, *149*, 2161–2177. [CrossRef]
43. Guo, F.; Aryana, S.; Han, Y.; Jiao, Y. A Review of the Synthesis and Applications of Polymer-Nanoclay Composites. *Appl. Sci.* **2018**, *8*, 1696. [CrossRef]
44. Kotal, M.; Bhowmick, A.K. Polymer Nanocomposites from Modified Clays: Recent Advances and Challenges. *Prog. Polym. Sci.* **2015**, *51*, 127–187. [CrossRef]
45. Orta, M.d.M.; Martín, J.; Santos, J.L.; Aparicio, I.; Medina-Carrasco, S.; Alonso, E. Biopolymer-Clay Nanocomposites as Novel and Ecofriendly Adsorbents for Environmental Remediation. *Appl. Clay Sci.* **2020**, *198*, 105838. [CrossRef]
46. Xu, H.; Hao, Z.; Wang, C.; Deng, J.; Wang, T.; Zhang, J. In Situ Emulsion Polymerization to Multifunctional Polymer Nanocomposites: A Review. *Macromol. Chem. Phys.* **2023**, *224*, 2300185. [CrossRef]
47. Anjum, A.; Gupta, D.; Singh, B.; Garg, R.; Pani, B.; Kashif, M.; Jain, S. Clay–Polymer Nanocomposites for Effective Water Treatment: Opportunities, Challenges, and Future Prospects. *Environ. Monit. Assess.* **2024**, *196*, 666. [CrossRef] [PubMed]
48. Azeez, A.A.; Rhee, K.Y.; Park, S.J.; Hui, D. Epoxy Clay Nanocomposites—Processing, Properties and Applications: A Review. *Compos. Part B Eng.* **2013**, *45*, 308–320. [CrossRef]

96. Pıhtılı, G.; Tanyol, M. Modelling and Optimization of Indigo Carmine Adsorption onto P(NIPAM-Co-AN)/Clay Composite Using Response Surface Methodology. *Int. J. Pure Appl. Sci.* **2022**, *8*, 71–80. [CrossRef]
97. Jawad, A.H.; Abdulhameed, A.S.; Kashi, E.; Yaseen, Z.M.; ALOthman, Z.A.; Khan, M.R. Cross-Linked Chitosan-Glyoxal/Kaolin Clay Composite: Parametric Optimization for Color Removal and COD Reduction of Remazol Brilliant Blue R Dye. *J. Polym. Environ.* **2022**, *30*, 164–178. [CrossRef]
98. A pH-Sensitive Surface of Chitosan/Sepiolite Clay/Algae Biocomposite for the Removal of Malachite Green and Remazol Brilliant Blue R Dyes: Optimization and Adsorption Mechanism Study | Journal of Polymers and the Environment. Available online: https://link.springer.com/article/10.1007/s10924-022-02614-y?fromPaywallRec=true (accessed on 28 August 2024).
99. Reghioua, A.; Atia, D.; Hamidi, A.; Jawad, A.H.; Abdulhameed, A.S.; Mbuvi, H.M. Production of Eco-Friendly Adsorbent of Kaolin Clay and Cellulose Extracted from Peanut Shells for Removal of Methylene Blue and Congo Red Removal Dyes. *Int. J. Biol. Macromol.* **2024**, *263*, 130304. [CrossRef]
100. Junceda-Mena, I.; García-Junceda, E.; Revuelta, J. From the Problem to the Solution: Chitosan Valorization Cycle. *Carbohydr. Polym.* **2023**, *309*, 120674. [CrossRef] [PubMed]
101. Riofrio, A.; Alcivar, T.; Baykara, H. Environmental and Economic Viability of Chitosan Production in Guayas-Ecuador: A Robust Investment and Life Cycle Analysis. *ACS Omega* **2021**, *6*, 23038–23051. [CrossRef] [PubMed]
102. Babel, S.; Kurniawan, T.A. Low-Cost Adsorbents for Heavy Metals Uptake from Contaminated Water: A Review. *J. Hazard. Mater.* **2003**, *97*, 219–243. [CrossRef] [PubMed]
103. Singh, R.P.; Khait, M.; Zunjarrao, S.C.; Korach, C.S.; Pandey, G. Environmental Degradation and Durability of Epoxy-Clay Nanocomposites. *J. Nanomater.* **2010**, *2010*, 352746. [CrossRef]

minerals

MDPI

Review

An Overview of Soil Pollution and Remediation Strategies in Coal Mining Regions

Abdulmannan Rouhani [1,*], Jeff Skousen [2] and Filip M. G. Tack [3]

[1] Department of Environment, Faculty of Environment, The University of Jan Evangelista in Ústí nad Labem (UJEP), Pasteurova 3632/15, 400 96 Ústí nad Labem, Czech Republic
[2] Division of Plant and Soil Science, West Virginia University, Morgantown, WV 26506, USA; jeff.skousen@mail.wvu.edu
[3] Department of Green Chemistry and Technology, Ghent University, Frieda Saeysstraat 1, 9052 Ghent, Belgium; filip.tack@ugent.be
* Correspondence: a.rohani70@gmail.com

Abstract: Coal remains a very important source of energy for the global economy. Surface and underground coal mining are the two major methods of coal extraction, and both have benefits and drawbacks. Surface coal mining can have a variety of environmental impacts including ecosystem losses, landscape alteration, soil destruction, and changes to surface and groundwater quality and quantity. In addition, toxic compounds such as heavy metals, radioactive elements, polycyclic aromatic hydrocarbons (PAHs), and other organic contaminants are released in the environment, ultimately affecting the health of ecosystems and the general population. Underground mining has large impacts on underground water supplies and water quality, but generally has less visual surface impacts such as leaving waste and tailings on the surface and subsidence problems. In response to the concern about these environmental issues, many strategies have been developed by scientists and practitioners to minimize land degradation and soil pollution due to mining. Reclamation laws passed in numerous countries during the past 50 years have instituted practices to reduce the impacts of soil pollution including burying toxic materials, saving and replacing topsoil, and vegetating the land surface. While modern mining practices have decreased the environmental impacts, many sites are inadequately reclaimed and present long-lasting soil pollution problems. The current review summarizes progress in comprehending (1) coal mining impacts on soil pollution, (2) the potential risks of soil pollution associated with coal mining, and (3) different types of strategies for remediating these contaminated soils. Research and prospective directions of soil pollution in coal mining regions include refinements in assessing pollutant levels, the use of biochars and other amendments, phytoremediation of contaminated soils, and the release of toxic elements such as mercury and thallium.

Keywords: coal mining; soil pollution; heavy metals; PAHs; remediation methods

1. Introduction

Coal is the second-most-significant global energy source, contributing 27.1% to the main energy generation [1]. At the beginning of the 18th Century, coal became the main energy source, eventually leading to the Industrial Revolution. From the late Eighteenth Century onwards, the techniques of underground mining were developed, mainly in Britain [2]. Despite its positive impact on human development, coal mining has had adverse effects on the environment. Its activities have resulted in significant waste generation (tailings) and long-term environmental destruction, such as polluting the water systems, soil erosion, mine subsidence, ecosystem and biodiversity destruction, and land surface disruption [3]. Furthermore, mining has resulted in significant societal costs, including fatal accidents, health risks, and community displacement [4].

Historic mining practices have caused various hazards to the environment and humans. Some of those traditional practices are continued today in many parts of the world. In industrialized cultures, mining and reclamation laws have mitigated the degradation by stronger environmental requirements and requiring reclamation once mining has ended. Current mining and reclamation standards regulate site preparation, blasting, road building, overburden removal and placement, water control and maintenance, topsoil replacement, and vegetation establishment. Together, these practices greatly reduce the potential for land and water degradation during and after mining [5]. The largest problem in many countries is reclaiming sites that were abandoned and thus remain exposed to continued erosion and degradation of the surrounding environment. This problem often occurs where funding is unavailable for reclaiming abandoned lands or where laws have not been passed to deal with these abandoned sites. This paper deals with pollution in areas caused by wastes and tailings from surface and underground mining that have been left on the surface.

Coal mining can be classified into two types: surface and underground mining. Large quantities of waste are generated in either case due to the coalification process and other impacts associated with mining and processing technologies [6]. Based on the geological and mining circumstances, a total of 0.4–0.7 t of waste is produced for every tonne of coal that is mined [7]. Compared to underground mines, opencast (surface) coal mines have higher environmental impacts on the soil, water, and air, whereas underground mining entails the risk of land subsidence and groundwater impacts below the land [8]. Heavy metals from dumps of spoil or refuse can be mobilized by surface runoff, which can leach into the subsurface soil or into nearby water resources [9]. The disturbed areas in coal mining regions can, therefore, be environmentally compromised, exposing organisms in the food chain to pollutants by direct contact, along with the inhalation of toxic metals, which can translate into human health effects [10].

Soil contamination studies conducted in coal mining areas have shown that mining and the associated operations release toxic elements into the soil, which then affect nearby riparian environments, surface water, and farmland [11,12]. The elevated levels of heavy metals in materials left on the surface pose a potential risk to the environment and human health, because of their bioaccumulation, toxicity, and release to organisms in the environment [13,14]. This elevated metal pollution may be absorbed by plants and crops, which leads to inferior growth and plant tissue contamination [15]. Polycyclic aromatic hydrocarbons (PAHs) are the primary organic toxic components found in coal mining areas [16]. The emissions from the combustion of small coal particles and coal gangue (also called coal refuse or coal waste) are the primary sources of PAHs in coal mine districts. Depositional environments and coal rank are two primary factors that determine the PAH content in coal gangues [17]. Excessive deposition of coal gangue on the surface can lead to spontaneous combustion under the right conditions, resulting in atmospheric pollution. At the same time, leaching due to precipitation events will result in heavy metals and PAHs contained in the gauge being released into the soil and surface water [18].

Coal has been used for millennia as a fuel source. Archaeologists discovered in 2005 what appeared to be a 120,000-year-old Stone Age opencast coal mine in Germany. According to the China Coal Information Institute, coal in China was utilized as a fuel source during the Neolithic Era, approximately 10,000 years ago. In the Americas, the Aztecs were the first people to use coal. For Aztecs, coal was employed not just as a heating source, but also as a decorative element. In the 18th Century, colonists in eastern North America discovered coal reserves. Worldwide coal reserves are estimated at over 1.1 trillion t. Coal deposits can be found in more than 70 countries across all continents, with the largest reserves located in the United States, Russia, China, and India [19]. In the modern world, the global demand for low-cost energy, iron, steel, and cement has resulted in the continued growth of coal extraction. Coal reserves are projected to last longer than conventional oil and gas reserves by 115 years, given current extraction rates. Ninety percent of the world's coal production originates from ten countries [20].

Studies regarding soil pollution around coal mining areas have significantly increased in recent decades. The source, distribution, and content of organic and inorganic pollutants in soils around abandoned or active coal mine areas have been the subject of many investigations so far. Currently, the only existing reviews are limited to the broad evaluation of worldwide metal pollution in soils near coal mines and their health risk assessment [21,22] and also changes in soil characteristics caused by surface coal mining and subsequent reclamation [23,24]. The objectives of the study were: (1) to evaluate the contamination levels of organic and inorganic pollutants in soils around both surface and underground coal mining areas and (2) to identify the most-common remediation strategies used for the removal of pollutants from coal mining areas.

2. Methodology

We searched for articles from around the world that described soil pollution in mining areas and tried to sort them into categories and summarize them. Based on the aim of this study, we performed a search on the Scopus and Web of Science databases with the terms (coal mining) AND (soil) AND (PAHs) AND (heavy metal) AND (Remediation Strategies). Finally, in accordance with studies published from 2000 to 2023, a global analysis was undertaken in this study.

3. Contents and Chemical Forms of Trace Metals and Metalloids in Coal

Coal is a highly complex geological material that contains over 200 minerals and nearly all naturally occurring elements. It also has an intricate chemical structure [25]. According to Finkelman [26], the modes of occurrence of metals in coal refer to their chemical bonding and distribution within the coal. Ultimately, this knowledge is extremely important in foreseeing their potential impact on human health and the environment [25]. Many scientific publications have addressed the detrimental environmental effects of the metals contained in coal [27–29]. The majority of these studies were concerned with the volume of elements released into the environment due to coal mining, disposal, and combustion [25]. In the following sections, the occurrence and chemical forms of selected trace elements in coal are briefly detailed.

3.1. Cr

According to Ketris and Yudovich [30], common low-rank coals typically have 15 mg/kg and hard coals contain 17 mg/kg of Cr on average. Considering the toxicity and carcinogenicity of, in particular, the Cr(VI) form, the chemical form of occurrence is very important. Goodarzi et al. [31] found that Cr is associated with illite and occurs in coal primarily as Cr(III) rather than the hazardous Cr(VI) in the corresponding fly ashes. Several studies have indicated that Cr in coal is predominantly linked to organic matter, especially in low-rank coals [32], clay minerals, spinel-group minerals [26,33], and other mineral phases including carbonates [34]. In summary, the main association of Cr in coal is with organic matter and aluminosilicates [25].

3.2. Ni

The common low-rank coals have a mean Ni concentration of 9 mg/kg, which is approximately half of the worldwide concentration in hard coals at 17 mg/kg [30]. Dai et al. [35] revealed that the mean content of Ni in Chinese coals is 13.7 mg/kg, while Orem and Finkelman [33] discovered that the mean content of Ni in U.S. coals is 14 mg/kg. In coal, Ni is linked to clays, sulfides (mostly pyrite and other minor sulfides), and organic matter [26,32,36]. During a density fraction study, Kolker et al. [37] discovered that Ni is enriched in both the heaviest- and lightest-density fractions, suggesting that Ni in coal is bound with both organic and pyritic components. In conclusion, Ni in coal is mostly connected to clays, organic matter, and sulfides [25].

3.3. Cu

The mean contents of Cu in low-rank and hard coals around the world are similar, at 15 mg/kg and 16 mg/kg, respectively [30]. These values were comparable to those reported for U.S. and Chinese coals, which were 16 mg/kg and 17.5 mg/kg, respectively [33,35]. In most coals, Cu is often found as chalcopyrite and pyrite and can be organically bound, particularly in low-rank coals [26,32]. Finkelman et al. [38] found that Cu can be found in low-rank coals mainly in organic fractions (20%), clays (20%), chalcopyrite (30%), and pyrite (30%). Similar forms of occurrence can be found in bituminous coals, but in different percentages, with 5% organic, 20% clays, 30% chalcopyrite, and 45% pyrite. In conclusion, Cu in coal is mainly found in pyrite and chalcopyrite. In some low-rank coals, it could be found in the organic fractions [25].

3.4. Zn and Cd

Zinc and Cd have a similar chemical behavior. They are both commonly associated with sphalerite and usually behave in the same way in coal [39]. Zinc in coal was found at 28 mg/kg in hard coal and 18 mg/kg in common low-rank coal across the world [30]. According to Orem and Finkelman [33] and Dai et al. [35], the mean Zn concentration in U.S. and China coals is 53 mg/kg and 41.4 mg/kg, respectively. These values are significantly higher than the mean values in coals from other parts of the world. However, the concentrations of Cd are much lower in common low-rank coal with 0.24 mg/kg and hard coal with 0.20 mg/kg globally [30]. Mean Cd contents in U.S. and Chinese common coals are 0.47 mg/kg and 0.25 mg/kg, respectively [33,35].

Many studies have shown that sphalerite is the primary host for Zn in coal [40,41]. Although sphalerite is the most-common mineral containing Cd [42], other minerals including silicate and pyrite may also be associated with Cd [43]. Zinc in coal is often bound with sulfide and aluminosilicate, as well as organic matter [44,45]. In conclusion, sphalerite is the primary host of Zn and Cd in coal. Zn in low-rank coal is bound with the organic fraction [25].

3.5. As

Ketris and Yudovich [30] reported that the contents of As in low-rank and hard coals worldwide are 7.6 mg/kg and 9.0 mg/kg, respectively. However, Dai et al. [35] found that a lower mean concentration in Chinese coals of 3.8 mg/kg. According to Swaine [42], As is most commonly associated with pyrite; however, it can also be bound with phosphates, clay minerals, and organic matter. By employing numerous approaches, some researchers demonstrated that As has an organic association in some sub-bituminous and lignite coals [43,46]. In summary, As in coal is mainly found in pyrite and has an association with organic matter. Some As-bearing minerals, such as Ti-As sulfide, zeunerite, orpiment, realgar, getchellite, and arsenopyrite, have been found in coal; however, these minerals are neither common, nor unique hosts for As in coal [25].

3.6. Pb

On a worldwide basis, the mean Pb contents in hard and low-rank coals are 9.0 mg/kg and 6.6 mg/kg, respectively [30]. The mean Pb contents in U.S. and Chinese coals are higher than those found in common coals across the world, at 11 mg/kg and 15.1 mg/kg, respectively [33,35]. In general, Pb in low-rank coals is bound with organic matter [32,42]. Using selective leaching, Finkelman et al. [43] found that Pb in bituminous and low-rank coals is associated with various phases, including silicate, pyrite, monosulfides (galena), and other unknown phases. In conclusion, pyrite is the most-common mineral form of Pb in coal, while it can also be found in organic matter and trace minerals (including, clausthalite and galena) [25].

4. Levels of Soil Pollution in Coal Mining Areas

4.1. Heavy Metals and Metalloids

Around the mine area, a significant quantity of heavy metal dust and polluted wastewater may be produced during the many phases of the coal mining process, including transportation, wastewater treatment, tailings' placement and slurry pumping, topsoil removal, mineral extraction, and ore concentration [47]. These heavy metals potentially can migrate to ground water and aquatic environments via surface runoff. The local natural ecosystem can be severely disrupted by the long-term effects of mining disturbances, which include groundwater quality impacts, soil erosion, and land subsidence [48]. Coal gangue and fly ash are sources of toxic heavy metals, which can be emitted into the environment during coal transport, smoke and dust emissions from coal-fired power plants, stormwater discharges from mining areas, and other operations associated with the coal industry [49].

The primary source of pollution in coal mines is attributed to the process of leaching acidic mine tailings or the creation and release of acid mine drainage (AMD) [50]. AMD is formed because of the presence of pyrite and sulfide minerals within the coal seam and associated rocks. Exposure to water and air causes these sulfide minerals to be oxidized. Free sulfuric acid is one of the oxidation products [51,52]. Metal elements from sulfide oxidation, including Zn, Al, and Ni, as well as toxic materials containing Cr, Pb, and Cu, are released after the pH is decreased [53]. Metal pollutants resulting from mining activities can be transported over long distances by water and wind erosion or remain in tailings for later release with continued weathering [54]. In mining areas, these pollutants in soil may have adverse impacts on nutrient availability and microbial activity [55].

Currently, research on heavy metals in soil has primarily been concentrated on analyzing the distribution pattern, total content, and ecological and human health risks [56]. The source of heavy metals in the soil from the mining area, however, is more complicated given that it may have arisen from not only emissions during coal extraction, but also from atmospheric deposition from coal-fired power plants and the dispersion during the process of transporting coal as well [57]. Studies regarding the concentration of trace metals and metalloids in coal mining areas worldwide are provided in Table 1, and some of those studies are shown below.

According to the literature reviews, during the last ten years, China has conducted numerous studies on soil heavy metal pollution caused by coal mining activities. For example, Yan et al. [58] conducted a study in a coal mining district in Tai'an City and found that Hg and Cd were the primary pollutants in the area. Additionally, Cd exhibited the highest rate of contamination. This result was associated with roads that were employed for transporting coal from the mines. Sun et al. [59] reported similar results in the Tangshan Coal Industrial City, where a majority of soil samples exhibited slight to moderate contamination by Hg and Cd, mainly due to coal-related industrial activities, vehicle emissions, agrochemical application, and sewage irrigation. Zhang et al. [60] discovered that the primary sources of As and Hg in soils from major coal mining regions of Xinjiang were atmospheric dust emissions resulting from coal combustion. Additionally, the high volume of traffic and frequent transportation associated with coal delivery in the study area were the main source of soil Pb pollution. According to Li et al. [61], the increased concentrations of Cu and Cd in paddy soils in the vicinity of the Zhangji coal mine in Huainan City were associated with mining activities, i.e., from rainfall leaching. According to Dong et al. [62], the main cause of soil pollution in the Liuxin mining area of Xuzhou was attributed to coal mining activities. However, there were also other contributing factors such as industrial pollution, sludge addition, and untreated water irrigation.

Kou et al. [63] revealed that the soils in Shengli and Baorixile coal mines located in Inner Mongolia showed elevated levels of pH and heavy metals such as Hg, As, Ni, Mn, Zn, Pb, and Cu, surpassing the mean soil background values, as a result of high coal mining and burning volume. In Mongolia's main coal mining town (Sharyn Gol), Pecina et al. [64] found high levels of Zn, Pb, Cu, and Cd in the soil, resulting from the deposition of mining waste heaps. Due to AMD, moderate soil pollution by Ni, Zn, and Cu was found by Hossen et al. [65] in the area surrounding the Barapukuria coal mine in Bangladesh. Siddiqui et al. [11] provided a comprehensive analysis of the effect of elements in soils from the aforementioned mining area. Zinc exceeded the world normal limit [66] among the evaluated elements. The study revealed that Ag showed the highest degree of enrichment, followed by As. In this study, Zn and As were potentially derived from minerals such as mispickel, melnikovite, and sphalerite, which are commonly found in association with coal deposits.

A study conducted In Neturia, West Bengal, showed that opencast mining has led to an elevated level of toxic heavy metals in the soil, leading to heavy metal contamination and associated non-carcinogenic health risks to humans. Moreover, this study found that topsoil was exposed at the surface, which made it more susceptible to heavy metal pollution [10]. According to a study conducted by Siddiqui et al. [67], soils collected from the Jharia coalfield in India exhibited a moderate level of pollution risk in terms of trace elements such as Cu, Pb, Zn, Cr, Ni, Co, and Cd, released from mining spoils. Raj et al. [68] in their investigation of soils near an opencast coal mine in Eastern India found that coal mining activities including coal dust deposition and transport activities were the predominant source of metal pollution in the area. Reza et al. [69] indicated that agricultural soils in the Ledo coal mining area of India were significantly contaminated with Cd and Pb due to mine drainage. Ahmad et al. [70] found that the levels of Zn, Cu, Ni, and Cr in the soil of the Dukki and Sharigh coal mines in Pakistan exceeded the acceptable limits.

Marove et al. [71] found that the type of leaching solution strongly impacted the hazardous elements from soils around open-pit coal mines in Mozambique. They also discovered that some of these elements were highly bio-accessible, which could lead to health and ecological risks in the area. Zerizghi et al. [72] conducted a study on the soil in the surrounding area of the Greenside coal mine in South Africa. They found that the levels of heavy metals examined in the soil surpassed the local background value. Moreover, the predominant health and ecological hazards associated with Cr were discovered. Ameh and Aina [73] found that the Cr, Cd, and As concentrations in soil samples obtained from the polluted coal mine in Okaba, Nigeria, were within the standard limits for unpolluted soils. However, the Cu concentration in the soil samples exceeded the permissible limits. Galunin et al. [74] reported findings from a coal mining region in Brazil showing that the contents of Mn, Ca, Al, Mg, and pH separately had a considerable impact on the Cd sorption behavior, and similarly, these factors also impacted the desorption behavior.

Alekseenko et al. [75] found that coal mining waste disposal sites in Rostov Oblast, Russia, had a direct impact on the soil, which resulted in increased levels of Ba, Mo, Pb, Zn, Cr, V, Cu, and Mn. In Poland, Pietrzykowski et al. [76] found that areas impacted by the Smolnica hard coal mine that had been reclaimed and reforested did not pose any risk of hazardous levels or the bioavailability of metals. The study conducted by Fiket et al. [77] in Croatia demonstrated that the soils in the Labin City area exhibited distinct metal fingerprints as a result of coal mining activities, particularly coal combustion or transport infrastructure. Consequently, soils showed varying degrees of enrichment in Zn, Sr, Sb, Pb, Mo, Cu, and Co, ranging from moderate to extremely high. Boahen et al. [78] found that the levels of Zn, Cd, Be, and As in the soil samples collected from the North Bohemian Brown Coal Basin (Czech Republic) surpassed the recommended thresholds. The agricultural soils from the Northern Czech Republic were found to be contaminated by anthropogenic material associated with past mining activities, such as lignite mining in North Bohemia and hard coal mining and heavy industry in North Moravia [79]. Potential environmental implications from AMD and associated dissolved metals discharging in Portugal's Douro coalfield have been reported by Ribeiro et al. [80]. In Spain, Boente et al. [81] revealed that coal mining and processing history in the municipality of Langreo had a strong association with the increased levels of Cu in the soil.

Table 1. Concentration of trace metals and metalloids in coal mining areas worldwide (mg/kg).

Country		Mine Type	Status	HMs (mg/kg)								References
				Cd	Hg	Cu	Ni	Pb	Zn	As	Cr	
South Korea: Gangreung coalfield		unknown	abandoned	1.10	n.a.	41.0	42.6	32.9	87.3	n.a.	35.8	[82]
China: Linhuan coal mining area		underground	active	0.17	n.a.	n.a.	n.a.	n.a.	n.a.	n.a.	n.a.	[12]
China: Coal mining region of Tai'an		underground	active	0.20	0.03	26.5	29.61	27.6	66.7	9	56.4	[58]
China: Zhundong mining		surface	active	n.a.	0.01	n.a.	n.a.	16.3	0.0067	9.0	53.0	[60]
China: Kaili City		unknown	abandoned	0.52	0.11	13.3	n.a.	30.7	56.2	28.2	75.6	[83]
China: Yulin		underground	active	n.a.	n.a.	15.0	17.9	21.5	57.6	5.3	45.1	[84]
China: Xilinguole League		surface	active	0.11	n.a.	15.0	n.a.	20.0	49.3	9.1	52.9	[85]
China: Xuzhou coal mining region		unknown	active	0.88	n.a.	22.9	n.a.	19.0	128.3	17.1	58.1	[86]
China: Tangshan City		underground	active	0.15	0.06	22.4	16.8	22.9	70.3	5.9	37	[59]
China: Weibei coalfield		underground	active	0.28	n.a.	58.1	88.1	11.49	1139.8	n.a.	143.1	[87]
China: Zhangji coal mine		underground	active	0.54	n.a.	64.4	27.7	20.3	118.0	23.4	85.3	[61]
China: Yanzhou		underground	active	0.14	n.a.	23.1	29.5	23.7	66.3	n.a.	72.1	[88]
China: Eastern Junggar coal mine		surface	active	n.a.	n.a.	19.3	n.a.	16.7	47.8	33.5	68.5	[89]
China: Xinzhuangzi		unknown	unknown	0.09	n.a.	36.8	28.4	25.4	62.0	3.7	n.a.	[48]
China: Datong coal mine		underground	abandoned	n.a.	n.a.	26.4	34.3	50.2	76.1	n.a.	188.6	[90]
China: Xingren coal mine		unknown	active	1.09	0.71	134.1	64. 9	240.6	n.a.	477.5	173.6	[91]
China: Liuxin mining area		unknown	abandoned	0.40	0.03	31.8	n.a.	12.9	91.3	24.3	44.5	[62]
Mongolia: Sharyn Gol		surface	active	1.46	n.a.	22.1	n.a.	69.0	87.6	n.a.	n.a.	[64]
Mongolia	Shengli	surface	active	n.a.	0.03	19.6	25.0	16.8	60.7	9.0	n.a.	[63]
	Baorixile	surface	active	n.a.	0.08	19.7	20.8	26.3	64.2	8.6	n.a.	
Bangladesh: Barapukuria coal mine		underground	active	n.a.	n.a.	31.6	56.5	n.a.	101.9	n.a.	82.3	[65]
				0.25	n.a.	27.4	48.0	11.2	128.0	5.1	17.5	[11]
				n.a.	n.a.	n.a.	n.a.	433.0	296.0	17.5	n.a.	[66]
				n.a.	n.a.	29.9	98.5	188.6	160.0	22.4	107.3	[92]

Table 1. *Cont.*

Country		Mine Type	Status	HMs (mg/kg)								References
				Cd	Hg	Cu	Ni	Pb	Zn	As	Cr	
Bangladesh: Farmland around Barapukuria coal mine		underground	active	0.09	n.a.	2.1	n.a.	0.9	8.2	n.a.	n.a.	[93]
India: Neturia block		surface	active	n.a.	n.a.	677.0	811.5	265.2	893.7	33.7	851.7	[10]
India: Jharia coalfield		underground and surface	active	0.80	n.a.	11.3	11.3	11.4	19.9	n.a.	23.4	[67]
				0.40	n.a.	66.3	64.1	27.8	127.0	n.a.	43.0	[94]
India: Rohini		surface	active	1.11	0.58	n.a.	n.a.	16.9	n.a.	3.1	17.5	[68]
India: Korba coalfield		surface	active	n.a.	n.a.	218.0	n.a.	311.0	426.0	n.a.	567.0	[95]
India: Raniganj coalfield		underground and surface	active	n.a.	n.a.	256.6	347.3	88.2	369.1	15.4	306.3	[96]
India: Eastern part of Jharia coalfield		underground and surface	active	2.49	1.19	26.0	n.a.	15.7	55	2.7	55.0	[97]
India: Ledo coal mining area		surface	active	2.60	n.a.	n.a.	87.5	183.1	n.a.	n.a.	112.3	[69]
India: Surat district		unknown	active	0.77	n.a.	57.0	16.0	9.0	60.0	n.a.	8.0	[98]
Pakistan	Dukki coal mines	underground	active	n.a.	n.a.	421.6	125.1	n.a.	131.2	n.a.	149.1	[70]
	Sharigh coal mines	underground	active	n.a.	n.a.	53.7	115.4	n.a.	68.5	n.a.	95.7	
Iran: Aghdarband coal mine		underground	active	n.a.	n.a.	33.0	14.0	52.0	89.0	33.3	39.0	[99]
Iran: Tazareh coal mine area		underground	active	0.31	n.a.	27.1	43.1	19.2	78.8	0.2	91.5	[100]
Turkey: Oltu coal mine district		underground	active	0.05	1.15	28.0	97.0	49.4	36.8	n.a.	180.2	[101]
Turkey: Ovacik–Yaprakli		underground and surface	abandoned	n.a.	n.a.	58.0	296.8	n.a.	101.1	n.a.	348.0	[102]
Australia	Glenbawn	surface	active	n.a.	n.a.	59.0	n.a.	12.8	137.0	14.2	n.a.	[103]
	Traralgon	surface	active	n.a.	n.a.	22.6	n.a.	22.1	132.8	5.6	n.a.	
Mozambique: Moatize district		surface	active	n.a.	n.a.	34.0	31.0	26.0	78.0	3.0	95.0	[71]
South Africa: Witbank coalfield		underground	active	0.40	n.a.	33.1	98.3	30.7	110.6	19.4	653.3	[72]
South Africa: Emalahleni		underground	active	n.a.	n.a.	22.0	20.0	19.0	36.0	0.5	419.0	[104]

Table 1. *Cont.*

Country	Mine Type	Status	HMs (mg/kg)								References
			Cd	Hg	Cu	Ni	Pb	Zn	As	Cr	
Nigeria: Okaba	surface	active	1.05	n.a.	5.9	n.a.	n.a.	n.a.	1.4	5.2	[73]
Morocco: Jerada coal mine	underground	abandoned	n.a.	n.a.	32.6	n.a.	60.6	144.3	24.2	n.a.	[105]
Botswana: Morupule coal mine area	underground	active	n.a.	n.a.	40.3	60.5	36.6	304.5	13.2	155.6	[106]
Russia: Kizel Coal Basin	unknown	abandoned	n.a.	n.a.	n.a.	69.0	n.a.	80.0	10.0	178.0	[107]
			0.93	2.59	33.0	41.9	36.5	61.4	2.9	1653.0	[108]
Russia: Lipovtsy coalfield mine	surface	abandoned	0.03	n.a.	20.7	34.2	12. 8	187.8	n.a.	8.2	[109]
Russia: Rostov Oblast	underground	active	n.a.	n.a.	57.0	41.0	29.0	65.0	n.a.	90.0	[75]
Russia: Vorkuta coal mining area	underground	active	1.0	n.a.	27.0	31.0	38.0	380.0	5.6	27.0	[110]
Russia: Tula coal mining region	unknown	abandoned	n.a.	n.a.	n.a.	n.a.	n.a.	164.5	17.4	n.a.	[111]
Greece: Ptolemais lignite basin	surface	abandoned	n.a.	n.a.	n.a.	10.1	n.a.	n.a.	12.3	17.5	[112]
			0.4	n.a.	44.4	65	62.1	110.0	11.7	876.0	[113]
Poland: Upper Silesian Coal Basin	underground	active	2.55	n.a.	16.5	7.1	106.0	273.9	n.a.	67.1	[114]
			0.80	0.07	8.5	6.3	39.8	62.5	8.8	34.4	[115]
Poland: Smolnica hard coal mine	underground	abandoned	n.a.	n.a.	39.5	n.a.	61.5	80.0	n.a.	n.a.	[76]
Czechia: Litvínov City area	surface and	active	0.46	n.a.	31.1	26.0	48.9	175.0	41.7	45.6	[116]
	underground	abandoned									
Czechia: Sokolov Coal Basin	surface	active	0.98	n.a.	36.1	19.1	40.7	118.0	33.7	40.0	[117]
Czechia: Ostrava City	underground	active	0.21	0.19	21.1	n.a.	37.7	204.5	n.a.	17.4	[118]
Czechia: North Bohemian Region	surface	active	0.33	0.17	35.7	32.6	51.5	107.9	33.8	n.a.	[79]
Croatia: Raša coal mine	underground	abandoned	1.00	0.09	55.0	74.0	45.0	169.0	17.0	123.0	[119]
Croatia: Labin City area	underground	abandoned	2.01	n.a.	2226.0	176.0	484.4	2778.0	21.9	684.6	[77]
Spain: Langreo	underground	abandoned	0.60	0.40	39.0	18.3	91.6	136.2	21.8	18.9	[81]
Portugal: São Pedro da Cova	underground	abandoned	0.11	n.a.	50.2	24.3	50.2	97.0	22.6	74.1	[120]
Portugal: Douro coalfield	underground	abandoned	0.20	n.a.	36.5	21.4	30.8	57.0	38.3	92.3	[80]
Norway: Svalbard	underground	abandoned	n.a.	n.a.	n.a.	5.1	2.0	29.3	0.3	5.6	[52]
Wales (U.K.): Varteg mine	surface	abandoned	1.16	n.a.	63.9	n.a.	224.0	192.8	n.a.	n.a.	[121]

Note: n.a. = not analyzed.

This review revealed that the majority of publications in the analyzed areas originate from Asian countries, specifically China (17 articles) and India (9 articles). There is a research gap in this field of study from countries having high amounts of coal reserves, especially the U.S., Germany, Indonesia, Ukraine, and Kazakhstan. However, there are some limited studies from Australia, Poland, and Russia.

As shown in Figure 1, the most-analyzed element in coal mining areas is Zn followed by Cu and Pb, while Hg received much less attention compared to other metals. Coal mining activities often result in elevated levels of Hg in the surrounding environment [122]. Furthermore, as the extraction of coal increases, the concentration of Hg in regional soil also tends to rise [122]. Coal gangue particles emit Hg into the environment upon spontaneous combustion and leaching. Then, it migrates, transforms, and accumulates in ecological systems, ultimately entering the human body via multiple pathways and impacting human health [123]. Coal processing generates soot, dust, and wastewater, which can lead to the release of Hg into the environment. Additionally, the combustion of coal-preparation byproducts can lead to secondary emissions, which can contribute to local environmental contamination of Hg [124].

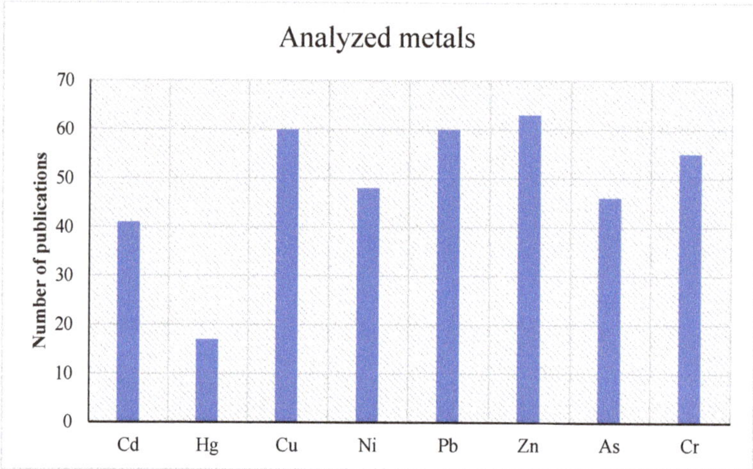

Figure 1. Most-analyzed metals around coal mining areas (mg/kg).

Thallium (Tl) is another element that is ignored in studies conducted around coal mining areas. This metal is classified as 1 of the 13 priority pollutants, which are considered to be more hazardous than Hg and Cd [125]. This metal is heavy, volatile, and highly toxic by exposure to contaminated soils and inhalation. While it is not commonly found in natural systems, it is significant in anthropogenic systems owing to its toxicity. Even at low concentrations in the environment, it has the potential to cause severe ecological risks [126]. Coal mining activities can lead to the release of Tl and its compounds into the environment, resulting in elevated levels of Tl in crops, soils, and water. This can pose a direct threat to human health through food chains and contact [127]. Zhang et al. [127] found that surface soils in the surrounding area of coal mines in Henan Province (China) had higher contents (0.77 mg/kg) of Tl than background values (0.42 mg/kg).

4.2. PAHs

Polycyclic aromatic hydrocarbons are a group of hydrophobic substances that contain at least two annelated aromatic rings and are categorized into subgroups based on the number of rings they contain [128]. They occur in coal and are generated through the process of the incomplete combustion of carbon-based fuels, including wood or coal. Additionally, they can also originate from natural sources such as petroleum input in oil seeps and volcanic eruptions [129]. These pollutants can also be generated through anthropogenic processes such as biomass burning, car exhausts, waste incineration, industrial emissions, and energy production [130]. The coal industry is another major contributor to the release of PAHs into the environment. This is due to the consumption of coal as a source of energy, as well as the accumulation of coal waste and the processing of coal, which have all led to the contamination of soil with PAHs [131]. In general, areas with heavy industrial activities in which coal facilities and metal industries are found are subject to higher levels of PAH emissions [132]. The European Union and the U.S. Environmental Protection Agency (USEPA) have both assigned PAHs as priority contaminants [133]. Wang et al. [134] conducted a comparative analysis of PAH concentrations in coals from different studies. Their findings revealed that the 16 priority PAHs as listed by the USEPA ranged from <0.1 to 260 mg/kg. Based on the coal rank and provenance, the total extractable PAHs in coals can range from a few tens to hundreds, in some cases thousands, of milligrams/kilogram [134,135].

In a study conducted by Xu et al. [136], it was found that the average polycyclic aromatic compounds (PACs) in the coal gangue, coal, and topsoil of typical coal mining regions in the Huaibei area of China were 10,908, 274,815, and 1528 mg/kg, respectively. In this study, alkyl PAHs were the main source of PACs, while coal and coal gangue exhibited notably higher levels of oxygenated PAHs compared to topsoil. Furthermore, the highest levels of PAH pollution were primarily identified in the vicinity of the coal mine region and nearby the road leading to the coal gangue landfill site. Liang et al. [137] conducted a study on soil in the Wuda coalfield area of China, wherein they observed the occurrence of coal fire sponges, which are polluted soil protrusions with a sponge-like appearance, in the Suhaitu mining area. This study showed a high level of contamination at 1000 mg/kg. Additionally, it was observed that low-molecular-weight PAHs were the predominant compounds, constituting over 50% of the total PAHs. In the topsoils from an extensive coal mine in Huainan, China, Zhang et al. [138] evaluated the presence, hazards, and influencing variables of PAHs. The concentration of PAHs in the soil was found to be comparatively higher than that of industrial, urban, and agricultural soils, while being lower than in some regions associated with coal mines and coal-fired power plants across the globe. Earlier work on this site by Wang et al. [139] revealed that the pollution level of PAHs in the soil profiles was higher in the upper layers of soil due to the gob pile and coal preparation plant. In this study, the mean PAH concentration measured in soil samples collected from coal mine areas was 0.84 mg/kg. This concentration is significantly higher than the natural PAH content in soil and also exceeds the Dutch standards. These findings suggest that coal mine areas have a higher carcinogenic burden.

According to a study conducted by Fan et al. [17], the levels of PAHs were found to increase as the distance from the coal gangue dump in the Gequan coal mine (China) decreased. These findings suggest that the organic matter present in the samples originates from coal particles found in the coal gangue dump. In a study from the Tiefa coal mine (Northeast China), Liu et al. [140] found that the surface soil had a total content of 16 PAHs within the range of 5.1 to 5642 mg/kg, with an average of 1118.3 mg/kg. The Tiefa coal mine's activities including coal gangue, unburned coal particulates, and coal combustion have resulted in soil pollution not only within the mining area, but also in the surrounding agricultural and lake bank soils. Moreover, the results indicate that the soils in the study area are heavily contaminated with PAHs and pose a significant health risk. According to Sun et al. [141], the coal gangue dump located at Jiulong coal mine is rich in deleterious organic compounds. Over the course of 15 years, wind and rain have contributed to the transportation of total organic carbon, sulfur, and aromatic compounds from the coal gangue dump into the surrounding soil and riverbed. Mizwar et al. [142] investigated surface soils of three distinct sites in South Kalimantan Province (Indonesia), namely coal stockpile, agricultural, and residential areas. Compared to agricultural and residential soil, they found that coal stockpile locations exhibited higher concentrations of PAHs in their soil. In this study, PAHs were a combination of pyrolytic and pyrogenic sources. Masto et al. [143] reported higher PAH contents in underground mine soils compared to surface coal mining in the Raniganj coalfield area (India), which may be attributed to natural coal burning at these sites. Yakovleva et al. [144] found that light polyarenes constituted a significant proportion (82–91%) of the total soil PAHs in both control and polluted sites in a study on soils affected by coal mining in Russia.

Jakovljević et al. [145] found that the soil in the Labin City area (West Croatia) had significantly increased levels of PAHs due to pollution from the Raša coal mining area, which has been ongoing for centuries. The findings showed the existence of pyrolytic PAHs formed by the burning of Raša coal at high temperatures in power plants, as well as unburned coal-derived PAHs resulting from Raša coal carbonization. In Spain, Boente et al. [146] found an elevated level of high-molecular-weight PAHs in an urban area that has been affected by coal mining activity such as coal combustion. They also noticed that benzo[a]pyrene levels exceeded the threshold in almost all of the assessed soil samples. In Germany, Hindersmann and Achten [147] found high levels of PAHs in urban soils affected by coal mining tailings. These PAHs were typically associated with non-point pyrogenic carbon sources such as particulate matter and soot. In all soil samples taken from the Saar River and the Mosel River downstream (Germany), Pies et al. [148] observed higher PAH concentrations as a result of past coal mining activities in this area. According to a report by Ugwu and Ukoha [149], the plant and soil samples taken from the Nigerian Okobo coal mine area showed low levels of toxic PAHs.

Similar to the heavy metals' and metalloids' section, the majority of conducted studies on PAHs' analysis around coal mining areas are from China. Furthermore, this field of study has been ignored or has limited attention in other countries having high levels of coal reserves.

5. Applied Soil Remediation Strategies in Coal Mining Areas

5.1. Bioaugmentation

The acidic environment and high levels of heavy metals at coal mine sites lead to a decrease in microbial abundance and diversity. In addition, coexisting abiotic parameters including dissolved oxygen, organic carbon, temperature, and ionic composition also contribute to the composition of microorganisms that survive in such an environment [150]. AMD-tolerant microorganisms include archaea, bacteria, and even certain eukaryotes, such as algae and fungi [151]. The adverse environmental conditions in soil and water at coal mine sites restrict the survival of bacteria to a few phylogenetic groups such as *Thiobacillus* spp. and *Methanogen* spp. [152]. The first microorganism that has been successfully separated from the acidic environment is *Thiobacillus ferrooxidans* [153]. The acid and metal tolerance of *Rhodococcus* spp., *Acinetobacter* spp., *Enterobacter* spp., *Staphylococcus* spp., *Klebsiella* spp., and *Bacillus* spp., which were obtained from coal mines and eventually cultured, have been reported [154,155]. According to Shylla et al. [156], indigenous bacteria present in soils with elevated levels of heavy metals could potentially be considered viable options for the bioremediation of heavy metals in polluted areas. Due to the fact that these bacteria are native to the system, it is possible that their ongoing interaction with the conditions of low pH and AMD enabled them to develop resistance. The differences in their ability to tolerate metals may be due to variations of in vitro mechanisms or cell wall composition [157].

Indigenous strains of *Thiobacillus* and *Bacillus* that are tolerant to metals have been discovered in mining environments. These strains have been proven to assist in the bio-reduction of toxic metals, including vanadium. This suggests that they could be a favorable method for metal bio-remediation [158]. For example, metal-resistant *Bacillus* spp. as described by Gupta et al. [159] can thrive in soils with Fe concentrations of 400–550 mg/kg. Furthermore, Upadhyay et al. [160] found that *Bacillus* spp. taken from polluted soils in coal mine areas exhibited tolerance to Cr (VI) and showed an efficient reduction of Cr (VI) to Cr (III). Several explanations have been provided regarding the mechanism underlying this conversion. It may happen when chromate functions as the final electron acceptor to obtain energy, or when bacteria produce waste products, or when multiple enzymes interact with Cr(VI) to mitigate its toxicity by turning it into Cr(III). Syed and Chinthala [161] found that *B. subtilis, B. cereus,* and *B. licheniformis* were effective at reducing the content of Pb in soils by 86%, 87%, and 78%, respectively. A study on bacteria in mine soils polluted with metals found that *Bacillus* was the most-predominant species identified. These organisms demonstrated notable levels of tolerance, as evidenced by their ability to endure the highest levels of 207–414 mg/kg Pb and 65–196 mg/kg Zn [162]. In this study, producing metal-chelating substances, including siderophores and organic acids, has been crucial in facilitating metal detoxification and enhancing metal tolerance in bacteria. Singh and Tiwary [163] conducted a study at the Chirimiri coal mines in India and discovered that the *Pseudomonas stutzeri* bacteria collected from the mines exhibited the ability to degrade high concentrations of pyrene and phenanthrene in polluted soil. In this study, several dioxygenase enzymes, namely protocatechuate, catechol, and gentisic acid, were produced by Pseudomonas stutzeri bacteria in order to degrade PAHs.

Despite all the reported advantages of this remediation strategy, Kurniawan et al. [164] claim that, since heavy metals are not effectively separated from the treated medium using bioaugmentation alone, this method cannot be used to treat heavy-metal-polluted soil in real-world settings at large scales. The majority of studies that have documented the effective removal of heavy metals from contaminated soil were primarily conducted in laboratory settings, where environmental conditions are carefully regulated.

5.2. Phytoremediation

The cultivation of hyperaccumulator plants, which are either naturally occurring or created as a result of genetic modification, is one of phytoremediation strategies for removing metals from the soil and accumulating them in plant biomass [165]. It has the potential to not only strengthen the ecological environment of the mining region, but also facilitate the recovery of vegetation [166]. Phytoremediation is a comprehensive remediation strategy that encompasses a range of techniques, including rhizodegradation, rhizofiltration, phytoevaporation, phytostabilization, and phytoextraction, all of which are mediated by microbial interactions with roots and soil [13]. In addition, it has economic benefits and aesthetic values, both of which can result in a wide range of social advantages [167,168]. As an illustration, Pandey and Bajpai [169] suggested in their study that abandoned mining sites have the potential to be converted into public spaces, such as parks, in order to offer ecosystem services and products. Additionally, this finding demonstrates the various advantages of phytoremediation for promoting sustainable ecosystem development [170]. Some concern has been expressed about the disposal of the biomass of hyperaccumulator plants, but there is evidence that metals bound into plant biomass can be stabilized and less available for release over time.

In coal gangue sites that experienced a decade of phytoremediation with single/mixed elm/poplar, Bai et al. [171] found that the soil contents of Th, U, Bi, Co, Ni, Pb, and Cu were considerably decreased. In their study, Feng et al. [172] applied herbaceous plants for performing vegetation restoration efforts in coal mines located in Hulunbuir (Inner Mongolia). The findings of this study indicated that the restoration of vegetation has had a positive impact on the overall restoration of the ecosystem within the studied area. According to a report by Ameh and Aina [73], the only plants that exhibited high potential as phytostabilizers of Cd in a coal mine soil in Nigeria contaminated with toxic metals were *Fuirena umbrellata* and *Selaginella myosurus*. Furthermore, *Hyptis Suaveolens* exhibited hyperaccumulation potential for Cu (>1000 mg/kg accumulation). Furthermore, they concluded that it is possible for native plants to naturally eliminate toxic elements from soils that have been contaminated with metal pollutants.

In Australia, successful Se extraction (48% and 28%) was achieved from phytoremediated post-mining coal wastes area by *Brassica juncea* [173]. Further Se extraction may have been accomplished in this study; however, *B. juncea* crops should be harvested immediately after they reach maturity. If the biomass of *B. juncea* is not harvested and dried, it has a tendency to break into chip-like particles that become easily dispersed by wind, potentially leading to the dispersal of accumulated heavy metals. Matanzas et al. [174] found that herbaceous species including *Lotus hispidus* and *Medicago lupulina* have the ability to translocate As and Pb in the polluted soils of a brownfield site from Spain. Consequently, these two plants demonstrate their capacity to serve as bioindicators for the presence of As and/or Pb pollution. Moreover, they could potentially function as phytoextractors or accumulators under varying conditions, such as when there is a higher concentration of potentially toxic elements in the soil compared to the study site. In a more-recent study at this site, Fernández-Braña et al. [175] found that *Buddleja davidii, Betula celtiberica*, and *Acer pseudoplatanus* were effective for phytostabilization in regions with elevated pollutant concentrations. However, these plant species were only suitable for phytoextraction in soils with low-to-moderate pollution levels.

Desai et al. [121] found that using *Alnus glutinosa* and *Betula pendula* for forestation was successful in remediating metals (Mn, Pb, Zn, Cu, and Cd) on moderately polluted lands resulting from opencast coal mining in South Wales. Additionally, this study examined soil measurements taken from various points along a 14-year forestation chronosequence. The results consistently showed that, as the age of the tree plantation increased, the level of soil metal contaminants decreased. This finding provides further evidence that forestation has a positive impact on reducing soil metal loadings. After 25 years of phytoremediation by the *Eucalyptus* hybrid tree in the Jharia coalfield (India), Bandyopadhyay et al. [176] found a considerable reduction in the soil content of Cu, Zn, and Pb. They suggested the application of metal-tolerant woody trees, specifically the *Eucalyptus* hybrid, with high biomass enhanced phytoremediation of coal mine overburden dumps contaminated with metals. These trees have the ability to accumulate considerable quantities of metals in their tissues and decrease metal contents in the soil. Niu et al. [90] reported that particular plant species, such as *Weigela hortensis* and *Ligustrum lucidum*, were found to have a significant potential for reducing the content of Pb, Ni, and Cr in soil from a reclaimed coal mining area in China. According to Mellem et al. [177], *Amaranthus dubius* is capable of hyperaccumulating As in regions associated with coal mining, but has limited capacity for the bioaccumulation of Ni, Cu, Pb, Hg, and Cr. This plant can extract and transport As to its aerial parts and is able to tolerate high levels of this metal.

Phytoremediation is a strategy that has various advantages and disadvantages. It is important to take these into account when considering the application of this strategy. Although cost-effectiveness is a favorable aspect, the duration required to observe the outcomes may be prolonged. It is imperative that the concentration of pollutants and the presence of other toxins should not exceed the tolerance thresholds of the plant species under consideration for utilization. The process of identifying plants with the capability to remediate multiple contaminants concurrently is a challenging operation. It is necessary to consider the limitations and potential for these plants to become part of the food web when implementing this strategy [178].

5.3. Biochar

Biochar is an appropriate substance for environmental applications, particularly in the remediation of contaminated soils, due to its low production cost and availability [179]. It is typically alkaline and possesses a significant surface area along with various active functional carbon groups capable of binding numerous cations [180]. Thus, biochar has the ability to reduce the availability, leachability, and mobility of toxic elements in polluted soils. Additionally, it can also decrease the absorption of these elements by plants [181]. Furthermore, biochar exhibited significant efficacy in mitigating the levels of PAHs in soil. Biochar generated at temperatures exceeding 400 °C indicated a notable propensity for PAHs owing to surface oxidation–reduction reactions. In addition, they show a higher level of aromatic clusters in a condensed form and possess a high degree of porosity characterized by the presence of well-developed nano- and micro-pores, which enable the sorption of low-molecular-weight PAHs [182].

Heavy metal remediation mechanisms in biochar involve precipitation, cation exchange, electrostatic attraction, reduction–oxidation, and physical adsorption. Biochar has been noticed to possess an interesting characteristic, whereby it can have an impact on the behavior of metals. The porous structures found in biochar have the capability to facilitate the transformation of metals into stable forms [183]. The -COOR and -OH functional groups in biochar surfaces contribute to the sorption of many heavy metals, making them unavailable [184]. According to studies, a competitive condition arises among metal ions in their interaction with functional groups on biochar surfaces due to the cationic nature of most metals. This leads to an enhanced immobilization of potentially hazardous elements in polluted soil. In general, the application of biochar treatment results in a significant increase in the levels of reducible and oxidizable heavy metals [185]. Biochar contains graphene moieties that serve as sites for both redox and adsorption reactions, resulting in a high affinity for metal ions and the capacity to transfer electrons to adsorbed reactants [184].

Reclamation principles are generally similar across different mine environments, despite differences in pollution type and level. The general acceptance of biochar for coal mine reclamation has been limited by lacking knowledge regarding its advantages, its accessibility at reclamation areas, the techniques required for its application, its cost, and the long-term effects of its field implementation. In addition, the study of biochar for the purpose of coal mine reclamation is a novel field of study that needs patience to assess its long-term effects when used in degraded areas [186].

The impact of applying chemical fertilizer and *Eucalyptus* wood biochar simultaneously to remediate Co-, Ni-, Zn-, and Cr-polluted soils from an operational coal mine dump in India was evaluated by Chandra et al. [187]. This study revealed that enhancing the mixing ratio of biochar from 0.5 to 5% (w/w) resulted in a considerable reduction of extractable heavy metal concentrations in the soil. The study conducted by Mujtaba Munir et al. [188] assessed the potential synergistic impacts of hydrothermally treated coal (HTC), raw coal (RC), and biochar (BC) on the accumulation, transformation, speciation, and immobilization of Pb, Cr, and Cd in soil polluted by the Huainan coalfield in Anhui, China. The findings showed that the co-application of BC-2% and BC-HTC amendments proved to be more efficient in mitigating Cd, Cr, and Pb contents in comparison to the singular application of RC or HTC amendments. This was achieved by increasing the organic carbon content and pH in the soil. Additionally, the application of BC-2% and BC-HTC amendments resulted in a respective increase of 1.5 and 2.5 units in soil pH. This led to the reduction of Pb, Cr, and Cd to more-stabilized forms in the soil.

Using soil from farms in the Huainan coal mine district, Dai et al. [189] conducted a laboratory experiment to investigate the effects of biochar amendments on the bioavailability and speciation of heavy metals. In this study, the concentrations of Cd, As, Zn, and Cu in polluted soil decreased by 42%, 7%, 51%, and 57%, respectively, when rice–straw-derived biochar was added to the soil. In a study conducted by Jain et al. [190], *Lemongrass* (*Cymbopogon ciatratus*) -derived biochar was applied as a soil amendment for spoil samples taken from coal deposits characterized by elevated levels of sulfur. According to the results, the application of biochar had a positive impact on the Palmarosa plant's metal tolerance index, increasing it by 54%. Additionally, it led to a reduction in acid generation from acidic mine waste. In an earlier study, Jain et al. [191] assessed the impact of Lemongrass-derived biochar application on the heavy metal contents of *Bacopa monnieri* plants growing in acidic coal mine spoil. In this study, using biochar in acidic mine spoil led to reduced levels of heavy metals; Pb decreased by 93%, Fe by 50%, Cu by 42%, Cr by 65%, and Al by 60%. After the application of algal-derived biochar on coal mine stockpiles, Roberts et al. [192] found that the concentrations of Cr, Ni, and Zn were reduced by 49%, 2%, and 55%, respectively. In this study, remediated soils with this biochar had lower or, in some cases, equal contents of metals compared to soil without biochar remediation.

6. Conclusions and Future Perspectives

Heavy metal soil pollution in coal mining areas has been known for many years. These pollutants are related not directly to the excavation of coal to the surface, but mainly to the deposition of dust from coal combustion in power plants located close to coal mines, coal transport activities, and the deposition of mining waste or coal combustion waste (ash coal). As long as post-mining areas remain unremediated and coal is used as a fuel or raw material for chemicals and other beneficial products, pollutants such as heavy metals can be released into the soil from coal consumption products and can be potentially hazardous. The release of these pollutants can be controlled by pre-mine planning and implementing mining and reclamation techniques to control, minimize, or eliminate the problems. The use of soil replacement on the surface can help mitigate pollutant releases by sorbing and stabilizing, depending on the pH, the organic matter content, the surface properties, and the crystalline structure. Soil remediation technologies have been developed and have shown effective mitigation of pollution levels and release. Methods to identify and analyze geologic materials that can cause problems, such as determining the sulfide contents of rocks, can help to determine the extent and minimize soil and water pollution. Reclamation activities using topsoil, amendments such as fertilizer and organic matter, and revegetation are important practices that need to be more widely adopted worldwide. Such techniques improve microbial interactions and processes that can decrease the availability of heavy metals. Phytoremediation is another process that has demonstrated good success in mitigating pollution. Biochar is capable of improving soil properties in an environmentally sustainable and cost-effective way, making it an excellent strategy for reclamation purposes.

Potential areas for future research include the following:

(1) Countries should identify and categorize the extent of mining-related disturbances, evaluate the pollution type and extent, and follow established guidelines for assessments and ecological remediation in mining regions.
(2) Coal consumption and transportation activities within the mining area can result in the release of other elements of concern into the environment. Two elements of concern, Hg and Tl, can have detrimental effects on ecosystems and pose a potential risk to human health through bioaccumulation in the food web. However, these hazardous metals, as well as others have not received attention in studies conducted around coal mining areas. Therefore, it is recommended to consider these elements in particular in future studies.
(3) Only a few studies focused on the remediation of coal-mine-degraded lands using the application of biochar. Further knowledge is needed to increase its use for reclamation and soil pollution remediation.

Author Contributions: A.R.: conceptualization, methodology, investigation, visualization, writing—original draft, writing—review and editing. J.S.: conceptualization, writing—review and editing. F.M.G.T.: conceptualization, writing—review and editing. All authors have read and agreed to the published version of the manuscript.

Funding: This research received no external funding.

Data Availability Statement: The datasets generated during this study are presented in the table in this manuscript.

Conflicts of Interest: The authors have no conflict of interest to declare that are relevant to the content of this article.

References

1. Zocche, J.J.; Sehn, L.M.; Pillon, J.G.; Schneider, C.H.; Olivo, E.F.; Raupp-Pereira, F. Technosols in coal mining areas: Viability of combined use of agro-industry waste and synthetic gypsum in the restoration of areas degraded. *Clean. Eng. Technol.* **2023**, *13*, 100618. [CrossRef]
2. Bian, Z.; Dong, J.; Lei, S.; Leng, H.; Mu, S.; Wang, H. The impact of disposal and treatment of coal mining wastes on environment and farmland. *Environ. Geol.* **2009**, *58*, 625–634. [CrossRef]

3. Byrne, C.F.; Stormont, J.C.; Stone, M.C. Soil water balance dynamics on reclaimed mine land in the southwestern United States. *J. Arid. Environ.* **2017**, *136*, 28–37. [CrossRef]
4. Yu, X. Coal mining and environmental development in southwest China. *Environ. Dev.* **2017**, *21*, 77–86. [CrossRef]
5. Skousen, J.; Zipper, C. Post-mining policies and practices in the Eastern USA coal region. *Int. J. Coal Sci. Technol.* **2014**, *1*, 135–151. [CrossRef]
6. Weiler, J.; Firpo, B.A.; Schneider, I.A.H. Technosol as an integrated management tool for turning urban and coal mining waste into a resource. *Miner. Eng.* **2020**, *147*, 106179. [CrossRef]
7. Maiti, S.K. *Ecorestoration of the Coalmine Degraded Lands*; Springer Science & Business Media: Berlin, Germany, 2012. [CrossRef]
8. Daozhong, C.H.; Qingli, Z.H.; Jie, W.A.; Xiaozhi, Z.H. Comparative analysis of ecological rucksack between open-pit and underground coal mine. *Energy Procedia* **2011**, *5*, 1116–1120. [CrossRef]
9. De, S.; Mitra, A.K. Mobilization of heavy metals from mine spoils in a part of Raniganj coalfield, India: Causes and effects. *Environ. Geosci.* **2004**, *11*, 65–76. [CrossRef]
10. Chakraborty, B.; Bera, B.; Roy, S.; Adhikary, P.P.; Sengupta, D.; Shit, P.K. Assessment of non-carcinogenic health risk of heavy metal pollution: Evidences from coal mining region of eastern India. *Environ. Sci. Pollut. Res.* **2021**, *28*, 47275–47293. [CrossRef]
11. Siddique, M.A.; Alam, M.K.; Islam, S.; Diganta, M.T.; Akbor, M.A.; Bithi, U.H.; Chowdhury, A.I.; Ullah, A.A. Apportionment of some chemical elements in soils around the coal mining area in northern Bangladesh and associated health risk assessment. *Environ. Nanotechnol. Monit. Manag.* **2020**, *14*, 100366. [CrossRef]
12. Tang, Q.; Chang, L.; Wang, Q.J.; Miao, C.; Zhang, Q.; Zheng, L.; Zhou, Z.; Ji, Q.; Chen, L.; Zhang, H. Distribution and accumulation of cadmium in soil under wheat-cultivation system and human health risk assessment in coal mining area of China. *Ecotoxicol. Environ. Saf.* **2023**, *253*, 114688. [CrossRef] [PubMed]
13. Mahar, A.; Wang, P.; Ali, A.; Awasthi, M.K.; Lahori, A.H.; Wang, Q.; Li, R.; Zhang, Z. Challenges and opportunities in the phytoremediation of heavy metals contaminated soils: A review. *Ecotoxicol. Environ. Saf.* **2016**, *126*, 111–121. [CrossRef] [PubMed]
14. Rouhani, A.; Makki, M.; Hejcman, M.; Shirzad, R.; Gusiatin, M.Z. Risk Assessment and Spatial Distribution of Heavy Metals with an Emphasis on Antimony (Sb) in Urban Soil in Bojnourd, Iran. *Sustainability* **2023**, *15*, 3495. [CrossRef]
15. Chojnacka, K.; Chojnacki, A.; Gorecka, H.; Gorecki, H. Bioavailability of heavy metals from polluted soils to plants. *Sci. Total Environ.* **2005**, *337*, 175–182. [CrossRef] [PubMed]
16. Ren, D.Y.; Zhao, F.H.; Dai, S.F.; Zhang, J.Y.; Luo, K.L. *Geochemistry of Trace Elements in Coal*; Science Press: Beijing, China, 2006.
17. Fan, J.; Sun, Y.; Li, X.; Zhao, C.; Tian, D.; Shao, L.; Wang, J. Pollution of organic compounds and heavy metals in a coal gangue dump of the Gequan Coal Mine, China. *Chin. J. Geochem.* **2013**, *32*, 241–247. [CrossRef]
18. Qi, C.; Fourie, A. Cemented paste backfill for mineral tailings management: Review and future perspectives. *Miner. Eng.* **2019**, *144*, 106025. [CrossRef]
19. Dave, S.R.; Tipre, D.R. Coal Mine Drainage Pollution and Its Remediation. In *Microorganisms in Environmental Management: Microbes and Environment*; Springer: Berlin/Heidelberg, Germany, 2012; pp. 719–743. [CrossRef]
20. Wagner, N.J. *Geology of Coal. Encyclopedia of Geology*, 2nd ed.; Elsevier: Amsterdam, The Netherlands, 2021. [CrossRef]
21. Sahoo, P.K.; Equeenuddin, S.M.; Powell, M.A. Trace elements in soils around coal mines: Current scenario, impact and available techniques for management. *Curr. Pollut. Rep.* **2016**, *2*, 1–14. [CrossRef]
22. Xiao, X.; Zhang, J.; Wang, H.; Han, X.; Ma, J.; Ma, Y.; Luan, H. Distribution and health risk assessment of potentially toxic elements in soils around coal industrial areas: A global meta-analysis. *Sci. Total Environ.* **2020**, *713*, 135292. [CrossRef]
23. Feng, Y.; Wang, J.; Bai, Z.; Reading, L. Effects of surface coal mining and land reclamation on soil properties: A review. *Earth-Sci. Rev.* **2019**, *191*, 12–25. [CrossRef]
24. Rouhani, A.; Gusiatin, M.Z.; Hejcman, M. An overview of the impacts of coal mining and processing on soil: Assessment, monitoring, and challenges in the Czech Republic. *Environ. Geochem. Health* **2023**, *2023*, 1–32. [CrossRef]
25. Dai, S.; Finkelman, R.B.; French, D.M.; Hower, J.C.; Graham, I.T.; Zhao, F. Modes of occurrence of elements in coal: A critical evaluation. *Earth-Sci. Rev.* **2021**, *222*, 103815. [CrossRef]
26. Finkelman, R.B. The Origin, Occurrence, and Distribution of the Inorganic Constituents in Low-Rank Coals. In Proceedings of the Basic Coal Science Workshop; US Department of Energy: Houston, TX, USA, 1981; pp. 69–90.
27. Finkelman, R.B.; Greb, S.F. Environmental and Health Impacts. In *Applied Coal Petrology*; Suarez-Ruiz, I., Crelling, J.C., Eds.; Academic Press: Amsterdam, The Netherlands, 2008; pp. 263–287, Chapter 10.
28. Ribeiro, J.; Suarez-Ruiz, I.; Ward, C.R.; Flores, D. Petrography and mineralogy of self-burning coal wastes from anthracite mining in the El Bierzo Coalfield (NW Spain). *Int. J. Coal Geol.* **2016**, *154*, 92–106. [CrossRef]
29. Medunić, G.; Grigore, M.; Dai, S.; Berti, D.; Hochella, M.F.; Mastalerz, M.; Valentim, B.; Guedes, A.; Hower, J.C. Characterization of superhigh-organic-sulfur Raša coal, Istria, Croatia, and its environmental implication. *Int. J. Coal Geol.* **2020**, *217*, 103344. [CrossRef]
30. Ketris, M.P.; Yudovich, Y.E. Estimations of Clarkes for Carbonaceous biolithes: World averages for trace element contents in black shales and coals. *Int. J. Coal Geol.* **2009**, *78*, 135–148. [CrossRef]
31. Goodarzi, F.; Huggins, F.E.; Sanei, H. Assessment of elements, speciation of As, Cr, Ni and emitted Hg for a Canadian power plant burning bituminous coal. *Int. J. Coal Geol.* **2008**, *74*, 1–12. [CrossRef]
32. Mukherjee, K.N.; Dutta, N.R.; Chandra, D.; Singh, M.P. Geochemistry of trace elements of Tertiary coals of India. *Int. J. Coal Geol.* **1992**, *20*, 99–113. [CrossRef]

33. Orem, W.H.; Finkelman, R.B. Coal Formation and Geochemistry. In *Sediments, Diagenesis, and Sedimentary Rocks: Treatise on Geochemistry*; Mackenzie, F.T., Ed.; Elsevier-Pergamon: Oxford, UK, 2003; Volume 7, pp. 191–222.
34. Goodarzi, F.; Sanei, H.; Stasiuk, L.D.; Bagheri-Sadeghi, H.; Reyes, J. A preliminary study of mineralogy and geochemistry of four coal samples from northern Iran. *Int. J. Coal Geol.* **2006**, *65*, 35–50. [CrossRef]
35. Dai, S.; Ren, D.; Chou, C.-L.; Finkelman, R.B.; Seredin, V.V.; Zhou, Y. Geochemistry of trace elements in Chinese coals: A review of abundances, genetic types, impacts on human health, and industrial utilization. *Int. J. Coal Geol.* **2012**, *94*, 3–21. [CrossRef]
36. Karayiğit, A.I.; Bircan, C.; Oskay, R.G.; Türkmen, I.; Querol, X. The geology, mineralogy, petrography, and geochemistry of the Miocene Dursunbey coal within fluvio-lacustrine deposits, Balıkesir (Western Turkey). *Int. J. Coal Geol.* **2020**, *228*, 103548. [CrossRef]
37. Kolker, A.; Senior, C.; van Alphen, C.; Koenig, A.; Geboy, N. Mercury and trace element distribution in density separates of a South African Highveld (# 4) coal: Implications for mercury reduction and preparation of export coal. *Int. J. Coal Geol.* **2017**, *170*, 7–13. [CrossRef]
38. Finkelman, R.B.; Dai, S.; French, D. The importance of minerals in coal as the hosts of chemical elements: A review. *Int. J. Coal Geol.* **2019**, *212*, 103251. [CrossRef]
39. Riley, K.W.; French, D.H.; Farrell, O.P.; Wood, R.A.; Huggins, F.E. Modes of occurrence of trace and minor elements in some Australian coals. *Int. J. Coal Geol.* **2012**, *94*, 214–224. [CrossRef]
40. Hower, J.C.; Campbell, J.L.; Teesdale, W.J.; Nejedly, Z.; Robertson, J.D. Scanning proton microprobe analysis of mercury and other trace elements in Fe-sulfides from a Kentucky coal. *Int. J. Coal Geol.* **2008**, *75*, 88–92. [CrossRef]
41. Mastalerz, M.; Drobniak, A. Arsenic, cadmium, lead, and zinc in the Danville and Springfield coal members (Pennsylvanian) from Indiana. *Int. J. Coal Geol.* **2007**, *71*, 37–53. [CrossRef]
42. Swaine, D.J. *Trace Elements in Coal*; Butterworth: London, UK, 1990; 278p.
43. Finkelman, R.B.; Palmer, C.A.; Wang, P. Quantification of the modes of occurrence of 42 elements in coal. *Int. J. Coal Geol.* **2018**, *185*, 138–160. [CrossRef]
44. Querol, X.; Alastuey, A.; Zhuang, X.G.; Hower, J.C.; Lopez-Soler, A.; Plana, F.; Zeng, R.S. Petrology, mineralogy and geochemistry of the Permian and Triassic coals in the Leping area, Jiangxi Province, southeast China. *Int. J. Coal Geol.* **2001**, *48*, 23–45. [CrossRef]
45. Wang, J.; Yamada, O.; Nakazato, T.; Zhang, Z.G.; Suzuki, Y.; Sakanishi, K. Statistical analysis of the concentrations of trace elements in a wide diversity of coals and its implications for understanding elemental modes of occurrence. *Fuel* **2008**, *87*, 2211–2222. [CrossRef]
46. Xie, P.; Li, Q.; Liu, J.; Song, H.; Wei, J. Geochemistry of arsenic and selenium in a Ge-poor coal from the Wulantuga coal-hosted Ge ore deposit, Inner Mongolia, North China. *Int. J. Coal Sci. Technol.* **2014**, *1*, 383–389. [CrossRef]
47. Qiao, M.; Cai, C.; Huang, Y.; Liu, Y.; Lin, A.; Zheng, Y. Characterization of soil heavy metal contamination and potential health risk in metropolitan region of northern China. *Environ. Monit. Assess.* **2011**, *172*, 353–365. [CrossRef]
48. Chen, Y.; Yuan, L.; Xu, C. Accumulation behavior of toxic elements in the soil and plant from Xinzhuangzi reclaimed mining areas, China. *Environ. Earth Sci.* **2017**, *76*, 226. [CrossRef]
49. Candeias, C.; da Silva, E.A.; Salgueiro, A.R.; Pereira, H.G.; Reis, A.P.; Patinha, C.; Matos, J.X.; Ávila, P.H. Assessment of soil contamination by potentially toxic elements in the aljustrel mining area in order to implement soil reclamation strategies. *Land Degrad. Dev.* **2011**, *22*, 565–585. [CrossRef]
50. Skousen, J.; Ziemkiewicz, P.; McDonald, L. Acid mine drainage formation, control and treatment: Approaches and strategies. *Extr. Ind. Soc.* **2019**, *6*, 241–249. [CrossRef]
51. Zhou, J.; Dang, Z.; Cai, M.F.; Liu, C.Q. Soil heavy metal pollution around the Dabaoshan mine, Guangdong province, China. *Pedosphere* **2007**, *17*, 588–594. [CrossRef]
52. Askaer, L.; Schmidt, L.B.; Elberling, B.; Asmund, G.; Jónsdóttir, I.S. Environmental impact on an Arctic soil–plant system resulting from metals released from coal mine waste in Svalbard (78 N). *Water Air Soil Pollut.* **2008**, *195*, 99–114. [CrossRef]
53. Larsen, D.; Mann, R. Origin of high manganese concentrations in coal mine drainage, eastern Tennessee. *J. Geochem. Explor.* **2005**, *86*, 143–163. [CrossRef]
54. Cui, X.; Geng, Y.; Sun, R.; Xie, M.; Feng, X.; Li, X.; Cui, Z. Distribution, speciation and ecological risk assessment of heavy metals in Jinan Iron & Steel Group soils from China. *J. Clean. Prod.* **2021**, *295*, 126504. [CrossRef]
55. Zhang, X.; Zhong, T.; Liu, L.; Ouyang, X. Impact of soil heavy metal pollution on food safety in China. *PLoS ONE* **2015**, *10*, e0135182. [CrossRef]
56. Rouhani, A.; Shadloo, S.; Naqibzadeh, A.; Hejcman, M.; Derakhsh, M. Pollution and Health Risk Assessment of Heavy Metals in the Soil Around an Open Landfill Site in a Developing Country (Kazerun, Iran). *Chem. Afr.* **2023**, *6*, 2139–2149. [CrossRef]
57. Luo, L.; Ma, Y.; Zhang, S.; Wei, D.; Zhu, Y.-G. An inventory of trace element inputs to agricultural soils in China. *J. Environ. Manag.* **2009**, *90*, 2524–2530. [CrossRef]
58. Yan, T.; Zhao, W.; Yu, X.; Li, H.; Gao, Z.; Ding, M.; Yue, J. Evaluating heavy metal pollution and potential risk of soil around a coal mining region of Tai'an City, China. *Alex. Eng. J.* **2022**, *61*, 2156–2165. [CrossRef]
59. Sun, L.; Guo, D.; Liu, K.; Meng, H.; Zheng, Y.; Yuan, F.; Zhu, G. Levels, sources, and spatial distribution of heavy metals in soils from a typical coal industrial city of Tangshan, China. *Catena* **2019**, *175*, 101–109. [CrossRef]
60. Zhang, H.; Zhang, F.; Song, J.; Tan, M.L.; Kung, H.; Johnson, V.C. Pollutant source, ecological and human health risks assessment of heavy metals in soils from coal mining areas in Xinjiang, China. *Environ. Res.* **2021**, *202*, 111702. [CrossRef]

61. Li, H.; Xu, W.; Dai, M.; Wang, Z.; Dong, X.; Fang, T. Assessing heavy metal pollution in paddy soil from coal mining area, Anhui, China. *Environ. Monit. Assess.* **2019**, *19*, 518. [CrossRef] [PubMed]
62. Dong, J.; Yang, Q.-W.; Sun, L.-N.; Zeng, Q.; Liu, S.-J.; Pan, J.; Liu, X.L. Assessing the concentration and potential dietary risk of heavy metals in vegetables at a Pb/Zn mine site, China. *Environ. Earth Sci.* **2011**, *64*, 1317–1321. [CrossRef]
63. Kou, J.; Gan, Y.; Lei, S.; Meng, W.; Feng, C.; Xiao, H. Soil health and ecological risk assessment in the typical coal mines on the Mongolian Plateau. *Ecol. Indic.* **2022**, *142*, 109189. [CrossRef]
64. Pecina, V.; Juřička, D.; Hedbávný, J.; Klimánek, M.; Kynický, J.; Brtnický, M.; Komendova, R. The impacts of mining on soil pollution with metal(loid)s in resource-rich Mongolia. *Sci. Rep.* **2023**, *13*, 2763. [CrossRef]
65. Hossen, M.A.; Chowdhury, A.I.; Mullick, M.R.; Hoque, A. Heavy metal pollution status and health risk assessment vicinity to Barapukuria coal mine area of Bangladesh. *Environ. Nanotechnol. Monit. Manag.* **2021**, *16*, 100469. [CrossRef]
66. Bhuiyan, M.A.; Parvez, L.; Islam, M.A.; Dampare, S.B.; Suzuki, S. Heavy metal pollution of coal mine-affected agricultural soils in the northern part of Bangladesh. *J. Hazard. Mater.* **2010**, *173*, 384–392. [CrossRef]
67. Siddiqui, A.U.; Jain, M.K.; Masto, R.E. Pollution evaluation, spatial distribution, and source apportionment of trace metals around coal mines soil: The case study of eastern India. *Environ. Sci. Pollut. Res.* **2020**, *27*, 10822–10834. [CrossRef]
68. Raj, D.; Kumar, A.; Maiti, S.K. Evaluation of toxic metal(loid)s concentration in soils around an open-cast coal mine (Eastern India). *Environ. Earth Sci.* **2019**, *78*, 645. [CrossRef]
69. Reza, S.K.; Baruah, U.; Singh, S.K.; Das, T.H. Geostatistical and multivariate analysis of soil heavy metal contamination near coal mining area, Northeastern India. *Environ. Earth Sci.* **2014**, *73*, 5425–5433. [CrossRef]
70. Ahmad, N.; Niamatullah; Hussain, J.I.; Ahmad, I.; Asif, M. Estimation of health risk to humans from heavy metals in soil of coal mines in Harnai, Balochistan. *Int. J. Environ. Anal. Chem.* **2020**, *102*, 3894–3905. [CrossRef]
71. Marove, C.A.; Sotozono, R.; Tangviroon, P.; Tabelin, C.B.; Igarashi, T. Assessment of soil, sediment and water contaminations around open-pit coal mines in Moatize, Tete province, Mozambique. *Environ. Adv.* **2022**, *8*, 100215. [CrossRef]
72. Zerizghi, T.; Guo, Q.; Tian, L.; Wei, R.; Zhao, C. An integrated approach to quantify ecological and human health risks of soil heavy metal contamination around coal mining area. *Sci. Total Environ.* **2022**, *814*, 152653. [CrossRef] [PubMed]
73. Ameh, E.G.; Aina, D.O. Search for autochthonous plants as accumulators and translocators in a toxic metal-polluted coal mine soil in Okaba, Nigeria. *Sci. Afr.* **2020**, *10*, e00630. [CrossRef]
74. Galunin, E.V.; Ferreti, J.; Zapelini, I.W.; Vieira, I.; Ricardo Teixeira Tarley, C.; Abrão, T.; Santos, M.J. Cadmium mobility in sediments and soils from a coal mining area on Tibagi River watershed: Environmental risk assessment. *J. Hazard. Mater.* **2014**, *265*, 280–287. [CrossRef] [PubMed]
75. Alekseenko, V.; Bech, J.; Alekseenko, A.V.; Shvydkaya, N.; Roca, N. Environmental impact of disposal of coal mining wastes on soils and plants in Rostov Oblast, Russia. *J. Geochem. Explor.* **2018**, *184*, 261–270. [CrossRef]
76. Pietrzykowski, M.; Socha, J.; van Doorn, N.S. Linking heavy metal bioavailability (Cd, Cu, Zn and Pb) in Scots pine needles to soil properties in reclaimed mine areas. *Sci. Total Environ.* **2014**, *470–471*, 501–510. [CrossRef]
77. Fiket, Ž.; Medunić, G.; Vidaković-Cifrek, Ž.; Jezidžić, P.; Cvjetko, P. Effect of coal mining activities and related industry on composition, cytotoxicity and genotoxicity of surrounding soils. *Environ. Sci. Pollut. Res.* **2019**, *27*, 6613–6627. [CrossRef]
78. Boahen, F.A.; Száková, J.; Kališová, A.; Najmanová, J.; Tlustoš, P. The assessment of the soil–plant-animal transport of the risk elements at the locations affected by brown coal mining. *Environ. Sci. Pollut. Res.* **2022**, *30*, 337–351. [CrossRef]
79. Vácha, R.; Skála, J.; Čechmánková, J.; Horváthová, V.; Hladík, J. Toxic elements and persistent organic pollutants derived from industrial emissions in agricultural soils of the Northern Czech Republic. *J. Soils Sediments* **2015**, *15*, 1813–1824. [CrossRef]
80. Ribeiro, J.C.; Silva, E.; Li, Z.; Ward, C.R.; Flores, D. Petrographic, mineralogical and geochemical characterization of the Serrinha coal waste pile (Douro Coalfield, Portugal) and the potential environmental impacts on soil, sediments and surface waters. *Int. J. Coal Geol.* **2010**, *83*, 456–466. [CrossRef]
81. Boente, C.; Albuquerque, M.T.; Fernández-Braña, A.; Gerassis, S.; Sierra, C.; Gallego, J.P. Combining raw and compositional data to determine the spatial patterns of Potentially Toxic Elements in soils. *Sci. Total Environ.* **2018**, *631*, 1117–1126. [CrossRef] [PubMed]
82. Kim, J.Y.; Chon, H.T. Pollution of a water course impacted by acid mine drainage in the Imgok creek of the Gangreung coal field, Korea. *Appl. Geochem.* **2001**, *16*, 1387–1396. [CrossRef]
83. Chen, D.; Feng, Q.; Liang, H. Effects of long-term discharge of acid mine drainage from abandoned coal mines on soil microorganisms: Microbial community structure, interaction patterns, and metabolic functions. *Environ. Sci. Pollut. Res.* **2021**, *28*, 53936–53952. [CrossRef] [PubMed]
84. He, A.; Li, X.; Ai, Y.; Li, X.; Li, X.; Zhang, Y.; Gao, Y.; Liu, B.; Zhang, X.; Zhang, M.; et al. Potentially toxic metals and the risk to children's health in a coal mining city: An investigation of soil and dust levels, bioaccessibility and blood lead levels. *Environ. Int.* **2020**, *141*, 105788. [CrossRef] [PubMed]
85. Cheng, W.; Lei, S.; Bian, Z.; Zhao, Y.; Li, Y.; Gan, Y. Geographic distribution of heavy metals and identification of their sources in soils near large, open-pit coal mines using positive matrix factorization. *J. Hazard. Mater.* **2020**, *387*, 121666. [CrossRef]
86. Maqbool, A.; Bian, Z.; Akram, M.W. Bioassessment of heavy metals in wheat crop from soil and dust in a coal mining area. *Pollution* **2019**, *5*, 323–337. [CrossRef]
87. Hussain, R.; Luo, K.; Liang, H.; Hong, X. Impact of the coal mining-contaminated soil on the food safety in Shaanxi, China. *Environ. Geochem. Health* **2019**, *41*, 1521–1544. [CrossRef]

88. Li, F.; Li, X.; Hou, L.; Shao, A. Impact of the Coal Mining on the Spatial Distribution of Potentially Toxic Metals in Farmland Tillage Soil. *Sci. Rep.* **2018**, *8*, 14925. [CrossRef]
89. Abliz, A.; Shi, Q.; Keyimu, M.; Sawut, R. Spatial distribution, source, and risk assessment of soil toxic metals in the coal-mining region of northwestern China. *Arab. J. Geosci.* **2018**, *11*, 793. [CrossRef]
90. Niu, S.; Gao, L.; Zhao, J. Heavy metals in the soils and plants from a typical restored coal-mining area of Huainan coalfield, China. *Environ. Monit. Assess.* **2017**, *189*, 484. [CrossRef] [PubMed]
91. Qin, F.; Wei, C.; Zhong, S.; Huang, X.; Pang, W.; Jiang, X. Soil heavy metal(loid)s and risk assessment in vicinity of a coal mining area from southwest Guizhou, China. *J. Cent. South Univ.* **2016**, *23*, 2205–2213. [CrossRef]
92. Halim, M.A.; Majumder, R.K.; Zaman, M.N. Paddy soil heavy metal contamination and uptake in rice plants from the adjacent area of Barapukuria coal mine, northwest Bangladesh. *Arab. J. Geosci.* **2015**, *8*, 3391–3401. [CrossRef]
93. Hossain, M.N.; Paul, S.K.; Hasan, M.M. Environmental impacts of coal mine and thermal power plant to the surroundings of Barapukuria, Dinajpur, Bangladesh. *Environ. Monit. Assess.* **2015**, *187*, 187. [CrossRef] [PubMed]
94. Pandey, B.; Agrawal, M.; Singh, S. Ecological risk assessment of soil contamination by trace elements around coal mining area. *J. Soils Sediments* **2015**, *16*, 159–168. [CrossRef]
95. Das, A.K.; Patel, S.S.; Kumar, R.R.; Krishna, K.S.; Dutta, S.; Saha, M.C.; Sengupta, S.; Guha, D. Geochemical sources of metal contamination in a coal mining area in Chhattisgarh, India using lead isotopic ratios. *Chemosphere* **2018**, *197*, 152–164. [CrossRef]
96. Manna, A.; Maiti, R. Geochemical contamination in the mine affected soil of Raniganj Coalfield—A river basin scale assessment. *Geosci. Front.* **2017**, *9*, 1577–1590. [CrossRef]
97. Raj, D.; Chowdhury, A.; Maiti, S.K. Ecological risk assessment of mercury and other heavy metals in soils of coal mining area: A case study from the eastern part of a Jharia coal field, India. *Hum. Ecol. Risk Assess. Int. J.* **2017**, *23*, 767–787. [CrossRef]
98. Ladwani, K.D.; Ladwani, K.D.; Manik, V.S.; Ramteke, D.S. Assessment of heavy metal contaminated soil near coal mining area in Gujarat by toxicity characteristics leaching procedure. *Int. J. Life Sci. Biotechnol. Pharma Res.* **2012**, *1*, 73–80.
99. Dabiri, R.; Adli, F.; Javanbakht, M. Environmental impacts of Aghdarband coal mine: Pollution by heavy metals. *Geopersia* **2017**, *7*, 311–321. [CrossRef]
100. Sakizadeh, M.; Mirzaei, R.; Ghorbani, H. Support vector machine and artificial neural network to model soil pollution: A case study in Semnan Province, Iran. *Neural Comput. Appl.* **2017**, *28*, 3229–3238. [CrossRef]
101. Tozsin, G. Hazardous elements in soil and coal from the Oltu coal mine district, Turkey. *Int. J. Coal Geol.* **2014**, *131*, 1–6. [CrossRef]
102. Yenilmez, F.; Kuter, N.; Emil, M.K.; Aksoy, A. Evaluation of pollution levels at an abandoned coal mine site in Turkey with the aid of GIS. *Int. J. Coal Geol.* **2011**, *86*, 12–19. [CrossRef]
103. Schneider, L.; Rose, N.L.; Lintern, A.; Sinclair, D.; Zawadzki, A.; Holley, C.; Aquino-López, M.A.; Haberle, S. Assessing environmental contamination from metal emission and relevant regulations in major areas of coal mining and electricity generation in Australia. *Sci. Total Environ.* **2020**, *728*, 137398. [CrossRef] [PubMed]
104. Maya, M.; Musekiwa, C.; Mthembi, P.; Crowley, M. Remote sensing and geochemistry techniques for the assessment of coal mining pollution, Emalahleni (Witbank), Mpumalanga. *S. Afr. J. Geomat.* **2015**, *4*, 174–188. [CrossRef]
105. Khalil, A.; Taha, Y.; Benzaazoua, M.; Hakkou, R. Applied Methodological Approach for the Assessment of Soil Contamination by Trace Elements around Abandoned Coal Mines—A Case Study of the Jerada Coal Mine, Morocco. *Minerals* **2023**, *13*, 181. [CrossRef]
106. Zhai, M.; Totolo, O.; Modisi, M.P.; Finkelman, R.B.; Kelesitse, S.M.; Menyatso, M. Heavy metal distribution in soils near Palapye, Botswana: An evaluation of the environmental impact of coal mining and combustion on soils in a semi-arid region. *Environ. Geochem. Health* **2009**, *31*, 759–777. [CrossRef] [PubMed]
107. Dziuba, E.; Buzmakov, S.; Khotyanovskaya, Y. Study of geochemical features of soils on the territory of an abandoned coal mining area using geoinformation technologies. *Environ. Geochem. Health* **2023**, *2023*, 1–21. [CrossRef]
108. Ushakova, E.; Menshikova, E.; Blinov, S.; Osovetsky, B.; Belkin, P. Environmental assessment impact of acid mine drainage from Kizel coal basin on the Kosva bay of the Kama Reservoir (Perm Krai, Russia). *Water* **2022**, *14*, 727. [CrossRef]
109. Arefieva, O.D.; Tregubova, V.G.; Gruschakova, N.V.; Starozhilov, V.T. Properties of Soils of Abandoned Coal Mine Industrial Areas (Primorsky Krai, Russia). *J. Geosci. Environ. Prot.* **2018**, *6*, 78. [CrossRef]
110. Dymov, A.A.; Kaverin, D.A.; Gabov, D.N. Properties of soils and soil-like bodies in the Vorkuta area. *Eurasian Soil Sci.* **2013**, *46*, 217–224. [CrossRef]
111. Komnitsas, K.; Modis, K. Soil risk assessment of As and Zn contamination in a coal mining region using geostatisretics. *Sci. Total Environ.* **2006**, *371*, 190–196. [CrossRef] [PubMed]
112. Modis, K.; Vatalis, K.I. Assessing the risk of soil pollution around an industrialized mining region using a geostatistical approach. *Soil Sediment Contam. Int. J.* **2014**, *23*, 63–75. [CrossRef]
113. Pentari, D.; Typou, J.; Goodarzi, F.; Foscolos, A.E. Comparison of elements of environmental concern in regular and reclaimed soils, near abandoned coal mines Ptolemais–Amynteon, northern Greece: Impact on wheat crops. *Int. J. Coal Geol.* **2006**, *65*, 51–58. [CrossRef]
114. Sutkowska, K.; Teper, L.; Czech, T.; Walker, A. Assessment of the Condition of Soils before Planned Hard Coal Mining in Southern Poland: A Starting Point for Sustainable Management of Fossil Fuel Resources. *Energies* **2023**, *16*, 737. [CrossRef]
115. Loska, K.; Wiechuła, D.; Korus, I. Metal contamination of farming soils affected by industry. *Environ. Int.* **2004**, *30*, 159–165. [CrossRef]

116. Hanousková, B.; Száková, J.; Rychlíková, E.; Najmanová, J.; Košnář, Z.; Tlustoš, P. The risk assessment of inorganic and organic pollutant levels in an urban area affected by intensive industry. *Environ. Monit. Assess.* **2021**, *193*, 68. [CrossRef]
117. Zádrapová, D.; Titěra, A.; Száková, J.; Čadková, Z.; Cudlín, O.; Najmanová, J.; Tlustoš, P. Mobility and bioaccessibility of risk elements in the area affected by the long-term opencast coal mining. *J. Environ. Sci. Health* **2019**, *54*, 1159–1169. [CrossRef]
118. Doležalová Weissmannová, H.; Mihočová, S.; Chovanec, P.; Pavlovský, J. Potential ecological risk and human health risk assessment of heavy metal pollution in industrial affected soils by coal mining and metallurgy in Ostrava, Czech Republic. *Int. J. Environ. Res. Public Health* **2019**, *16*, 4495. [CrossRef]
119. Lieberman, N.R.; Izquierdo, M.T.; Muñoz-Quirós, C.; Cohen, H.; Chenery, S.R. Geochemical signature of superhigh organic sulphur Raša coals and the mobility of toxic trace elements from combustion products and polluted soils near the Plomin coal-fired power station in Croatia. *Appl. Geochem.* **2020**, *114*, 104472. [CrossRef]
120. Santos, P.; Espinha Marques, J.; Ribeiro, J.; Mansilha, C.; Melo, A.; Fonseca, R.; Sant'Ovaia, H.; Flores, D. Geochemistry of Soils from the Surrounding Area of a Coal Mine Waste Pile Affected by Self-Burning (Northern Portugal). *Minerals* **2023**, *13*, 28. [CrossRef]
121. Desai, M.; Haigh, M.; Walkington, H. Phytoremediation: Metal decontamination of soils after the sequential forestation of former opencast coal land. *Sci. Total Environ.* **2019**, *656*, 670–680. [CrossRef] [PubMed]
122. Osipova, N.A.; Tkacheva, E.V.; Arbuzov, S.I.; Yazikov, E.G.; Matveenko, I.A. Mercury in coals and soils from coal-mining regions. *Solid Fuel Chem.* **2019**, *53*, 411–417. [CrossRef]
123. Miller, C.L.; Watson, D.B.; Lester, B.P.; Lowe, K.A.; Pierce, E.M.; Liang, L. Characterization of soils from an industrial complex contaminated with elemental mercury. *Environ. Res.* **2013**, *125*, 20–29. [CrossRef]
124. Ouyang, D.; Liu, K.; Wu, Q.; Wang, S.; Tang, Y.; Li, Z.; Liu, T.; Han, L.; Cui, Y.; Li, G.; et al. Effect of the coal preparation process on mercury flows and emissions in coal combustion systems. *Environ. Sci. Technol.* **2021**, *55*, 13687–13696. [CrossRef]
125. Peter, A.J.; Viraraghavan, T. Thallium: A review of public health and environmental concerns. *Environ. Int.* **2005**, *31*, 493–501. [CrossRef]
126. Gao, B.; Sun, K.; Ren, M.Z.; Liang, X.R.; Peng, P.A.; Sheng, G.Y.; Fu, J.M. Ecological risk assessment of thallium pollution in the surface sediment of Beijiang River. *Ecol. Environ.* **2008**, *17*, 528–532.
127. Zhang, C.; Ren, S.; Cheng, H.; Zhang, W.; Ma, J.; Zhang, C.; Guo, Z. Thallium pollution and potential ecological risk in the vicinity of coal mines in Henan Province, China. *Chem. Speciat. Bioavailab.* **2018**, *30*, 107–111. [CrossRef]
128. Yan, D.; Wu, S.; Zhou, S.; Tong, G.; Li, F.; Wang, Y.; Li, B. Characteristics, sources and health risk assessment of airborne particulate PAHs in Chinese cities: A review. *Environ. Pollut.* **2019**, *248*, 804–814. [CrossRef]
129. Wilcke, W. Global patterns of polycyclic aromatic hydrocarbons (PAHs) in soil. *Geoderma* **2007**, *141*, 157–166. [CrossRef]
130. Coxon, T.; Goldstein, L.; Odhiambo, B.K. Analysis of spatial distribution of trace metals, PCB, and PAH and their potential impact on human health in Virginian Counties and independent cities, USA. *Environ. Geochem. Health* **2019**, *41*, 783–801. [CrossRef]
131. Hendryx, M.; Wang, S.; Romanak, K.A.; Salamova, A.; Venier, M. Personal exposure to polycyclic aromatic hydrocarbons in Appalachian mining communities. *Environ. Pollut.* **2020**, *257*, 113501. [CrossRef]
132. Boente, C.; Matanzas, N.; García-González, N.; Rodríguez-Valdés, E.; Gallego, J.R. Trace elements of concern affecting urban agriculture in industrialized areas: A multivariate approach. *Chemosphere* **2017**, *183*, 546–556. [CrossRef] [PubMed]
133. Bandowe, B.A.M.; Nkansah, M.A. Occurrence, distribution and health risk from polycyclic aromatic compounds (PAHs, oxygenated-PAHs and azaarenes) in street dust from a major West African Metropolis. *Sci. Total Environ.* **2016**, *553*, 439–449. [CrossRef]
134. Wang, R.; Sun, R.; Liu, G.; Yousaf, B.; Wu, D.; Chen, J.; Zhang, H. A review of the biogeochemical controls on the occurrence and distribution of polycyclic aromatic compounds (PACs) in coals. *Earth-Sci. Rev.* **2017**, *171*, 400–418. [CrossRef]
135. Achten, C.; Hofmann, T. Native polycyclic aromatic hydrocarbons (PAH) in coals–a hardly recognized source of environmental contamination. *Sci. Total Environ.* **2009**, *407*, 2461–2473. [CrossRef] [PubMed]
136. Xu, Z.; Qian, Y.; Hong, X.; Luo, Z.Y.; Gao, X.; Liang, H. Contamination characteristics of polycyclic aromatic compounds from coal sources in typical coal mining areas in Huaibei area, China. *Sci. Total Environ.* **2023**, *873*, 162311. [CrossRef]
137. Liang, M.; Liang, H.; Gao, P.; Rao, Z.; Liang, Y. Characterization and risk assessment of polycyclic aromatic hydrocarbon emissions by coal fire in northern China. *Environ. Geochem. Health* **2021**, *44*, 933–942. [CrossRef]
138. Zhang, J.; Liu, F.; Huang, H.; Wang, R.; Xu, B. Occurrence, risk and influencing factors of polycyclic aromatic hydrocarbons in surface soils from a large-scale coal mine, Huainan, China. *Ecotoxicol. Environ. Saf.* **2020**, *192*, 110269. [CrossRef] [PubMed]
139. Wang, R.; Liu, G.; Chou, C.; Liu, J.; Zhang, J. Environmental Assessment of PAHs in Soils Around the Anhui Coal District, China. *Arch. Environ. Contam. Toxicol.* **2010**, *59*, 62–70. [CrossRef] [PubMed]
140. Liu, J.; Liu, G.; Zhang, J.; Yin, H.; Wang, R. Occurrence and risk assessment of polycyclic aromatic hydrocarbons in soil from the Tiefa coal mine district, Liaoning, China. *J. Environ. Monit.* **2012**, *14*, 2634–2642. [CrossRef]
141. Sun, Y.Z.; Fan, J.S.; Qin, P.; Niu, H. Pollution extents of organic substances from a coal gangue dump of Jiulong Coal Mine, China. *Environ. Geochem. Health* **2009**, *31*, 81–89. [CrossRef] [PubMed]
142. Mizwar, A.; Priatmadi, B.J.; Abdi, C.; Trihadiningrum, Y. Assessment of polycyclic aromatic hydrocarbons (PAHs) contamination in surface soil of coal stockpile sites in South Kalimantan, Indonesia. *Environ. Monit. Assess.* **2016**, *188*, 152. [CrossRef] [PubMed]
143. Masto, R.E.; Sheik, S.; Nehru, G.; Selvi, V.A.; George, J.C.; Ram, L.C. Assessment of environmental soil quality around Sonepur Bazari mine of Raniganj coalfield, India. *Solid Earth* **2015**, *6*, 811–821. [CrossRef]

144. Yakovleva, E.V.; Gabov, D.; Beznosikov, V.A.; Kondratenok, B.M.; Dubrovskiy, Y.A. Accumulation of PAHs in Tundra Plants and Soils under the Influence of Coal Mining. *Polycycl. Aromat. Compd.* **2017**, *37*, 203–218. [CrossRef]
145. Jakovljević, I.; Mešić, I.; Pehnec, G. Soil pollution of the Labin city area with polycyclic aromatic hydrocarbons derived from Raša coal mining and associated industries. *Rud.-Geološko-Naft. Zb.* **2022**, *37*, 139–150. [CrossRef]
146. Boente, C.; Baragaño, D.; Gallego, J.L. Benzo[a]pyrene sourcing and abundance in a coal region in transition reveals historical pollution, rendering soil screening levels impractical. *Environ. Pollut.* **2020**, *266*, 115341. [CrossRef]
147. Hindersmann, B.; Achten, C. Urban soils impacted by tailings from coal mining: PAH source identification by 59 PAHs, BPCA and alkylated PAHs. *Environ. Pollut.* **2018**, *242*, 1217–1225. [CrossRef]
148. Pies, C.; Yang, Y.; Hofmann, T. Distribution of polycyclic aromatic hydrocarbons (PAHs) in floodplain soils of the Mosel and Saar River. *J. Soils Sediments* **2007**, *7*, 216–222. [CrossRef]
149. Ugwu, K.E.; Ukoha, P.O. Analysis and sources of polycyclic aromatic hydrocarbons in soil and plant samples of a coal mining area in Nigeria. *Bull. Environ. Contam. Toxicol.* **2016**, *96*, 383–387. [CrossRef]
150. Tan, G.L.; Shu, W.S.; Zhou, W.H.; Li, X.L.; Lan, C.Y.; Huang, L.N. Seasonal and spatial variations in microbial community structure and diversity in the acid stream draining across an ongoing surface mining site. *FEMS Microbiol. Ecol.* **2009**, *70*, 277–285. [CrossRef] [PubMed]
151. Méndez-García, C.; Peláez, A.I.; Mesa, V.; Sánchez, J.; Golyshina, O.V.; Ferrer, M. Microbial diversity and metabolic networks in acid mine drainage habitats. *Front. Microbiol.* **2015**, *6*, 475. [CrossRef]
152. Gogoi, J.; Pathak, N.; Dowrah, J.; Boruah, H.P.D.; Gogoi, J.; Pathak, N.; Dowrah, J.; Deka Boruah, H.P. In situ selection of tree species in environmental restoration of opencast coalmine wasteland. In *Proceedings of Int. Sem. on MPT*; Allied Publisher: Mumbai, India, 2007; pp. 678–681.
153. Williamson, J.C.; Johnson, D.B. Determination of the activity of soil microbial populations in stored and restored soils at opencast coal sites. *Soil Biol. Biochem.* **1990**, *22*, 671–675. [CrossRef]
154. Gandhi, V.; Priya, A.; Priya, S.; Daiya, V.; Kesari, J.; Prakash, K.; Kumar Jha, A.; Kumar, K.; Kumar, N. Isolation and molecular characterization of bacteria to heavy metals isolated from soil samples in Bokaro Coal Mines, India. *Pollution* **2015**, *1*, 287–295. [CrossRef]
155. Majumder, P.; Palit, D. Microbial diversity of soil in some coal mine generated wasteland of raniganj coalfield, West Bengal, India. *Int. J. Curr. Microbiol. Appl. Sci.* **2016**, *5*, 637–641. [CrossRef]
156. Shylla, L.; Barik, S.K.; Joshi, S.R. Characterization and bioremediation potential of native heavy-metal tolerant bacteria isolated from rat-hole coal mine environment. *Arch. Microbiol.* **2021**, *203*, 2379–2392. [CrossRef]
157. Murthy, S.; Bali, G.; Sarangi, S.K. Lead biosorption by a bacterium isolated from industrial effluents. *Int. J. Microbiol. Res.* **2012**, *4*, 196–200. [CrossRef]
158. He, C.; Zhang, B.; Lu, J.; Qiu, R. A newly discovered function of nitrate reductase in chemoautotrophic vanadate transformation by natural mackinawite in aquifer. *Water Res.* **2020**, *189*, 116664. [CrossRef] [PubMed]
159. Gupta, K.; Chatterjee, C.; Gupta, B. Isolation and characterization of heavy metal tolerant Gram-positive bacteria with bioremedial properties from municipal waste rich soil of Kestopur canal (Kolkata), West Bengal, India. *Biologia* **2012**, *67*, 827–836. [CrossRef]
160. Upadhyay, N.; Vishwakarma, K.; Singh, J.; Mishra, M.; Kumar, V.; Rani, R.; Mishra, R.K.; Chauhan, D.K.; Tripathi, D.K.; Sharma, S. Tolerance and reduction of chromium (VI) by Bacillus sp. MNU16 isolated from contaminated coal mining soil. *Front. Plant Sci.* **2017**, *8*, 778–790. [CrossRef] [PubMed]
161. Syed, S.; Chinthala, P. Heavy metal detoxification by different Bacillus species isolated from solar salterns. *Scientifica* **2015**, *2015*, 319760. [CrossRef] [PubMed]
162. Yahaghi, Z.; Shirvani, M.; Nourbakhsh, F.; de la Peña, T.C.; Pueyo, J.J.; Talebi, M. Isolation and characterization of Pb-solubilizing bacteria and their effects on pb uptake by Brassica juncea: Implications for microbe-assisted phytoremediation. *Environ. Microbiol. Biotechnol.* **2018**, *28*, 1156–1167. [CrossRef]
163. Singh, P.; Tiwary, B.N. Optimization of conditions for polycyclic aromatic hydrocarbons (PAHs) degradation by Pseudomonas stutzeri P2 isolated from Chirimiri coal mines. *Biocatal. Agric. Biotechnol.* **2017**, *10*, 20–29. [CrossRef]
164. Kurniawan, S.B.; Ramli, N.N.; Said, N.S.M.; Alias, J.; Imron, M.F.; Abdullah, S.R.S.; Othman, A.R.; Purwanti, I.F.; Hasan, H.A. Practical limitations of bioaugmentation in treating heavy metal contaminated soil and role of plant growth promoting bacteria in phytoremediation as a promising alternative approach. *Heliyon* **2022**, *8*, e08995. [CrossRef]
165. Vithanage, M.; Dabrowska, B.B.; Mukherjee, A.B.; Sandhi, A.; Bhattacharya, P. Arsenic uptake by plants and possible phytoremediation applications: A brief overview. *Environ. Chem. Lett.* **2012**, *10*, 217–224. [CrossRef]
166. Gadi, B.R.; Kumar, R.; Goswami, B.; Rankawat, R.; Rao, S.R. Recent developments in understanding fluoride accumulation, toxicity, and tolerance mechanisms in plants: An overview. *J. Soil Sci. Plant Nutr.* **2020**, *21*, 209–228. [CrossRef]
167. Pandey, V.C.; Bauddh, K. *Phytomanagement of Polluted Sites: Market Opportunities in Sustainable Phytoremediation*; Elsevier: Amsterdam, The Netherlands, 2018. [CrossRef]
168. Shojaee Barjoee, S.; Malverdi, E.; Kouhkan, M.; Alipourfard, I.; Rouhani, A.; Farokhi, H.; Khaledi, A.A. Health assessment of industrial ecosystems of Isfahan (Iran) using phytomonitoring: Chemometric, micromorphology, phytoremediation, air pollution tolerance and anticipated performance indices. *Urban Clim.* **2023**, *48*, 101394. [CrossRef]
169. Pandey, V.C.; Bajpai, O. Phytoremediation: From Theory toward Practice. In *Phytomanagement of Polluted Sites*; Elsevier: Amsterdam, The Netherlands, 2019; pp. 1–49. [CrossRef]

170. Sun, H.; Zhang, J.; Wang, R.; Li, Z.; Sun, S.; Qin, G.; Song, Y. Effects of Vegetation Restoration on Soil Enzyme Activity in Copper and Coal Mining Areas. *Environ. Manag.* **2021**, *68*, 366–376. [CrossRef]
171. Bai, D.S.; Yang, X.; Lai, J.L.; Wang, Y.W.; Zhang, Y.; Luo, X.G. In situ restoration of soil ecological function in a coal gangue reclamation area after 10 years of elm/poplar phytoremediation. *J. Environ. Manag.* **2022**, *305*, 114400. [CrossRef]
172. Feng, H.; Zhou, J.; Zhou, A.; Bai, G.; Li, Z.; Chen, H.; Su, D.; Han, X. Grassland ecological restoration based on the relationship between vegetation and its below-ground habitat analysis in steppe coal mine area. *Sci. Total Environ.* **2021**, *778*, 146221. [CrossRef] [PubMed]
173. Monei, N.L.; Puthiya Veetil, S.K.; Gao, J.; Hitch, M. Selective removal of selenium by phytoremediation from post/mining coal wastes: Practicality and implications. *Int. J. Min. Reclam. Environ.* **2021**, *35*, 69–77. [CrossRef]
174. Matanzas, N.; Afif, E.; Díaz, T.E.; Gallego, J.R. Phytoremediation potential of native herbaceous plant species growing on a paradigmatic brownfield site. *Water Air Soil Pollut.* **2021**, *232*, 290. [CrossRef]
175. Fernández-Braña, A.; Salgado, L.; Gallego, J.L.R.; Afif, E.; Boente, C.; Forján, R. Phytoremediation potential depends on the degree of soil pollution: A case study in an urban brownfield. *Environ. Sci. Pollut. Res.* **2023**, *30*, 67708–67719. [CrossRef]
176. Bandyopadhyay, S.; Rana, V.; Maiti, S.K. Chronological variation of metals in reclaimed coal mine soil and tissues of Eucalyptus hybrid tree after 25 years of reclamation, Jharia coal field (India). *Bull. Environ. Contam. Toxicol.* **2018**, *101*, 604–610. [CrossRef] [PubMed]
177. Mellem, J.J.; Baijnath, H.; Odhav, B. Bioaccumulation of Cr, Hg, As, Pb, Cu and Ni with the ability for hyperaccumulation by Amaranthus dubius. *Afr. J. Agric. Res.* **2012**, *7*, 591–596. [CrossRef]
178. Favas, P.J.; Pratas, J.; Varun, M.; D'Souza, R.; Paul, M.S. Phytoremediation of soils contaminated with metals and metalloids at mining areas: Potential of native flora. *Environ. Risk Assess. Soil Contam.* **2014**, *3*, 485–516. [CrossRef]
179. Zhang, A.; Li, X.; Xing, J.; Xu, G. Adsorption of potentially toxic elements in water by modified biochar: A review. *J. Environ. Chem. Eng.* **2020**, *8*, 104196. [CrossRef]
180. Liu, M.; Che, Y.; Wang, L.; Zhao, Z.; Zhang, Y.; Wei, L.; Xiao, Y. Rice straw biochar and phosphorus inputs have more positive effects on the yield and nutrient uptake of Lolium multiflorum than arbuscular mycorrhizal fungi in acidic Cd-contaminated soils. *Chemosphere* **2019**, *235*, 32–39. [CrossRef]
181. Beiyuan, J.; Awad, Y.M.; Beckers, F.; Wang, J.; Tsang, D.C.W.; Ok, Y.S.; Wang, S.-L.; Wang, H.; Rinklebe, J. (Im) mobilization and speciation of lead under dynamic redox conditions in a contaminated soil amended with pine sawdust biochar. *Environ. Int.* **2020**, *135*, 105376. [CrossRef]
182. Yang, W.; Qu, T.; Flury, M.; Zhang, X.; Gabriel, S.; Shang, J.; Li, B. PAHs sorption to biochar colloids changes their mobility over time. *J. Hydrol.* **2021**, *603*, 126839. [CrossRef]
183. Shaaban, M.; Van Zwieten, L.; Bashir, S.; Younas, A.; Nú~nez-Delgado, A.; Chhajro, M.A.; Kubar, K.A.; Ali, U.; Rana, M.S.; Mehmood, M.A.; et al. A concise review of biochar application to agricultural soils to improve soil conditions and fight pollution. *J. Environ. Manag.* **2018**, *228*, 429–440. [CrossRef]
184. Xie, T.; Reddy, K.R.; Wang, C.; Yargicoglu, E.; Spokas, K. Characteristics and applications of biochar for environmental remediation: A review. *Crit. Rev. Environ. Sci. Technol.* **2015**, *45*, 939–969. [CrossRef]
185. Mohamed, I.; Zhang, G.S.; Li, Z.G.; Liu, Y.; Chen, F.; Dai, K. Ecological restoration of an acidic Cd contaminated soil using bamboo biochar application. *Ecol. Eng.* **2015**, *84*, 67–76. [CrossRef]
186. Ghosh, D.; Maiti, S.K. Can biochar reclaim coal mine spoil? *J. Environ. Manag.* **2020**, *272*, 111097. [CrossRef]
187. Chandra, S.; Medha, I.; Bhattacharya, J.; Vanapalli, K.R.; Samal, B. Effect of the Co-Application of Eucalyptus Wood Biochar and Chemical Fertilizer for the Remediation of Multimetal (Cr, Zn, Ni, and Co) Contaminated Soil. *Sustainability* **2022**, *14*, 7266. [CrossRef]
188. Mujtaba Munir, M.A.; Liu, G.; Yousaf, B.; Ali, M.; Cheema, A.I.; Rashid, M.S.; Rehman, A. Bamboo-biochar and hydrothermally treated-coal mediated geochemical speciation, transformation and uptake of Cd, Cr, and Pb in a polymetal(iod)s-contaminated mine soil. *Environ. Pollut.* **2020**, *265*, 114816. [CrossRef]
189. Dai, S.; Li, H.; Yang, Z.; Dai, M.; Dong, X.; Ge, X.; Sun, M.; Shi, L. Effects of biochar amendments on speciation and bioavailability of heavy metals in coal-mine-contaminated soil. *Hum. Ecol. Risk Assess. Int. J.* **2018**, *24*, 1887–1900. [CrossRef]
190. Jain, S.; Khare, P.; Mishra, D.; Shanker, K.; Singh, P.; Singh, R.P.; Das, P.; Yadav, R.; Saikia, B.K.; Baruah, B.P. Biochar aided aromatic grass [Cymbopogon martini (Roxb.) Wats.] vegetation: A sustainable method for stabilization of highly acidic mine waste. *J. Hazard. Mater.* **2020**, *390*, 121799. [CrossRef]
191. Jain, S.; Singh, A.; Khare, P.; Chanda, D.; Mishra, D.; Shanker, K.; Karak, T. Toxicity assessment of Bacopa monnieri L. grown in biochar amended extremely acidic coal mine spoils. *Ecol. Eng.* **2017**, *108*, 211–219. [CrossRef]
192. Roberts, D.A.; Cole, A.J.; Paul, N.A.; De Nys, R. Algal biochar enhances the re-vegetation of stockpiled mine soils with native grass. *J. Environ. Manag.* **2015**, *161*, 173–180. [CrossRef] [PubMed]

MDPI AG
Grosspeteranlage 5
4052 Basel
Switzerland
Tel.: +41 61 683 77 34

Minerals Editorial Office
E-mail: minerals@mdpi.com
www.mdpi.com/journal/minerals

www.ingramcontent.com/pod-product-compliance
Lightning Source LLC
Chambersburg PA
CBHW061618210926
51850CB00009B/2366
* 9 7 8 3 7 2 5 8 5 2 9 9 4 *